Modern Programmable Interconnect Design

AF183787

Stefan Nikolić

Modern Programmable Interconnect Design

Springer

Stefan Nikolić
Faculty of Sciences
University of Novi Sad
Novi Sad, Serbia

ISBN 978-3-031-80631-5 ISBN 978-3-031-80629-2 (eBook)
https://doi.org/10.1007/978-3-031-80629-2

© The Editor(s) (if applicable) and The Author(s), under exclusive license to Springer Nature Switzerland AG 2025

This work is subject to copyright. All rights are solely and exclusively licensed by the Publisher, whether the whole or part of the material is concerned, specifically the rights of translation, reprinting, reuse of illustrations, recitation, broadcasting, reproduction on microfilms or in any other physical way, and transmission or information storage and retrieval, electronic adaptation, computer software, or by similar or dissimilar methodology now known or hereafter developed.
The use of general descriptive names, registered names, trademarks, service marks, etc. in this publication does not imply, even in the absence of a specific statement, that such names are exempt from the relevant protective laws and regulations and therefore free for general use.
The publisher, the authors and the editors are safe to assume that the advice and information in this book are believed to be true and accurate at the date of publication. Neither the publisher nor the authors or the editors give a warranty, expressed or implied, with respect to the material contained herein or for any errors or omissions that may have been made. The publisher remains neutral with regard to jurisdictional claims in published maps and institutional affiliations.

This Springer imprint is published by the registered company Springer Nature Switzerland AG
The registered company address is: Gewerbestrasse 11, 6330 Cham, Switzerland

If disposing of this product, please recycle the paper.

Any idea is better than none.
— László F. Nagy

To my late grandmother Mirjana.

Preface

With Moore's law coming to an end, today's rising demand for compute puts increasingly more hope in specialized hardware implemented on reconfigurable architectures such as *Field-Programmable Gate Arrays* (FPGAs). Yet, it is often neglected that these architectures themselves experience problems caused by technology scaling. In fact, due to their lower logic density and the need to provide interconnect programmability, rising wire resistance impacts the performance of FPGAs particularly negatively. This is further complicated by the traditional problem of reconfigurable architecture design: exact critical paths are not known at fabrication time.

For FPGAs to deliver what is expected from them, they must continue to adapt to new technological and application realities. Local optimization of programmable interconnect around a known prior solution, which, for the past two decades, has been ensuring that performance increase of moving to the next technology node is achieved on time, no longer yields the expected results. Since technology scaling has moved away from a predictable trajectory driven by geometric shrinking, into the realm of an ever-increasing number of one-off *scaling boosters*, it is now harder than ever to develop a lasting intuition about how to escape these local optima. Coupling this with an ever-increasing number of different applications that require FPGA acceleration in a variety of different device integration and use contexts, we believe that the only real solution to this issue is to use design automation to seek new local optima. The problem with this, however, is that design automation algorithms comparable to those that exist in a standard *Application-Specific Integrated Circuit* (ASIC) flow have so far not been developed—riding the wave of Moore's law in the past decades made tackling the hard combinatorial optimization problems that arise in designing programmable interconnect unattractive. Where ASIC *Computer-Aided Design* (CAD) algorithms, once developed, could be used to produce thousands of different circuits, basic economy behind FPGA's success meant that in every generation, only a handful of device families with shared programmable fabric architecture were produced. As a result, a large number of algorithms exist for transforming an ASIC described at a high level of abstraction using a hardware description language into a highly optimized netlist of gates, while programmable interconnect architectures still have to be described entirely by specifying every single wire and switch that comprise them—this would be equivalent to specifying an ASIC design by listing every single logic gate and all connections between them. Well, if such a specification has to be made only once per FPGA generation—or better yet, only once for a number of generations, as has largely been the case over the past two decades—maybe the effort of developing a set of algorithms that could do this automatically is not easily justified. The problem today, as we already mentioned, is that scaling time-tested designs with only minor modifications no longer yields the required results, while manually specifying a new design at this level of abstraction is hardly feasible after the stark increase in the number of variables that have to be taken into account due to technological and application changes.

In this book, we address the issue by developing new algorithms for automated design of several important aspects of programmable interconnect.

Who Should Read This Book?

As we have already mentioned, the main purpose of this book is to introduce novel algorithms that demonstrate both feasibility and utility of automating programmable interconnect design in the era of scaled fabrication technologies. Chapters dealing with this topic are contained in Part II of the book and will be of most interest to researchers and practitioners already working in the domains of FPGA architecture. Of course, given that evaluation of new architectures often requires new CAD algorithms as well, Part II, and especially Chap. 8, could be of interest to those involved in researching topics related to FPGA CAD. Finally, since the main topic of the book is design automation for a very specific but difficult and highly important problem, in Part II, practitioners and researchers from the broader CAD community may find inspiration for new ways to apply their expertise.

Part I provides a gentle introduction to the problem of programmable interconnect design, starting from the basic principles. Its intention is to allow a novice reader or a practitioner from another field to understand the subsequent chapters. Previous books on the topic were mostly published while the basic principles were still being developed or were very recent, which made it difficult to fully appreciate their significance and present them in a simple and consistent fashion. Part I attempts to rectify these shortcomings by profiting from the time distance available to the author. As such, it could be of interest to anyone trying to understand not only how FPGAs operate on a deeper level but also why they look the way they do.

Finally, Part III would mostly be of interest to a prospective researcher in the fields of FPGA architecture and CAD, looking for exciting problems to tackle.

After briefly introducing some related topics, it returns to drawing conclusions and providing a summary of what are in author's humble opinion the most important questions that remain open. The author's hope is precisely that the present book will inspire more people to enter these fascinating fields and quickly make this list of open questions obsolete, while discovering others that are still more important and intriguing.

Novi Sad, Serbia Stefan Nikolić

Acknowledgments

There is a very large number of people without whom my PhD thesis that formed the basis for this book would never have seen the light of day. A comprehensive list of all these individuals is given in the thesis itself [Nik23a]. I would like to thank them all here once more, but I fear that the aforementioned list is far too long and personal for a typical acknowledgments section in a monograph. Hence, I will not repeat it here, but will only extend my additional gratitude to the people without whom this book would not have come out of print even after the thesis had long been completed. I am deeply grateful to my thesis advisor Prof. Paolo Ienne, the book's editor Charles Glaser, Prof. Giovanni De Micheli, Dr. Dinesh Gaitonde, Prof. Vaughn Betz, Chirag Ravishankar, Dr. Herman Schmit, and Dr. Jonathan Greene for their support, encouragement, and constructive feedback.

Reference

[Nik23a] Nikolić, S. (2023). *Automating the Design of Programmable Interconnect for Reconfigurable Architectures.* PhD Thesis, École Polytechnique Fédérale de Lausanne.

Contents

Part III Additional Topics

Chapter 1
Introduction

Contents

As Moore's law is no longer delivering the performance improvements that it used to deliver over the past half a century, custom hardware architectures specifically tuned to the requirements of the algorithm that they are intended to implement are increasingly looked at as a promising way forward [Hen19, Lei20]. At the same time, however, attempts to keep Moore's law afloat and squeeze out further benefits from technology have made custom chip design prohibitively expensive, with costs reaching half a billion dollars at the 5 nm node [Bau20]. Hence, for a majority of custom hardware designs, actually producing an *Application-Specific Integrated Circuit* (ASIC) is simply not an option. Instead, they have to rely on reconfigurable integrated circuits such as *Field-Programmable Gate Arrays* (FPGAs).

For a long time, FPGAs have been considered a "poor man's ASIC," used mostly by those who could not afford design and fabrication of a custom chip. There were two notable exceptions, however: ASIC emulation and communications [Tri94, Tri15]. What made these domains special is that architectures (especially in ASIC emulation) and even algorithms (especially in communications) were so volatile that the ability to reconfigure the same physical chip at no cost and with close to zero delay was very important. Since perhaps the single most promising way of continuing to increase performance of computing systems is innovation at the algorithm level [Lei20], more and more application domains begin to profit from hardware reconfigurability that allows acceleration of the appropriate algorithm before it becomes obsolete [Ans23]; machine learning that requires model modifications on almost a daily basis [Jou21] is just one example. For this reason, traditionally fixed-functionality *Systems on Chip* (SoC) are now also increasingly incorporating reconfigurable fabrics [Jar22].

One of the main reasons why performance of integrated circuits is not increasing as significantly as it used to is poor scaling of interconnect delays [Tok22]. Interconnect has always been the Achilles' heel of FPGAs [Tri94]. Whereas in

© The Author(s), under exclusive license to Springer Nature Switzerland AG 2025

S. Nikolić, *Modern Programmable Interconnect Design*,
https://doi.org/10.1007/978-3-031-80629-2_1

an ASIC, connections are implemented by simply tracing the appropriate metal, in an FPGA, it is necessary to stitch together prefabricated wires by driving the right values onto select inputs of multiplexers between them. Superficially seen, these multiplexers are the greatest difference between a programmable wire of an FPGA and a fixed wire of an ASIC, and in the days when transistor delays dominated metal delays, they were also responsible for bulk of the FPGA's inferior performance [Tri94]. For decades, Moore's law brought increases in transistor performance, while the negative impact of diminishing crossectional area of wires on their delay was kept in check by technological innovation as well as simple reduction of wirelength through increased logic density. This enabled programmable interconnect architecture of FPGAs to gain performance benefits without major modifications of what was essentially a reasonably straightforward extension of *Mask-Programmed Gate Arrays* (MPGAs).

Around 7 nm, this situation changed. If ASIC performance suffered from a surge in wire resistance [Che14], FPGA performance suffered all the more, since the prefabricated wires were loaded by a much higher capacitive load [Tok16] of all the multiplexers that grant the interconnect architecture its reconfigurability. Hence, the latest generations of FPGAs from both major vendors—namely the 7 nm Xilinx/AMD Versal [Gai19] and Altera/Intel Agilex [Chr20, Cut21]—reported the most significant changes to the programmable interconnect topology in almost twenty years. Perhaps even more significantly, due to their highly regular structure, FPGAs have been used by foundries to test and develop new fabrication processes for a number of years [Tri15]; yet, they are now two technology nodes behind TSMC's latest offering [TSM23]. This may suggest that the long-delayed major modifications to programmable interconnect are difficult to pull off even for the industry leaders.

Why has custom hardware all of a sudden become so appealing? The answer is simple: General-purpose platforms are no longer delivering the required performance. It may then be logical to ask whether FPGAs themselves, which have traditionally also been general-purpose platforms, should follow suit, with different devices being customized for different application domains. In fact, this idea was pursued in academic research already decades ago [Bet95] and has recently received increased attention from industry as well [Lan21, Sch22a]. This positions it as a viable way to overcome the aforementioned technological issues.

Let us briefly see how customization of the reconfigurable fabric can help in this regard. We mentioned that the existence of multiplexers between prefabricated wires only superficially appears to be the largest difference between FPGA and ASIC connections. At the 7 nm node and beyond, high resistance of wires makes their delay very sensitive to their length [Che14]. Given that logic density of FPGAs is intrinsically much lower than that of ASICs—the gap even exceeds 20× in some designs [Abd14]—it should not be surprising that FPGAs struggle more than ASICs to keep the delays of their significantly longer wires in check. Traditionally, FPGAs have combated the density gap by hardening common functionality that is shared by a majority of designs as fixed-functionality hardware integrated with the reconfigurable fabric [Tri15]. The more general purpose an FPGA architecture is intended to be, the less likely it is to find functionalities that are common to a majority of designs that it needs to implement, resulting in reduced opportunities for density increase. By giving up on generality and moving toward application-domain-specific reconfigurable architectures instead, it becomes possible to reap the benefits of custom functionality hardening to a much greater extent. Apart from the immediate benefit of replacing the *Look-up Tables* (LUTs) with less general but considerably faster gates (or circuits) [Sch22a], this also produces the secondary effect of drastically increasing density and hence attenuating the wire delay scaling issue. Of course, specialization comes at a price, as it undermines the main economic principle driving FPGA adoption: The steep cost of new chip design is amortized over a large number of users that can later customize it to their own needs; the narrower this customer base becomes after specialization, the less room for cost amortization remains.

For the most part, recently published attempts at designing application-domain-specific FPGAs were limited to hardening new functionalities while inheriting the programmable interconnect architecture from existing general-purpose FPGAs [Aro21, Lan21, Sch22a], despite the obvious potential for also customizing the programmable interconnect architecture itself [Lan21]. Given that interconnect is the main performance bottleneck, this too suggests that designing good programmable interconnect architectures is much more difficult than it may seem at first—arguably even more so than designing new hardened fixed-functionality blocks.

One of the reasons why this could be the case is that functionality hardening is very similar to standard ASIC design and can largely rely on the same design methodologies and automated design flows [Yaz19]. On the other hand, programmable interconnect architecture design is a problem that is entirely specific to reconfigurable architectures. Since for most of the past two decades this problem has not been all that relevant, as existing solutions could be effectively scaled with only minor modifications, comparable design automation tools that could help solving it have not been developed.

As we have already mentioned, hardware reconfigurablity is experiencing a surge in popularity, from edge devices offering a few thousand LUTs for less than half a dollar [Ren21], through reconfigurable portions of traditionally fixed-functionality SoCs [Ach21], to high-end datacenter devices [AMD23]. This drives a need for designing efficient programmable interconnect architectures optimized for many different objectives and under many different constraints. With more and more companies and individuals becoming involved in designing reconfigurable hardware, it is possible that this will speed up innovation in programmable interconnect architecture as well, which is required for continued performance improvements at the latest technology nodes. After all, crowdsourcing human ingenuity has led to impressive success in other domains [Eib12]—far beyond what a small group of leading experts was capable of achieving alone.

Before raising too high the hopes that democratizing the effort of programmable interconnect design is going to solve all the challenges, it is worth mentioning that the *Versatile Place and Route* (VPR) project [Bet97] already democratized access to very flexible FPGA architecture evaluation tools, similar to those used at Altera/Intel [Lew03], more than 25 years ago. Arguably, VPR is less amusing for non-FPGA enthusiasts than *Foldit* [Cen23] is for nonbiologists and nonchemists. It certainly also has a much steeper learning curve and a much smaller user base. Hence, it is not possible to use it as a good measure of the effect that turning programmable interconnect design problems into engaging puzzle games with large following in the general public could have. Nevertheless, it provides the best available measure of what effect could be expected when more professionals are involved in the effort, and this does not seem all that promising: A vast majority of academic FPGA architecture work still largely relies on programmable interconnect design decisions that were reached toward the end of the previous century. Similar trends can be observed from the recent rise in access to highly flexible ready-for-fabrication FPGA synthesis tools [Tan19a, Koc21]. Although these tools claim to allow just about any programmable interconnect architecture to be implemented, the actual attempts at their use again mostly either rely on decades-old academic conclusions [Tan19], or architectures used in commercial FPGAs that are no longer under patent protection after 20 years of exploitation [Koc21].

Perhaps the problem of programmable interconnect design is too complex for humans to reason about it effectively. Or maybe it is just too uninteresting for people to want to be bothered by it when they can instead design blocks at a higher level of abstraction as computer architects do. Whatever be the reason, we argue that leaving programmable interconnect design to a manual effort is very likely to leave a lot of optimization potential untapped in many different applications of reconfigurable hardware. Moreover, spending human ingenuity, time, and effort on solving a problem that lags so much in terms of design automation when compared to other problems occurring in reconfigurable architecture design is simply unnecessarily wasteful. Hence, we task ourselves with developing missing algorithms for automating several fundamental problems of programmable interconnect design. While we cannot actually judge whether these problems are difficult for humans to reason about, it is well known that they are computationally difficult. Additionally, their solution space is very large, in any instance of practical relevance, and comparing any two solutions is also computationally very expensive. Hence, brute-force approaches, or those based on various meta-heuristics that have been successfully applied to other aspects of FPGA architecture design, are ineffective there. Fortunately, programmable interconnect possesses several specificities that can be leveraged to efficiently navigate these large design spaces, which is precisely what we do in this book.

1.1 Book Organization

Most of the content of this book is based on the author's PhD thesis, entitled "Automating the Design of Programmable Interconnect for Reconfigurable Architectures," which he defended at EPFL in 2023. We have already presented the general outline of the book's main parts and their intended readership in the preface. Here we give a more detailed presentation of the contents of each individual chapter.

In Chap. 2, we make a quick tour of historical evolution of FPGAs, introducing all of the fundamental principles of reconfigurable architectures along the way. This provides a deeper understanding of why FPGAs look as they do today, as well as how the fundamental technological changes occurring at the moment prevent FPGAs from simply continuing on their previous evolution trajectory.

In Chap. 3, we take a brief step back from the constraints of practical FPGAs and precisely define the problem of programmable interconnect architecture design in its most abstract sense. This will serve as a useful guide for understanding the nature and scope of the methods that will be presented in Part II of the book. We also provide more background information specific to *Island-Style* FPGAs [Bet99]—a class of FPGA architectures that encompasses the vast majority of modern commercial devices and that will thus also be the main focus of this book. Finally, we explain the details of a typical FPGA CAD flow.

To be able to design a programmable interconnect architecture that is appropriate for a given fabrication technology, it is first necessary to understand how this technology impacts the performance of different architectures. In Chap. 4, we introduce a new physical modeling framework that we developed to capture the effects of resistance scaling at advanced FinFET nodes going all the way down to 3 nm. The framework relies on state-of-the-art resistance models combined with realistic floorplans, both raised to a level of abstraction that is appropriate for rapid architecture modeling and evaluation. We use the framework to revisit one of the most fundamental, yet simplest architectural exploration experiments that have been performed in the previous decades: finding the optimal logic cluster size. The results demonstrate how changes in technological parameters can influence even such fundamental decisions and also help explain some of the modifications of the programmable interconnect architecture introduced in the Intel Agilex FPGAs.

Experiments conducted in Chap. 4 confirm the prior observation of Lin et al. that increased wire resistance in scaled technologies reduces the maximum feasible length of channel wires [Lin10]. This in turn dramatically reduces the search space for the problem of optimizing channel segmentation, even making exhaustive exploration possible. On the other hand, rising resistance increases the need to reduce the capacitive load on channel wires exerted by routing multiplexers and shorten wires at the lowest and most resistive metal layers [Chr20]. This shifts the focus to the problem of switch-pattern design. However, due to design space explosion, that problem is significantly more difficult to tackle through the traditional black-box approach where place and route algorithms are used to evaluate individually listed solutions. In Chap. 5, we present a novel method that overcomes this fundamental scalability barrier by avoiding explicit enumeration of individual solutions. Instead, it relies on implicitly representing the entire design space in the *Routing-Resource Graph*—a data structure used to represent the programmable interconnect architecture to the *PathFinder* routing algorithm [McM95], which is used in a majority of modern FPGA routers in academia and industry alike [Kap12]. By slightly modifying its cost function so that the negotiation principle normally used to drive signals away from congested regions of the FPGA can be applied in reverse to make signals converge on a common set of switches, we were able to essentially let PathFinder itself design the switch-pattern, thus entirely avoiding enumeration of individual solutions.

Commercial FPGAs have been able to leverage high levels of regularity of the architecture to enable highly optimized full-custom layouts with limited engineering effort. The ability to do this has been identified as key for preventing the gap between FPGAs and ASICs from deepening further, which would have been the case if FPGA vendors were to employ a standard cell approach instead [Kim17]. In Chap. 6, we extend the switch-pattern exploration technique presented in Chap. 5 so that it only produces solutions that satisfy any arbitrary definition of regularity that can be encoded as a set of constraints of an *Integer Linear Program* (ILP). We then analyze the impact of restricting solutions to those respecting several forms of regularity that are known to appear in commercial architectures and demonstrate that this does not significantly reduce the architecture's performance. With this extension, we obtain a complete algorithm for fully automated design of switch-patterns that are suitable for immediate fabrication, thus solving a fundamental problem in programmable interconnect design that has long been open. Nevertheless, the current implementation of the algorithm suffers from certain scalability limitations that still make it impractical for designing FPGAs with more than a few thousands of LUTs. Chapter 5 also includes a detailed analysis of the origins of these limitations and suggests potential ways to overcome them.

As we have mentioned, the important differences between reconfigurable connections of FPGAs and fixed connections of ASICs are no longer limited to the existence of routing multiplexers: The much lower logic density of FPGAs makes their wires significantly longer and slower. Nevertheless, this does not mean that multiplexers do not further significantly slow down a signal that has to traverse them. Already since the very first FPGA—the Xilinx XC2064 [Xil93]—FPGAs employed a small number of fixed connections that could directly connect neighboring LUTs, without passing through any multiplexers of the programmable interconnect. In scaled technologies, such wires have an additional benefit in that, unlike the wires of the programmable interconnect, they are not loaded by any routing multiplexers used for steering signals to other wires, which makes them much faster. Putting aside the low logic density that still significantly elongates them, these direct wires are as close as one could get to an ASIC connection in a reconfigurable setting. Nevertheless, for the fixed wires to be really useful, FPGA CAD algorithms must be able to map user circuits in such a way that their critical paths are implemented using them. Most commercial FPGAs have included only very simple direct connection patterns such as cascades, which limited their impact. In Chap. 7, we present an efficient algorithm for automated design of direct connection patterns with no restrictions other than the maximum connection span and preserved tilability that is required for island-style FPGAs. By adopting a "minimum harm" approach, in which direct connections are always an addition to the flexibility of programmable interconnect, never taking away from it, we were able to relate the problem to that of *Maximum coverage* and develop a very fast greedy algorithm for solving it.

Maximizing the utility of direct connections between LUTs requires appropriately aligning the endpoint LUTs of a connection of the circuit being implemented with the endpoint LUTs of a direct connection of the FPGA. In Chap. 8, we present a custom ILP-based detailed placement algorithm to solve this problem. It leverages the sparsity of optimized direct connection patterns to obtain an efficient ILP encoding. The new algorithm more than doubles the effect of direct connections on critical path delay reduction. Chapter 8 also concludes the part of the book that presents the original scientific contributions of the author.

Nevertheless, the present book would not be complete without the mention of hardened *Networks on Chip* (NoCs) and interconnect design for multi-die devices. These two important topics are briefly introduced in Chaps. 9 and 10, respectively. Lastly, in Chap. 11, we draw the final conclusions and comment on future work.

References

[Abd14] Abdelfattah, M. S., & Betz, V. (2014). The case for embedded networks on chip on field-programmable gate arrays. *IEEE Micro, 34*(1), 80–89. ISSN: 1937-4143.

[Ach21] Achronix Semiconductor Corporation (2021). *10 Millionth Achronix Speed-core eFPGA IP Core Shipped.* https://www.achronix.com/press-releases/10-millionth-achronix-speedcore-efpga-ip-core-shipped. Accessed on 06 April, 2023.

[AMD23] AMD (2023). *Adaptable Accelerator Cards for Data Center Workloads.* https://www.xilinx.com/products/boards-and-kits/alveo.html. Accessed on 06 April, 2023.

[Ans23] Ansell, T. (2023). Google investment in open source custom hardware development including no-cost shuttle program. In *Proceedings of the 2023 International Symposium on Physical Design*. Event, V.

[Aro21] Arora, A., Mehta, S., Betz, V., & John, L. K. (2021). Tensor slices to the rescue: Supercharging ML acceleration on FPGAs. In *The 2021 ACM/SIGDA International Symposium on Field-Programmable Gate Arrays*. Virtual Event (pp. 23–33).

[Bau20] Bauer, H., Burkacky, O., Kenevan, P., Lingemann, S., Pototzky, K., & Wiseman, B. (2020). *Semiconductor design and manufacturing: Achieving leading-edge capabilities.* Technical Report. McKinsey.

[Bet95] Betz, V., & Rose, J. (1995). Using architectural "Families" to increase FPGA speed and density. In *Proceedings of the 1995 ACM Third International Symposium on Field-Programmable Gate Arrays*. Monterey, CA (pp. 10–16).

[Bet97] Betz, V., & Rose, J. (1997). VPR: A new packing, placement and routing tool for FPGA research. In *Proceedings of the 7th International Workshop on Field-Programmable Logic and Applications*. London (pp. 213–22).

[Bet99] Betz, V., Rose, J., & Marquardt, A. (1999). *Architecture and CAD for Deep-Submicron FPGAs*. Kluwer Academic Publishers. ISBN: 0-7923-8460-1.

[Cen23] Center for Game Science (University of Washington), Institute for Protein Design (University of Washington), Cooper Lab (Northeastern University), Dartmouth Khatib Lab (University of Massachusetts, Davis), Siegel Lab (University of CA, Meiler Lab (Vanderbilt University), et al. (2023). *Foldit*. https://fold.it/. Accessed on 07 April, 2023.

[Che14] Chen, J. H.-C., Standaert, T., Alptekin, E., Spooner, T., & Paruchuri, V. (2014). Interconnect performance and scaling strategy at 7 nm node. In *IEEE International Interconnect Technology Conference*. San Jose, CA (pp. 93–96).

[Chr20] Chromczak, J., Wheeler, M., Chiasson, C., How, D., Langhammer, M., Vanderhoek, T., et al. (2020). Architectural enhancements in Intel®Agilex™FPGAs. In *Proceedings of the 2020 ACM/SIGDA International Symposium on Field-Programmable Gate Arrays*. Monterey, CA (pp. 140–149).

[Cut21] Cutress, I. (2021). *Intel's Process Roadmap to 2025: with 4nm, 3nm, 20A and 18A?!* AnandTech. https://www.anandtech.com/show/16823/intel-accelerated-offensive-process-roadmap-updates-to-10nm-7nm-4nm-3nm-20a-18a-packaging-foundry-emib-foveros. Accessed on 29 March, 2023.

[Eib12] Eiben, C. B., Siegel, J. B., Bale, J. B., Cooper, S., Khatib, F., Shen, B. W., et al. (2012). Increased diels-alderase activity through backbone remodeling guided by Foldit players. *Nature Biotechnology, 30*, 190–192. ISSN: 1546-1696.

[Gai19] Gaide, B., Gaitonde, D., Ravishankar, C., & Bauer, T. (2019). Xilinx adaptive compute acceleration platform: Versal architecture. In *Proceedings of the 2019 ACM/SIGDA International Symposium on Field-Programmable Gate Arrays*. Seaside, CA (pp. 84–93).

[Hen19] Hennessy, J. L., & Patterson, D. A. (2019). A new golden age for computer architecture. *Communications of the ACM, 62*(2), 48–60. ISSN: 0001-0782.

[Jar22] Jaros, A. (2022). *The future of embedded FPGAs — eFPGA: The proof is in the tape out*. Circuit Cellar. https://circuitcellar.com/insights/the-future-of-embedded-fpgas-efpga-the-proof-is-in-the-tape-out/. Accessed on 30 March, 2023.

[Jou21] Jouppi, N. P., Yoon, D. H., Ashcraft, M., Gottscho, M., Jablin, T. B., Kurian, G., et al. (2021). Ten lessons from three generations shaped Google's TPUv4i. In *Proceedings of the 48th Annual International Symposium on Computer Architecture*. Virtual Event (pp. 1–14).

[Kap12] Kaptanoglu, S. (2012). Pathfinder: A negotiation-based performance-driven router for FPGAs. In *FPGA and reconfigurable computing hall-of-fame endorsement*.

[Kim17] Kim, J. H., & Anderson, J. H. (2017). Synthesizable standard cell FPGA fabrics targetable by the verilog-to-routing CAD flow. *ACM Transactions on Reconfigurable Technology and Systems (TRETS), 10*(2), 1–23. ISSN: 1936-7406.

[Koc21] Koch, D., Dao, N., Healy, B., Yu, J., & Attwood, A. (2021). FABulous: An embedded FPGA framework. In *The 2021 ACM/SIGDA International Symposium on Field-Programmable Gate Arrays*. Virtual Event (pp. 45–56).

[Lan21] Langhammer, M., Nurvitadhi, E., Pasca, B., & Gribok, S. (2021). Stratix 10 NX architecture and applications. In *The 2021 ACM/SIGDA International Symposium on Field-Programmable Gate Arrays*. Virtual Event (pp. 57–67).

[Lei20] Leiserson, C. E., Thompson, N. C., Emer, J. S., Kuszmaul, B. C., Lampson, B. W., Sanchez, D., et al. (2020). There's plenty of room at the top: What will drive computer performance after Moore's law? *Science, 368*(6495), 1–7.

[Lew03] Lewis, D., Betz, V., Jefferson, D., Lee, A., Lane, C., Leventis, P., et al. (2003). The stratix routing and logic architecture. In *Proceedings of the 2003 ACM/SIGDA Eleventh International Symposium on Field Programmable Gate Arrays*. Monterey, Calif (pp. 12–20).

[Lin10] Lin, M., Wawrzynek, J., & Gamal, A. E. (2010). Exploring FPGA routing architecture stochastically. *IEEE Transactions on Computer-Aided Design of Integrated Circuits and Systems, 29*(10), 1509–1522.

[McM95] McMurchie, L., & Ebeling, C. (1995). PathFinder: A negotiation-based performance-driven router for FPGAs. In *Proceedings of the 1995 ACM Third International Symposium on Field-Programmable Gate Arrays*. Monterey, CA (pp. 111–117).

[Ren21] Renesas Electronics Corporation (2021). *Renesas Enters FPGA Market with the First Ultra-Low-Power, Low-Cost Family Addressing Low-Density, High-Volume Applications*. https://www.renesas.com/us/en/about/press-room/renesas-enters-fpga-market-first-ultra-low-power-low-cost-family-addressing-low-density-high-volume. Accessed on 06 April 2023.

[Sch22a] Schmit, H., & Denton, M. (2022). Multi-input serial adders for FPGA-like computational fabric. In *Proceedings of the 2022 ACM/SIGDA International Symposium on Field-Programmable Gate Arrays*. Virtual Event (pp. 35–41).

[Tan19] Tang, X., Giacomin, E., Alacchi, A., & Gaillardon, P. (2019). A study on switch block patterns for tileable FPGA routing architectures. In *2019 International Conference on Field-Programmable Technology (ICFPT)*. Tianjin (pp. 247–250).

[Tan19a] Tang, X., Giacomin, E., Alacchi, A., Chau-Viere, B., & Gaillardon, P.-E. (2019). OpenFPGA: An opensource framework enabling rapid prototyping of customizable FPGAs. In *Proceedings of the 29th International Conference on Field Programmable Logic and Applications (FPL)*. Barcelona (pp. 367–374).

[Tok16] Tőkei, Z., Ciofi, I., Roussel, P., Debacker, P., Raghavan, P., van der Veen, M. H., et al. (2016). On-chip interconnect trends, challenges and solutions: How to keep RC and reliability under control. In *2016 IEEE Symposium on VLSI Technology*. Honolulu, Hawaii (pp. 1–2).

[Tok22] Tőkei, Z. (2022). Logic scaling options for the next 10 years: From FinFet to CFET, from dual damascene to semi damascene. In *Proceedings of the 2022 ACM/SIGDA International Symposium on Field-Programmable Gate Arrays*. Virtual Event.

[Tri94] Trimberger, S. (1994). *Field-programmable gate array technology*. Springer Science+Business Media.

[Tri15] Trimberger, S. M. (2015). Three ages of FPGAs: A retrospective on the first thirty years of FPGA technology. *Proceedings of the IEEE, 103*(3), 318–331. ISSN: 1558-2256.

[TSM23] TSMC (2023). *Logic Technology*. https://www.tsmc.com/english/dedicatedFoundry/technology/logic. Accessed on 29 March 2023.

[Xil93] Xilinx Inc. (1993). *XC2000 Logic Cell Array Families*. San Jose, CA.

[Yaz19] Yazdanshenas, S., & Betz, V. (2019). COFFE 2: Automatic modelling and optimization of complex and heterogeneous FPGA architectures. *TRETS, 12*(11), 3:1–27.

Part I
The What, the Why, and the How of Programmable Interconnect

This part introduces the fundamental principles of programmable interconnect design. In Chap. 2 we expose these principles by following the evolution trajectory of real-world FPGAs, from their inception to the present day. Observing the major developments in several generations of real commercial devices and discussing the driving forces behind them, let us appreciate not only how FPGAs function, but also why they look the way they do. We also analyze some real-world applications in order to illustrate the benefits of reconfigurability, despite its high inherent cost.

In Chap. 3, we take a step back and introduce the problem of programmable interconnect design in its most abstract form, freed from the intricacies of commercial device design. This allows a much clearer presentation of the fundamental principles identified in Chap. 2, but it also provides a theoretical framework for the discussions in the subsequent chapters that will eventually converge back to the practical problems of designing a commercially viable architecture. Here we also present many relevant classical theoretical results that will be useful later, and we introduce the main parts of a typical CAD flow used to map user designs onto an FPGA.

Chapter 2
Where Did the FPGAs Come from and Where Are They Headed?

Contents

Field-Programmable Gate Arrays (FPGAs) emerged in the mid-1980s, providing a way to produce complex custom hardware at one's desk. The concept of hardware reconfigurability that was thus created is truly fascinating: An integrated circuit can be customized with no manufacturing steps whatsoever; all that is needed is loading an appropriate configuration into the FPGA's memory. The key principle that made this possible was the use of stored-select multiplexers for connecting prefabricated logic blocks. However, this opened two fundamental questions of reconfigurable architecture design: (1) which logic blocks should be prefabricated and (2) how should the stored-select multiplexer network be organized.

In this chapter, we give a brief overview of the historical development of FPGAs, from their precursors to the present day. Understanding the conditions that have been influencing the approach taken by FPGA architects in answering the two fundamental questions will enable us to reason about how the profound changes in these conditions that are happening today may influence the future of reconfigurable architecture development. In particular, we shall see that for

© The Author(s), under exclusive license to Springer Nature Switzerland AG 2025

S. Nikolić, *Modern Programmable Interconnect Design*,
https://doi.org/10.1007/978-3-031-80629-2_2

a long time, fabrication technology trends favored focusing on answering the first fundamental question—which logic blocks should be prefabricated—while leaving the attempts to answer the second fundamental question—how should the network of stored-select multiplexers be organized—to relatively minor incremental improvements. Decades of prioritizing functionality hardening, which can often leverage well-established VLSI design automation methods and algorithms, at the expense of programmable interconnect architecture design, meant that comparable design automation methods have not been developed for solving this difficult problem. It is the purpose of Part II of this book to explore some avenues toward rectifying that issue, while the purpose of this chapter is to convince the reader that reconfigurable architectures have finally reached a point when these methods are truly needed.

2.1 Mass Production at a Microscale

Soon after the integrated circuit was invented in 1958 [Sai17], entire logic gates became a single component that could be placed and connected to other gates on a printed circuit board in order to form a more complex circuit [Hal96]. Advances in technology quickly brought further miniaturization, which allowed more and more gates to be integrated on a common chip. Due to electrical limitations, these gates had a small number of inputs—typically up to four [Mor71]. Moreover, gates in the same technology were composed of the same elements in the pull-up and pull-down networks: for instance, resistors in the former and transistors in the later. Hence, it was natural to expect that different circuits would look similar when metal connections are neglected. If this similarity could be pushed to the extreme so that different integrated circuits would be identical with respect to the arrangement of the basic elements on the chip, they could share all of the *front-end-of-line* (FEOL) fabrication steps, greatly reducing the manufacturing costs. By late 1960s, this idea materialized in the so-called *Mask-Programmed Gate Arrays* (MPGAs) [Tri94].

Figure 2.1 shows one such gate array: the Ferranti ULA 2000. ULA 2000 was composed of a regular grid of cells, each containing a handful of transistors and resistors. Depending on how its elements were connected together, one cell could implement a 2-input NAND gate, a 3-input NOR gate, a 2-input XOR gate, or it could be combined with neighboring cells to create more complex blocks such as flip-flops [Fer77]. Spreading the rows and columns of the array left sufficient space for wires connecting different cells to be traced without intersecting with those that are traced on top of the cells to convert them into appropriate gates. Careful prefabrication of *crossunder* metal segments enabled completion of circuits on ULA 2000 using only one additional custom metal layer [Ram80, Fer84]. Requiring only one mask in turn meant that a customized chip could be obtained within a few weeks from completion of the design, and at a cost of merely several thousands of dollars [Tri94].

Mass production of the same cell across a large chip was possible because even in a custom *Application-Specific Integrated Circuit* (ASIC), different gates were

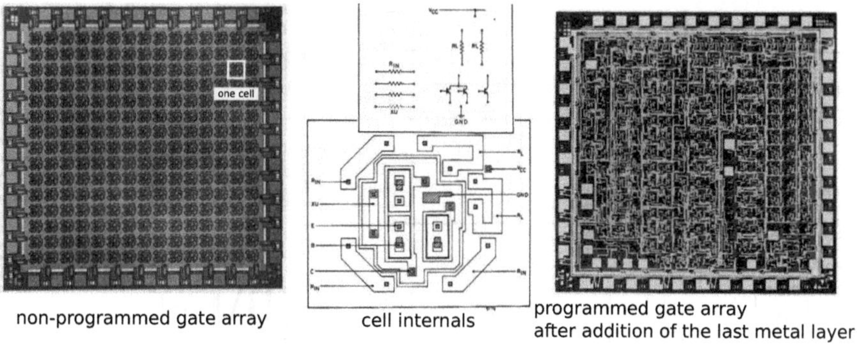

Fig. 2.1 Ferranti ULA 2000 Mask-Programmed Gate Array [Fer77]. Courtesy of Ferranti

composed of the same elements naturally occurring with comparable density across the entire chip, due to the typical gate's small input count. The ability to create any gate at any location on the MPGA that this mass production of the same generic cell created was is turnkey for limiting the gap between MPGAs and custom ASICs. If different, already specialized gates were prefabricated at different locations of the array instead, which would have introduced placement constraints when mapping a circuit onto the array, potentially drastically increasing the length of connections relative to an ASIC implementation with no such constraints. Unlike today, in 1970s this may not have been a particular concern for performance, because transistor delays dominated metal delays. However, longer connections would have called for wider routing channels, reducing the MPGA's density and increasing its cost.

2.2 Discarding the Last Mask

Removing the need for that last mask would mean that the prefabricated gate array would not have to reenter the factory for customization. This is precisely the problem that FPGAs solved. Since every Boolean function can be expressed using binary NAND operations, replacing the logic cell with a 2-input NAND gate—as some early FPGAs have done [Ros93]—would be sufficient, if suboptimal [Sin92]. The more challenging problem is how to connect the cells together to construct the required circuit, without using any fabrication steps. Hence, we first focus on how SRAM-based FPGAs [Hau07] solved this problem.

2.2.1 *Turns on a Grid*

Connecting together different cells of an MPGA amounts to tracing wires through channels between its rows and columns. Let us introduce a simplified model of

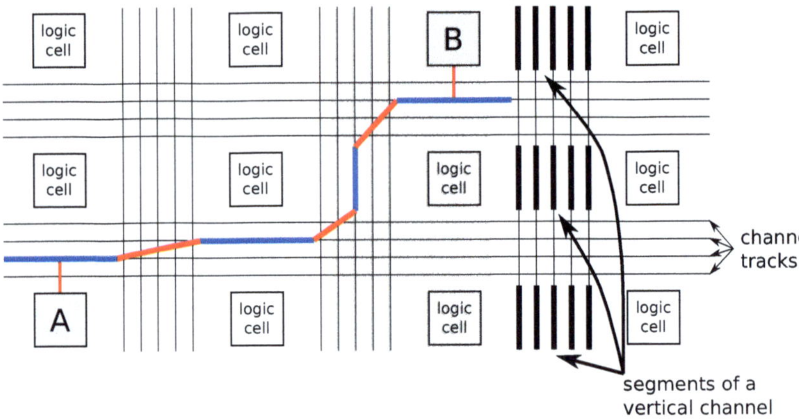

Fig. 2.2 Tracks and segments of an MPGA. Restrictions on the minimum metal pitch constrain the wires in the channels of an MPGA to *tracks*. When connections are forced to turn only at logic cell corners, routing channels are broken into straight *segments* [ElG81]. Specifying a connection between two cells then amounts to listing the consecutive track-segment pairs (blue) connected by additional metal (red)

this process, adopted from El Gamal [ElG81]: Each wire can only take a turn at a corner of a cell, but it can also continue in the same direction. Furthermore, wires are constrained to discrete *tracks*, which approximate the metal pitch constraints. This is illustrated in Fig. 2.2. Restricting wires to change tracks only at cell corners, where decisions about turning or continuing are made, splits each channel into straight *segments*. Then, connecting two cells together reduces to choosing consecutive track-segment pairs that will form the corresponding connection.

2.2.2 *Prefabricated Wires*

Spacing between rows and columns of logic cells in an MPGA is dimensioned in such a way that the routing channels can accommodate as many tracks as are expected to be required by the most demanding user circuits [ElG81]. For any particular circuit being implemented, not all tracks will be used in every segment. However, selectively fabricating wires in some track-segment pairs, according to the circuit's needs, is conceptually equivalent to fabricating wires in all track-segment pairs and leaving the unused ones hanging; this is precisely the approach taken by FPGAs, where wires cannot be selectively fabricated. For example, the very first FPGA—Xilinx XC2064—contained four prefabricated wire tracks in each horizontal channel and five tracks in each vertical channel [Xil93] (see Fig. 2.3).

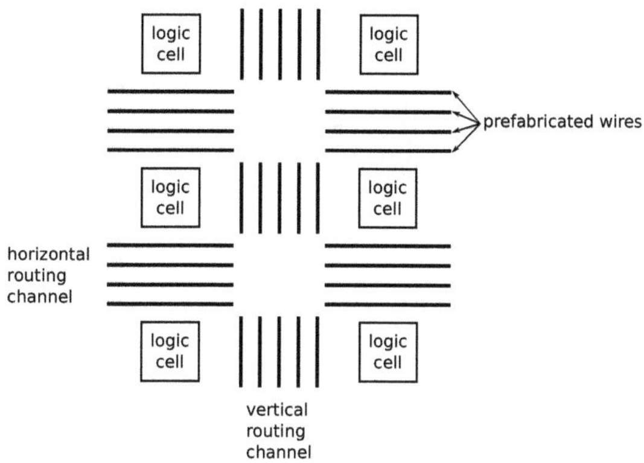

Fig. 2.3 Prefabricated wires in the routing channels of XC2064 [Xil93]. Determining how many wires to prefabricate is a similar problem to determining the spacing between rows and columns of an MPGA and is usually solved experimentally [Ros89, Tri97], although theoretical bounds have been derived [ElG81]

2.2.3 Stored-Select Multiplexers

What remains is to connect different prefabricated wires together to form connections between cells. Conceptually, when deciding which set of prefabricated wires is going to form the connection between two particular cells, we are selecting one out of a multitude of possible predecessors for each wire. In an MPGA, the possible predecessors are all wires in each segment that ends in the vicinity of the start of the segment in question; once the appropriate predecessor is selected, the connection is realized by a metal trace. Of course, FPGAs in which all resources are prefabricated do not have this possibility. Selecting one of the possible predecessors can instead be achieved dynamically by driving each prefabricated wire by a multiplexer, the data inputs of which are provided by the potential predecessors that should be selected from. Achieving the same flexibility of connectivity as is possible in an MPGA would require that the multiplexer at the start of each prefabricated wire takes inputs from all wires that end in its vicinity. This is not very practical, as it would make the multiplexer too large and slow, so typically, only a subset of the potential predecessors is connected to each multiplexer. Connectivity of XC2064 is shown in Fig. 2.4.

Multiplexers in an FPGA substitute solid metal in an MPGA. Hence, once the predecessor of each prefabricated wire has been determined for the particular customization of the FPGA, this selection should be made permanent until the next customization is applied. For this reason, the select inputs are provided by SRAM cells, in which the predecessor selection is stored, as shown in Fig. 2.5. In practice, multiplexers are implemented using pass transistors [Lew12] or transmission gates [Chi13, Chr20]. Since both of these components are bidirectional, if the multiplexer's output is not buffered, the entire multiplexer becomes bidirectional. Use of bidirectional, nonbuffered switches was common in early architectures, such as XC2064, but has since been abandoned [Lew03, Lem04].

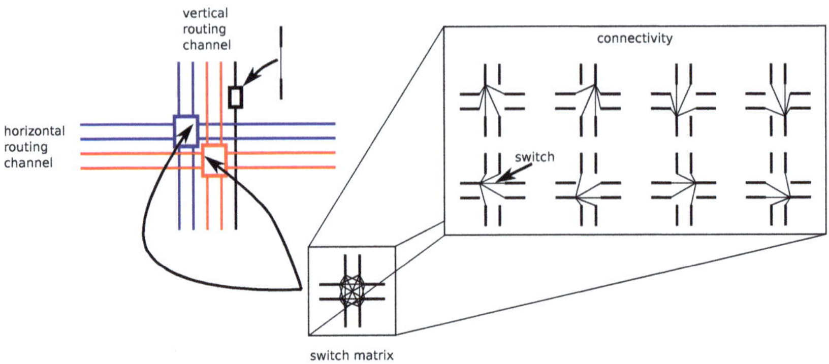

Fig. 2.4 Programmable switches of XC2064 [Xil93]. Forming connections using prefabricated wires amounts to selecting the appropriate predecessor of each wire. This can be done post-fabrication by driving each wire with a multiplexer. Storing select inputs of each multiplexer in SRAM cells retains the selection until the next customization of the chip

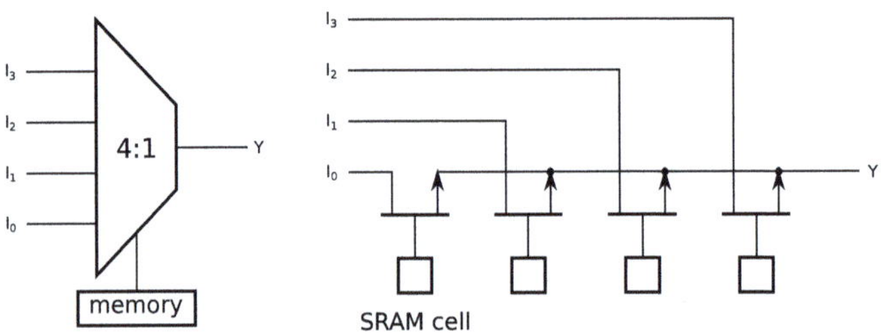

Fig. 2.5 Stored-select multiplexer. The basic building block of programmable interconnect is a multiplexer with select inputs stored in a memory. It is usually implemented using pass transistors or transmission gates, with gate terminals driven by SRAM cells [Lew12, Chr20]. If the output is not buffered, the switches that constitute the multiplexer are bidirectional. SRAM cell count can be reduced by breaking a larger multiplexer into a two-level tree, where multiplexers in the first level share SRAM cells [Lew16]. See Fig. 4.7 for an example of a two-level 4:1 multiplexer

2.2.4 What About the Logic Cells?

As we have already mentioned, a key feature of ULA 2000 was that all cells were identical and could be transformed into different gates. This reduced placement constraints compared to the situation in which different cells contain different prefabricated gates. In particular, each cell of ULA 2000 could implement one NAND-2, one NOR-3, or one XOR-2 [Fer77]. Other gates could be constructed by combining neighboring cells. Specialization of each generic cell was obtained by appropriately connecting the prefabricated elements. How can we achieve the same without using any additional masks? One idea could be to connect the prefabricated elements through programmable switches (stored-select multiplexers), as was done with intercell routing. Of course, the overhead of adding a pass transistor with

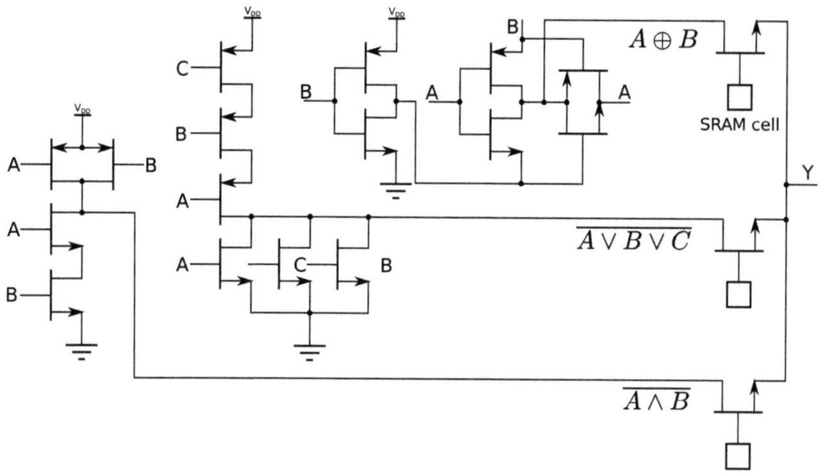

Fig. 2.6 Compound gate. The figure shows a 3-input cell composed of three gates that a single cell of ULA 2000 can implement. A stored-select multiplexer is used to select which gate is implemented by each cell, post-fabrication. Assuming that each SRAM cell is composed of five transistors [Koc13], the compound gate requires 34 transistors (not including buffers) and can implement three of the 256 possible 3-variable Boolean functions

the gate driven by an SRAM cell in order to form a programmable connection between two other prefabricated transistors is too large, although this idea has been pursued in antifuse-based one-time programmable FPGAs [Mar92]. Hence, we need to coarsen the granularity of components connected through programmable interconnect to better amortize its overhead. This is a general principle, as the coarser the granularity is, the less costly the stored-select multiplexers become in comparison with the logic blocks. Of course, there is a trade-off since blocks that are too large are not easy to efficiently utilize. Inefficient utilization does not only result in wasting the silicon area required for implementing the logic block itself, but also that of the routing multiplexers that provide inputs to it.

One idea for coarsening the granularity to approximately match the flexibility of a cell that exists in ULA 2000 could be to simply lay out in every cell the NAND-2, the NOR-3, and the XOR-2 and then use a stored-select multiplexer to decide post-fabrication which of the three gates will be implemented by the particular cell. This is illustrated in Fig. 2.6. The inefficiency of using the compound block is immediately clear: Three gates are fabricated, but only one of them can actually be used at a time. Moreover, transistors from neighboring cells of ULA 2000 can be combined to create much more efficient implementations of other 3-variable Boolean functions than what can be obtained by expressing the function in terms of the three basic gates prefabricated in the compound block of Fig. 2.6.

Fortunately, every Boolean function can be implemented using a tree of 2:1 multiplexers, where the select inputs are provided by the function's variables in both polarities, and the data inputs reduce to constants corresponding to the

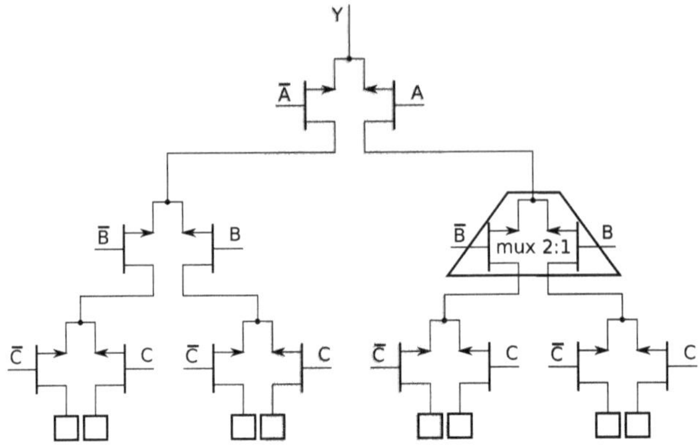

Fig. 2.7 Schematic of a 3-input *Look-up Table* (LUT). Every k-variable Boolean function can be implemented using a k-level full binary tree of 2:1 multiplexers [Sha49]. SRAM cells can be used for storing the truth table entries, which provide the data inputs to the multiplexer tree. Assuming that each SRAM again takes five transistors to implement, a 3-LUT can be realized using 60 transistors, not counting buffers. This is almost twice as many as for the compound gate of Fig. 2.6, but a 3-LUT can implement all 256 3-variable Boolean functions

appropriate entries of the function's truth table [Sha49]. The multiplexer tree is identical for all functions of the given number of variables (although minimization is possible [Bry86]), and the only difference between implementations of different functions are the constants driving the data inputs. Hence, this multiplexer tree can be used to form the basis of the reconfigurable cell. The only missing part is the ability to select either a 0 or a 1 at each data input. This is readily provided by SRAM cells, resulting in the so-called *Look-up Table* (LUT)—a generic programmable gate that forms the basis of most current FPGAs [Bou21].

Schematic of a 3-LUT is shown in Fig. 2.7. It requires almost twice the number of transistors as the compound cell of Fig. 2.6, but it can implement any of the 256 3-variable Boolean functions, as opposed to merely three. Doubling the 3-LUT and extending the multiplexer tree by one level create a 4-LUT. As we have mentioned, due to electrical constraints, gates rarely have more than four inputs. Hence, manually implementing any typical existing circuit on an FPGA based on a 4-LUT could be trivially achieved by swapping each gate for an LUT. This was likely very important for early adoption of FPGAs, before dedicated *technology mapping* algorithms were developed for them. Although such algorithms [Con94] enable much better implementations than this naive one, it was later experimentally determined that 4-LUTs in fact provide the best trade-off between efficient utilization of the logic cells and the reduction of the required programmable routing area [Ros89, Ahm00].

It is worth noting that the compound block of Fig. 2.6 does solve one important issue, however: Shorting the inputs of the three prefabricated gates makes them reuse the same stored-select multiplexers to bring the signals in from the routing channels. Hence, the aforementioned problem of silicon area taken by the stored-

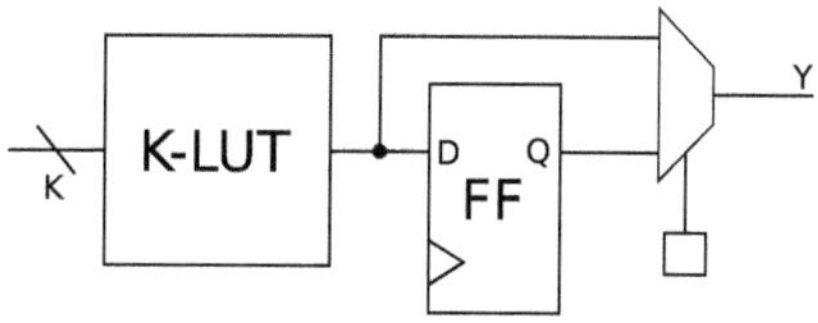

Fig. 2.8 *Basic Logic Element* (BLE) [Bet99]. Most FPGA architectures pair each LUT with a bypassable flip-flop (FF). Some newer commercial architectures allow the flip-flop to have an independent input and output, to better support heavily pipelined circuits [Lew13, Cha15]. In that case, a signal that has to access a flip-flop can do so without wasting its driving LUT

select multiplexers being wasted when logic blocks are underutilized will not occur. This principle of several blocks functioning as a *shadow* of one another has been demonstrated to bring great benefits when the multiplexer area dominates that of the logic circuitry itself [Jam06, Par13].

Most FPGAs, including the XC2064, also include bypassable hardened flip-flops, as very early on, it was determined that implementing flip-flops using LUTs connected through programmable routing is too costly [Ros89]. The simplest way to pair an LUT and a flip-flop that is often used in academic work [Luu14] is shown in Fig. 2.8.

2.3 Price of Removing the Last Mask

Removing the last mask of course comes at a price. LUTs are slower and larger than mask-specialized gates, and routing multiplexers considerably slow down the interconnect, already negatively impacted by the increased length of wires resulting from lower logic density. In the early 1990s, FPGAs were 2–3 times slower than MPGAs [Tri94]. The gap with ASICs has been even larger. Kuon and Rose identified that an FPGA implementation of a given circuit is on average 3.2× slower than that in standard cells [Kuo06]. Hennessy and Patterson state a 10× performance advantage of ASICs vs. FPGAs for modern high-performance designs [Hen19].

A logical question arises: Who would want to pay such a high penalty for hardware reconfigurability? Xuantie-910 CPU developed by Alibaba [Che20] provides one good illustration. The 64-bit CPU based on a custom-extended RISC-V ISA was capable of running at 200 MHz on a 16 nm Xilinx VU9P UltraScale+ FPGA, while an ASIC implementation on a 12 nm TSMC node[1] was capable of reaching clock frequencies of 2–2.5 GHz, resulting in a 10–12.5× performance gap, in line with the claims of Hennessy and Patterson. Yet, even the FPGA implementation was able to achieve a 20% higher per-core performance than a 14 nm Intel Xeon Platinum 8163 CPU on a particular set of relevant tasks [Che20]. All customers who cannot afford

[1] TSMC's 12 nm is a variant of the 16 nm process [TSM23].

to design and produce an ASIC could benefit from the 20% increase in performance offered by the FPGA implementation of the custom ISA extension. On the other hand, those who can afford a custom ASIC can still benefit from early deployment of the new custom hardware, while waiting for ASIC design and production, which can even take years in advanced technologies [Ans23].

For companies of the scale of Alibaba, such early deployment likely makes a very significant difference in profits. Hence, FPGAs provide the first stage in a two-stage custom hardware deployment pipeline. With cost of ASIC development reaching unprecedented heights in the most recent technologies [Bau20], for an increasing number of users, moving to the second stage is no longer even feasible and FPGAs remain their only choice. Similarly, in the fast-evolving fields such as machine learning, algorithms themselves can improve so much by the time a highly specialized ASIC is designed, produced, and deployed, that even if its development costs were affordable, it could well be rendered obsolete before it even appears [Ans23, Put23].

It is important to note that although the 20% performance increase is already considerable, benefits of FPGA-based custom hardware can be much higher, when the level of customization exceeds custom ISA extensions [Bec17]. Intuitively, FPGA and ASIC implementations of the same circuit can be compared analogously to a sculpture composed of Lego bricks and that cast in bronze (Fig. 2.9): A bronze sculpture of a human sitting in a certain pose will always be closer to the ideal than a sculpture of a human sitting in that same pose made out of Lego bricks. However, the Lego-brick sculpture will be much closer to the ideal than an equestrian sculpture cast in bronze. In general, provided that one has no access to a foundry that could cast their bronze sculpture, or time to wait for the mold to be made and casting completed, the more distant their design is from all of the bronze sculptures that can

(a) (b) (c)

Fig. 2.9 Sculpture analogy. The *Thinker* made of Lego bricks ((**b**); sculpture by Nathan Sawaya, photo courtesy of Simone Ramella) resembles less a person sitting and thinking than Rodin's bronze original ((**a**); courtesy of the National Gallery of Art, Washington). However, it is much closer to it than Meissonier's *Horseman in a Storm*, also in bronze ((**c**); courtesy of the National Gallery of Art, Washington). Much like Lego bricks, FPGAs cannot compete with ASICs that implement the same circuit, but they offer an easy way to implement a different circuit, potentially much closer to the requirements than any existing, or feasibly obtainable ASIC (due to budget or time-to-market constraints). (**a**) *Thinker* (bronze). (**b**) *Thinker* (Lego). (**c**) *Horseman in a Storm* (bronze)

be readily bought, the more advantageous having the Lego bricks will be. The same holds for FPGAs where LUTs and multiplexers produced in large quantities can be thought of as Lego bricks that are connected together as needed.

Further explanation of the benefits of reconfigurability goes beyond the scope of this book. We recommend that the interested readers refer to Putnam [Put23] and Alonso [Alo16] for a comprehensive overview of all the trade-offs between hardware accelerators—both implemented as ASICs and on top of FPGAs—and pure software implementations running on standard hardware. These discussions provide valuable insight into the difficulties of accelerating entire workloads in a complete system, where Amdahl's law can never be neglected, and where financial viability, maintainability, and manpower limitations have to be taken into account as well.

2.4 Race for the Latest Technology

As we have just mentioned, for many FPGA users, producing an ASIC is not an option. Nevertheless, there are users who could afford ASIC development but who may still opt for an FPGA implementation instead. Reduced development time, easier testing, possibility to apply post-fabrication bug fixes and design upgrades (as no fabrication exists as such) are the commonly listed inherent benefits of hardware reconfigurability that have no counterpart in ASICs [Tri94]. Yet, there is another appeal that FPGAs have traditionally been able to offer. Assume that a company can afford to design and produce an ASIC in technology X, whereas the latest available FPGA that they can purchase is produced in technology Y. If Y is significantly newer than X, the aforementioned performance gap between FPGA and ASIC implementations could be substantially reduced. This was especially true in the era of Dennard scaling in which FPGAs emerged, when clock frequency was increasing at a rate of about $1.7\times$ every 2 years [Lei20]. This meant that FPGAs could close a $4\times$ performance gap by being only three nodes ahead of the ASIC that the customer could afford. By spreading the cost of moving one and the same chip design to the next technology node over very large volumes sold to a large number of customers, FPGA vendors were traditionally able to remain at the forefront of technology and even help the foundries push it ahead [Tri15, Sta95]. While this was essential for enabling FPGAs to recover some of the inherent penalty of hardware reconfigurability, it also created an interesting situation for FPGA vendors themselves. In his interview at the Computer History Museum, Stephen Trimberger recalls how designing a new FPGA looked like at the time when Moore's law was in full swing [Com17]:

And so, you know, it was interesting because we did not have any time. A lot of decisions that we would like to have data for, my rule was... So here's the thought experiment: Assume Moore's Law is $2\times$ in 24 months. Call it 25 months. Call it 20. Okay, well, that is like four or five percent per month. If you work at a decision that is going to make less than a four or five percent difference, and it takes you more *than a month, it does not matter what your answer is. If you have a decision that is going to make one percent difference, if it is going to take you more than a week, it does not matter. And so we had a lot of these things that came back and said: "Gee this, you know, this sounds like a five percent thing. Can we get that done?" And so I squeeze it all down. Said, "Look, we can come up with an answer for that but really, the right thing to do is pick one. So we're going to pick one. We'll pick this one because it's power, or simple or whatever it was." And boom! Nothing clears up decision-making like realizing you do not need any of the data. <laughter> And yeah, oh yeah, we like to run all these test cases and so on and so forth. How long will that take? Four months. Is it going to make 25 percent difference? No. So we just flew through that, and we actually made a lot of really good decisions because we had a lot of people with expertise at that point, which we did not have five years earlier because there just was not that much expertise in the FPGA business.*

This story illustrates some very important points. First, it explains why for a long time there were relatively few and minor changes in programmable interconnect, as we will see shortly—anything but an incremental change of the previous architecture would have called for a significant amount of time, over which the potential benefits would have become available through simple scaling of the previous design. Hence, the most profitable decision that FPGA vendors could make in those days was to ensure that transition to the next technology node is never late. As Fig. 2.10 shows, for a long time, they were very successful at that. The second important point is that architectural development heavily relied on human intuition and accumulated expertise. As we shall also see shortly, in today's less predictable technologies, these are much more difficult to develop and retain.

2.5 What Happens When the Road Gets Bumpy?

When the benefits of moving to a new technology node can no longer be reaped by simple scaling of the previous design with some minor modifications, all of a sudden, all those architectural changes that promise an otherwise forgettable 4% performance increase get a different appeal. Naturally, time has to be afforded for coming up with promising modifications and testing them, which is likely to lead to a delay in an FPGA transitioning to the latest technology node. As explained by Dr. Trimberger, the process of proposing promising modifications can be greatly sped up by experience and intuition of the people involved in designing the next FPGA architecture. The problem arises, however, when technology changes so drastically that intuition developed through prior experience becomes largely inapplicable. Since the foundries themselves are increasingly struggling to keep performance and density increasing from node to node, they are relying on various one-time *scaling boosters* and *Design Technology Co-optimization* (DCTO) to an extent that is far greater than before [Tok22, Mor23]. This deepens the differences between consecutive technologies, making it difficult to develop intuition about which architectural modifications would pay off. Perhaps that explains why, for the first time in a number of decades, state-of-the-art FPGAs are several technology nodes behind the foundries' latest offering.

2.6 A New Age for FPGAs

Looking at the timeline of Fig. 2.10 may trigger a conclusion that the days of hardware reconfigurability are numbered. Not at all so! When FPGAs first gained popularity toward the end of the previous century, producing an ASIC was more than an order of magnitude cheaper than it is today [Bau20]. MPGAs also provided a 2–3× faster alternative, for merely a few thousands of dollars and with only a few weeks of delay. Yet, even then many customers needing lower production volumes chose to use FPGAs. Not only was custom hardware much more affordable then than it is today, but it was also much less needed: For most applications, it is suffice to wait for the next generation of general-purpose processors fabricated in the next technology, to obtain a drastic improvement in performance. Today, this possibility is long gone [Tho21], and hardware customization is considered one of the predominant avenues toward continued performance improvement [Hen19, Lei20]. Given the aforementioned prohibitive rise in cost of designing a new ASIC, reaching half a billion US dollars in latest technologies [Bau20], most often, FPGAs provide the only viable platform for this customization.

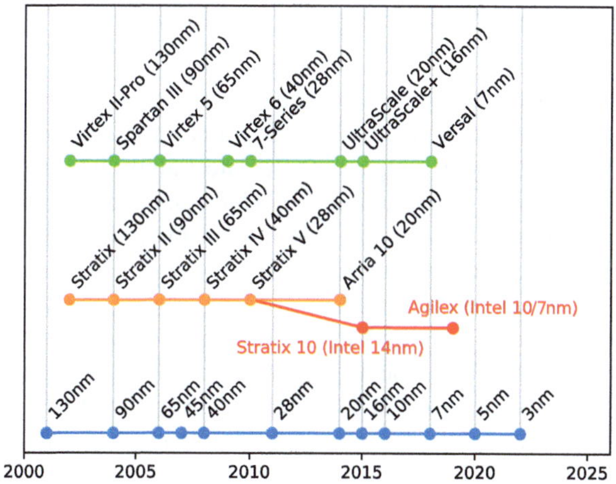

Fig. 2.10 Race for the latest technology. The plot shows the year of announcement of major FPGA families from Altera/Intel (orange and red) and Xilinx/AMD (green), compared to the start of production of different technology nodes by TSMC [TSM23] (blue). Until the 16 nm node, FPGAs were able to closely follow the technology development. Certain delays occurred at 7 nm, and as of 2024, there are yet to be any major architectures announced at 5 or 3 nm. Data about Stratix architectures is adopted from Intel's website [Int23], while the remaining points are adopted from *Design & Reuse* [Des04, Des05, Des06, Des09, Des10, Des14, Des14a, Des15, Des15a, Des18, Des19]. Intel's 10 nm node is considered to be equivalent to TSMC's 7 nm [Cut21]

Admittedly, many of the early adopters of FPGAs were using them for the inherent benefits of hardware reconfigurability: ability to modify the design very quickly and at no cost. This was and continues to be essential in the ASIC emulation domain, where the designs are under development or verification, making it unacceptable to wait several weeks for an MPGA to arrive, or pay the price even of a one-time field-programmable device, just to see the effect of modifying a single line of code [Tri94, Cad21]. These benefits have also been highly appealing in the communication domain, where protocols and algorithms may not be completely defined at the time of hardware development [Hil93, Tri15, Men+20].

Today, these quickly evolving application domains that were able to profit from rapid customization at no redeployment cost offered by FPGAs are supplemented by machine learning [Ans23]. For example, Denton and Schmit of Google Brain demonstrated that custom spatial implementation of specific machine learning models can achieve much lower latency compared to a general ASIC accelerator or a GPU [Den22]. Designing a new ASIC for each new machine learning model that may appear on a daily basis [Jou21] is far from feasible, but FPGAs provide a suitable alternative. The advantage over a general ASIC accelerator is increased further when LUTs of the FPGA are swapped for a custom bit-serial arithmetic block [Sch22a]. Note that with this change the FPGA is no longer general purpose, but it still essentially remains an FPGA, as we shall explain in more detail in Sect. 2.8.

2.7 FPGA or ASIC? No Longer a Question!

The bottom line is that reconfigurabile architectures are ideal for accelerating tasks for which the algorithms change at such a pace and bring such improvements that the ability to avoid a delay in adoption of the latest algorithm quickly offsets any inherent cost of hardware reconfigurability. On the other hand, for tasks with mature algorithmic and architectural solutions, where probable modifications have already entered the "diminishing returns" region of the performance increase curve, ASICs are more suitable, if they can be afforded.

Luckily, with current levels of VLSI miniaturization, and especially with the emergence of chiplets [San23], one no longer has to choose between reconfigurable and fixed platforms—it is possible to simultaneously have both. For example, a user of an FPGA may want to boot Linux for which they would need a CPU core, without any custom requirements. Of course, a general-purpose FPGA can easily implement a CPU, but as we have already mentioned, that CPU would have anywhere from 3.2 to 12.5× lower maximum clock frequency and as much as 8.7× larger area [Kuo06, Bou18] than if implemented in fixed silicon. So why not just harden a CPU next to an FPGA, on the same chip? This is exactly what FPGA architects have been doing at least since Virtex II Pro, introduced in 2002 [Xil11] and illustrated in Fig. 2.11a.

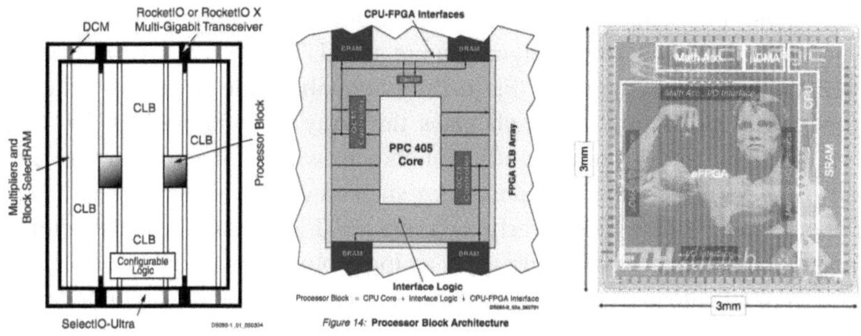

Fig. 2.11 Convergence of FPGAs and ASICs. (**a**) (courtesy of Xilinx/AMD) depicts the Virtex II Pro FPGA featuring two hardened PowerPC 405 CPU cores. (**b**) (courtesy of ETH Zürich) shows a die photo of the Arnold SoC consisting of a RISC-V CPU core and a QuickLogic eFPGA for custom accelerator implementation. Boundary between FPGAs and ASICs is getting increasingly blurred. (**a**) Virtex II Pro [Xil11]. (**b**) Arnold [Sch21]

In a typical Altera/Intel 28 nm FPGA with no hardened CPUs, the core reconfigurable fabric, composed of LUTs and programmable interconnect, was reported to account for a mere 30% of the entire die area [Lew12]. In more heterogeneous FPGAs that harden still more commonly used features than Virtex II Pro, such as the 7 nm Xilinx/AMD Versal, presented in 2019 [Gai19], the fraction of the chip dedicated to reconfigurable fabric is likely even much lower. On the other hand, in the Arnold chip shown in Fig. 2.11b—which its authors label an "eFPGA-augmented RISC-V SoC," rather than an FPGA with hardened functionality—the area ratios are completely inverted: eFPGA takes up almost 80% of the die area [Sch21].

Hence, it is possible to say that FPGAs and traditionally fixed-functionality ASICs have become increasingly alike, each combining the best of both worlds. This trend is expected to increase, with eFPGA-augmented SoCs predicted to reach $10 billion in annual sales in near future [Jar22]. With such proliferation of hardware reconfigurability, being able to design appropriate high-performance reconfigurable architectures is imperative. The purpose of this book is to make that significantly easier by enabling design automation of several aspects of programmable interconnect. Before entering the details of how this is achieved in subsequent chapters, it is important to reconsider what FPGAs actually are in their essence.

2.8 What Is an FPGA, Again?

Most likely, the first association to an FPGA is the LUT. Because they are the closest equivalent to logic gates in an ASIC, it has been stated that customers are interested in FPGA's LUT capacity and not the wires and multiplexers of the programmable interconnect [DeH99, Cha15, Com17]. To some extent, this is true, but what customers are actually paying for are not LUTs; it is a *Field-Programmable Gate Array*. Genus proximum of "Field-Programmable Gate Array" is "gate array," and this carries no notion of which gates the array is composed of. LUTs have been shown to be among the most efficient choices when a general-purpose array is needed [Sin92], but competing candidates such as And-Inverter Cones have also been proposed [Par12]. Some early commercial FPGA families also employed multiplexers or (networks of) logic gates as logic blocks instead of LUTs [Ros93]. The notion of a "gate array" also does not say anything about the architecture being general purpose. For specific applications, superior results can be obtained from bit-serial arithmetic units, for example [Sch22a].

On the other hand, differentia specifica of FPGA is "field programmable"—the ability to connect together the gates that form the gate array *after* the array has been fabricated, which is a quality afforded by programmable interconnect. Hence, it is really the programmable interconnect that forms the essence of FPGAs.

Despite migrating to the next technology node as soon as it was available being a key goal for FPGA architects between 2000 and 2018, development of FPGAs in that period was certainly far more exciting than shrinking a single unmodified design from one node to another. Let us briefly see how FPGAs developed in that period and how much of this development was dedicated to programmable interconnect.

2.9 FPGA Evolution: The Stratix Case

To understand how FPGAs have developed since the beginning of this century, it is of great interest to consider the Stratix architectures of Altera/Intel, because architectural decisions that entered the design of each generation of this family have been explained by members of the design teams themselves [Lew03, Lew05, Lew09, Lew13, Lew16]. Moreover, both the preceding and the succeeding families—Mercury [Hut02] and Agilex [Chr20]—have been presented in a similar manner, providing additional context to the evolution of Stratix.

2.9.1 Major Developments

2.9.1.1 Stratix: Island-Style Wins the Scaling Race

Not all early FPGAs adopted the *Island-Style* [Bet99] architecture of XC2064, where logic blocks conceptually (and in early days, when the number of available metal layers was very limited, quite literally) appeared as "islands" surrounded by segmented channels of prefabricated wires (Fig. 2.3). Although ingenious and with an effect lasting 40 years, island-style architectures could be seen as a relatively logical and straightforward step from MPGAs to field-programmable devices. It is yet to be demonstrated that a better architecture cannot be conceived, and it is thus unsurprising that different attempts at doing so have been made, especially when FPGAs were just starting to appear. Moreover, whether one style of a reconfigurable architecture is superior to another depends on both the applications that it is expected to support and the technology in which it is fabricated. Since both applications and technology evolve over time, so do the benefits of different architectural styles.

An example of very successful non-island-style architectures are the hierarchical FPGAs produced by Altera in the 1990s that were able to outcompete the contemporary island-style Xilinx FPGAs in both performance and capacity [Com17]. As these devices were a direct precursor to the Stratix series, we will briefly explain how they were organized. It is important to note that before entering the FPGA market, Altera had been designing PLAs [Alt90]. This may have been the reason why their pre-Stratix devices in many ways resemble more a PLA than an MPGA.

Altera APEX 20K FPGAs were based on *clusters* [*Logic Array Blocks* (LABs) in Altera terminology [Hut01]] of ten 4-LUT logic elements that were slightly more complex than the one of Fig. 2.8 [Alt04]. A cluster is a column of logic elements stacked on top of each other [Lew13] that share some local programmable interconnect. In particular, the local interconnect of APEX 20K implemented a complete graph, allowing any output of any logic element in the cluster, as well as any cluster input to be connected to any input of any logic element in the cluster [Hut01]. We will discuss the advantages of clustered organization in more detail in Chap. 3. The internal structure of an APEX 20K cluster is shown in the left part of Fig. 2.12.

Clusters in APEX 20K were in turn grouped into *superclusters* (*Mega LABs* in Altera terminology), containing 16 clusters [Hut01]. Superclusters also contained local interconnect (orange in Fig. 2.12), through which any signal entering the supercluster had to pass (red arrows in Fig. 2.12 [Alt04]). This local interconnect was sparser than that within the clusters, and it could be bypassed when signals had

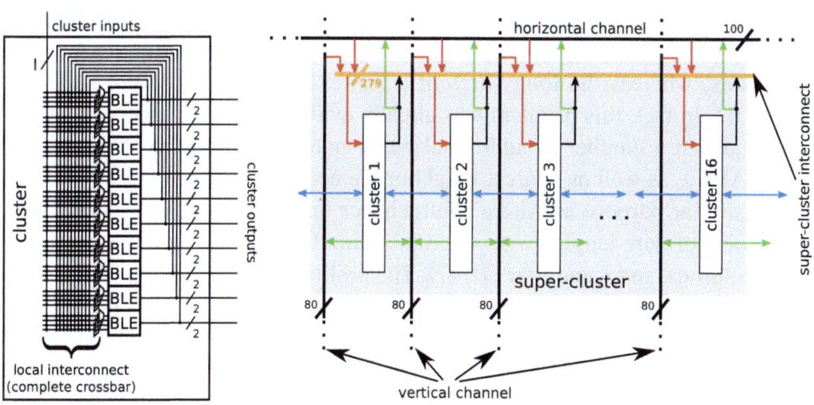

Fig. 2.12 Hierarchical logic block of an APEX 20K FPGA

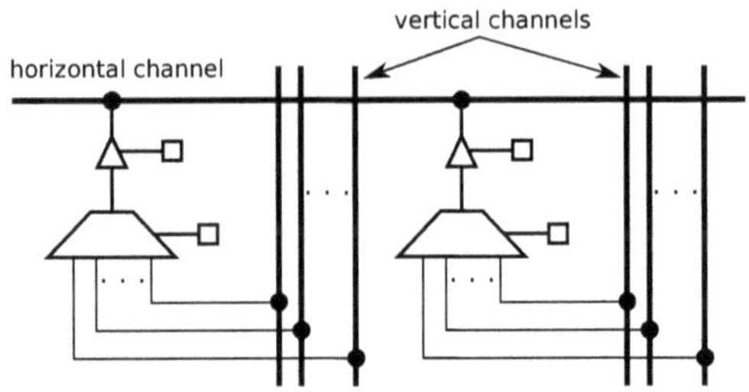

Fig. 2.13 Tri-stated routing multiplexer outputs allow wires that span the entire chip to be driven from multiple points [Alt01]

their origin and destination in neighboring clusters, by using the direct connections shown in blue in Fig. 2.12 [Alt04].

APEX 20K400 was a 2D array—much like XC2064—of superclusters, arranged in four columns and 26 rows [Hut01]. The main difference was in programmable interconnect. Namely, early Altera FPGAs had horizontal and vertical routing channels made up of wires that were spanning the entire width and height of the chip, respectively. Tristate buffers at the output of the store-select multiplexers driving channel wires allowed signals to switch between vertical and horizontal wires at intersections of two channels (Fig. 2.13) [Alt01]. Additionally, each of the channel wires could be programmably broken in half (by disengaging the corresponding programmable switch at the center of the chip [Hut01]). This created a hierarchical interconnect, where superclusters that were in the same half-row or in the same half-column were the fastest to connect, using only one half-wire; superclusters that were in the same row (column) required one horizontal (vertical) wire; superclusters that were in the same quadrant of the chip required one horizontal and one vertical half-wire; whereas all other superclusters took the longest to connect: one vertical and one horizontal full wire [Hut01]. A sketch of the APEX 20K400 hierarchical programmable interconnect architecture is shown in Fig. 2.14.

As we already mentioned, in the 1990s, hierarchical FPGAs were superior to the island-style ones [Com17]. As advertised in the FLEX 10K datasheet, segmented-channel routing of an island-style FPGA was slowed down by numerous multiplexers, whereas the long wires of a hierarchical device did not suffer from that [Alt01]. In fact, this problem was already evident to the designers of XC2064, who introduced a number of additional wires spanning the entire chip, much like those of Altera, as well as nearest-neighbor connections between logic blocks, that served a similar purpose as Altera's intracluster interconnect [Xil93]. Over time, a wider choice of wire lengths was added to Xilinx FPGAs, with some spanning two rows or columns, and some four [Tri97]. The problem of determining which lengths of prefabricated wires should appear in the routing channels, which is usually called *optimization of channel segmentation*, also received significant attention from the academic community [Bet99a, Cha01]. However, when these longer and faster wires were in high demand, some signals would have to be implemented using cascades of several shorter wires, incurring a multiplexer delay at each hop. This meant that the performance of a mapped design depended on wire demand and was thus

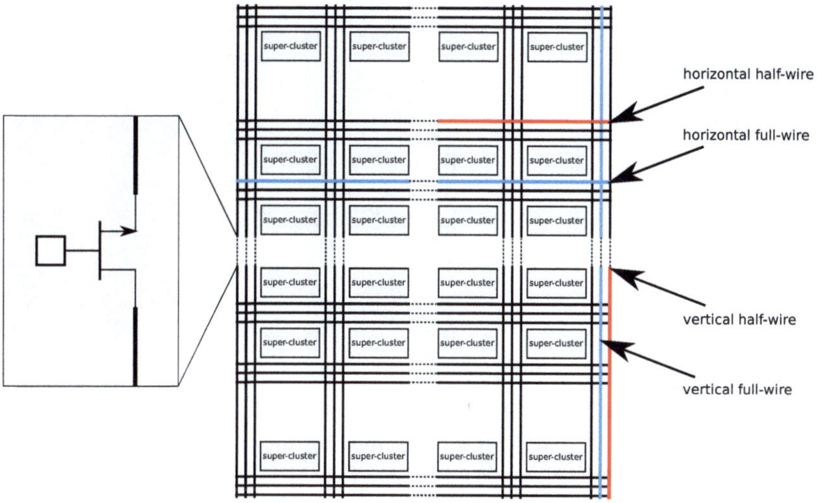

Fig. 2.14 Hierarchical interconnect of an APEX 20K device. Reproduced from Hutton et al. [Hut01]

less predictable than in a hierarchical device of the time, since multiplexer delays dominated those of metal [Alt01].

Despite these early advantages of hierarchical architectures, with scaling technology, the island-style approach became superior. First of all, increasing capacity of the device by increasing the size of the fully connected clusters was not scalable—local multiplexers would simply become too large and slow, as would the wires that provide their inputs (e.g., in APEX 20K architectures, supercluster connectivity was about as slow as traversing half of the chip in either direction [Hut01]). Neither was it possible to keep adding more channel wires spanning half the chip uninterrupted, in proportion with the number of cluster rows or columns added to increase the capacity of the device. Finally, reduction of cross-sectional area of the wires, caused by technology scaling, without appropriate reduction in length, caused by a need to increase device capacity, would quickly make the half-chip-spanning wires exceedingly slow. At the same time, transistor performance increased with technological progress making the argument about prohibitive cost of channel segmentation less compelling.

Some remedies for the above problems were attempted in the last of the Altera's hierarchical devices—Mercury [Hut02]—but the subsequent Stratix family marked a departure from the hierarchical approach altogether and adoption of the island-style one instead [Lew03]. Besides fixing the aforementioned issues, this brought several other benefits. First, the repetitive nature of island-style FPGAs made it possible to produce devices with different capacities by replicating an identical tile that needs to be laid out only once, drastically reducing development costs and time [Kuo04]. Second, the nonhierarchical nature of the interconnect made it easier to replace LUT clusters with other blocks, such as hardened multipliers and larger

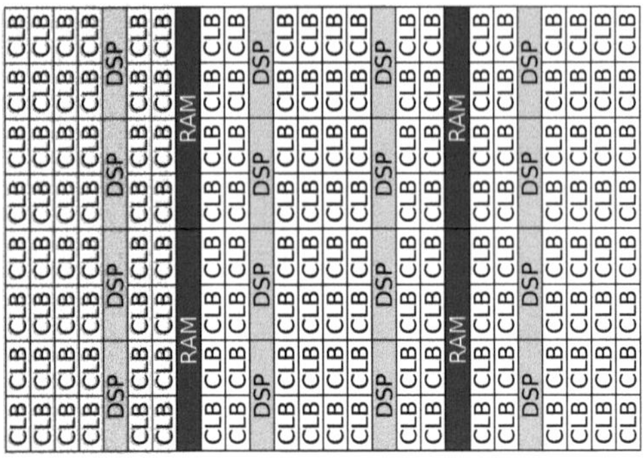

Fig. 2.15 Flat architecture of island-style FPGAs makes it easy to replace entire columns of LUTs by other kinds of logic blocks that are more efficient for certain application domains. The figure shows an example of the ASMBL columnar architecture pioneered by Xilinx [Bea08]

memories [Hut02, Lew03, Bea08]. This is illustrated in Fig. 2.15. Finally, reducing the gap between FPGAs and ASICs, by removing the hierarchy constraints, and moving back toward MPGAs made adoption of ASIC CAD algorithms much more straightforward. Given that FPGA device capacity was increasing exponentially at the start of this century, preventing place and route runtime from exploding was imperative [Bet20]. Luckily, ASIC CAD tools already had to cope with designs that were about an order of magnitude larger [Gor12], so the problem could be made much easier if the techniques developed there could be adapted to the specific constraints of FPGAs, of which island-style architectures had fewer than the hierarchical ones. Examples of successful adoption of ASIC algorithms are abundant, especially for placement [Gor12, Li19, Raj22], which suffers the most from problem size explosion: Unlike routing for which the runtime could be largely dictated by a relatively small bottleneck [Gor13], placement time is always directly related to the number of objects that need to be placed. We will review all relevant steps of the FPGA CAD flow in Chap. 3.

2.9.1.2 Stratix II: Fracturable LUTs Make a Comeback

The main difference between Stratix II and the original Stratix was the introduction of a *fracturable* 6-LUT in place of a 4-LUT, as the basis of the logic block. Prior results of Ahmed and Rose demonstrated that 6-LUTs provide higher performance than 4-LUTs, by allowing more levels of logic to be consumed by the LUT itself, thus avoiding the slow programmable interconnect [Ahm00]. However, all Boolean functions of $k < 6$ variables would waste 2^{6-k} memory bits of a 6-LUT, the corresponding fraction of its multiplexer tree, and the routing multiplexers driving the $6 - k$ inputs. Since most technology-mapped designs contain many functions of less than six variables [Hut04], Ahmed and Rose observed that this waste results in 6-LUTs having lower density than 4-LUTs and 5-LUTs [Ahm00].

To prevent this waste, some architectures like Virtex-II have provided 2:1 multiplexers that could be used to recursively combine outputs of pairs of 4-LUTs to create larger LUTs [Xil01]. This meant that whenever a function of four variables

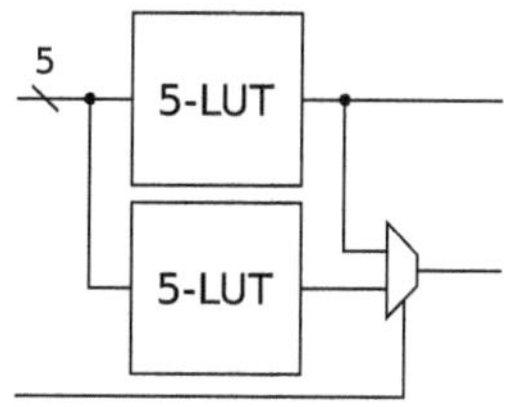

Fig. 2.16 Fracturable LUT [Ahm09]. Exposing an output of a prior stage of the multiplexer tree enables a 6-LUT to implement two different Boolean functions of the same five variables

had to be mapped onto an FPGA, it could simply use one of the available 4-LUTs and the only waste would be the small 2:1 combination multiplexers. However, Lewis et al. [Lew05] observed that on this kind of architecture, implementing a function of six variables is wasteful, because all of the inherent input sharing between the constituent 4-LUTs needs to be performed by programmable routing. In other words, whereas a 6-LUTs needs only six inputs, four independent 4-LUTs require sixteen, each driven by a (set of) stored-select multiplexer(s). This is another instance of the need to carefully balance the utilization of prefabricated logic blocks and programmable interconnect when designing an FPGA, which we discussed in Sect. 2.2.4.

Lewis et al. solved the problem by shorting some of the inputs among the two 5-LUTs that could be composed into a 6-LUT in Virtex-II [Lew05], giving rise to the concept of a *fracturable LUT*: In the limit, when all inputs are shared among the pair of composable LUTs, this amounts to exposing the intermediate outputs of the multiplexer tree of a larger LUT, as shown in Fig. 2.16. This extreme case was adopted by later generations of Virtex [Ahm09], although it was already previously in use in XC2064 [Xil93, Ros18]. Allowing some inputs to be independent, however, maximizes the chances that two Boolean functions of less than six variables can be successfully matched to a common 6-LUT, at the expense of some additional stored-select multiplexers to provide the independent inputs. This is the approach taken by Stratix II and later [Lew05, Lew16], as well as previously by AT&T ORCA [Hil93].

2.9.1.3 Stratix III and IV: It Is All About Power

In 65 nm Stratix III and 40 nm Stratix IV, power leakage became a major concern. Hence, programmable body bias functionality was introduced in these architectures, allowing different regions of the FPGA to switch between a low-power and a high-performance regime [Lew09]. Additionally, the fracturable LUTs introduced in Stratix II contained 64 bits of memory and eight independent inputs, which better amortized additional circuitry needed to allow LUTs to be used as read-write memories, like the ones that existed in the Xilinx 4000 series FPGAs [Xil98, Lew09]. Given that the Stratix III and IV architectural modifications were presented in the same paper [Lew09] and that they became available with a mere 16-month difference [Des07, Des08], it is very likely that Stratix IV is a minor redesign of Stratix III, migrated to 40 nm.

2.9.1.4 Stratix V: Skewing the Clock to Make it Faster

Modern, high-performance FPGA designs heavily rely on pipelining, which increases the demand on flip-flops. For this reason, Stratix V doubled the number of available flip-flops, increasing it to four per each 6-LUT [Lew13]. Pipelining can become ineffective when it is not possible to insert flip-flops in such a way as to break a path into sections of comparable delay [Cha15]. Due to various placement constraints, achieving a good balance among delays of different pipeline stages can often be difficult. To mitigate this issue, Stratix V introduced the possibility for the clock to be programmably delayed on some flip-flops [Lew13], allowing some pipeline stages to *borrow* time from the adjacent ones. This method of balancing delays between pipeline stages irrespective of placement constraints has been previously explored in academia [Sin02]. Lewis et al. report that when realized using pulse latches with programmable pulse width, the technique was able to improve the performance of designs implemented on Stratix V by up to 3% on average [Lew13]. Ganusov and Devlin report that a slightly different realization, using edge-triggered flip-flops, improved the performance of designs implemented on Xilinx/AMD UltraScale+ FPGAs by 5.5% on average [Gan16].

These two examples demonstrate how the perspective of the margin of performance improvement that justifies a relatively major design change in commercial FPGAs evolved since the 1990s: 3–5% are no longer considered insignificant.

2.9.1.5 Stratix 10: Interconnect Pipelining Pushed a Bit Too Far

Increasing the number of flip-flops available in a logic cluster allowed for better mapping of highly pipelined designs on Stratix V than the previous generations of Stratix. However, the ability to insert pipeline registers only inside the logic clusters limited the reduction of critical path delay to that of the longest section of the unpipelined path through programmable interconnect. As Fig. 2.17 adopted

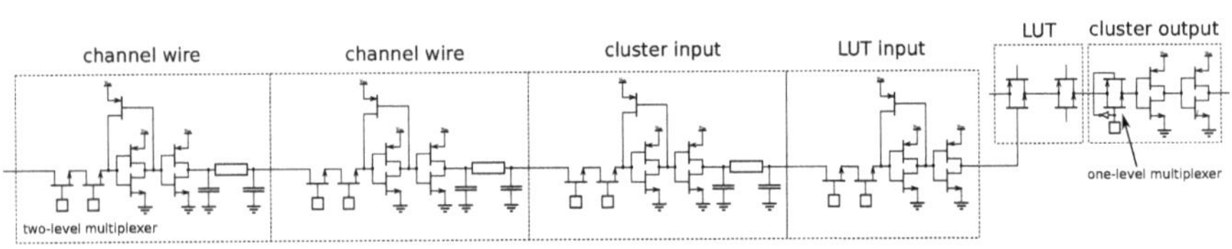

Fig. 2.17 Typical logic-interconnect stage on a critical path of an industrial circuit [Lew12]. A typical critical path in an industrial circuit contains three such stages [Lew12]. If the design is pipelined on an architecture like Stratix V, which only has flip-flops at LUT outputs, critical path delay cannot be reduced below one stage. However, as wire resistance increases with scaling technology, interconnect delay becomes increasingly dominant [Gai19]. This motivated insertion of interconnect pipeline registers in Stratix 10 [Lew16]. We note that because of permutability of LUT inputs, the router usually succeeds in using one of the faster LUT inputs (closer to the root of the multiplexer tree). However, the critical path typically passes through the second-fastest input [Lew12], which is the best option for a fracturable LUT implementing two 5-LUTs [Ahm09], since the fastest input has to be tied to a logic 1 (Fig. 2.16)

from Lewis and Chromczak [Lew12] shows, in a typical commercial design, this section is rather significant, especially taking into account that technology scaling causes wire delay to be increasingly dominant. To alleviate this issue, Stratix 10 introduced bypassable pipeline registers in all stored-select multiplexers that were part of the programmable routing architecture [Lew16]. This made it possible to reduce the critical path delay to as little as that of a single wire. Interconnect pipeline registers were subsequently also introduced in the Xilinx/AMD Versal architecture, but to a lesser extent, limiting the additional registers to isolation of the intracluster interconnect [Gai19] that, due to being traced at lower and more resistive metal layers, becomes a bottleneck in scaled technologies [Chr20, Nik21]. Interestingly enough, the latest Intel FPGA architecture, Agilex, also reduced the number of pipeline registers in multiplexers driving channel wires to 30% [Chr20].

2.9.2 What About Programmable Interconnect Topology?

As discussed in Sect. 2.8, programmable interconnect is the defining feature of FPGAs. Yet, apart from the original Stratix in which a major departure from hierarchical interconnect and adoption of island-style was made, and Stratix 10, which introduced interconnect pipeline registers, so far we have mostly discussed developments that were not related to the interconnect architecture itself. Figure 2.18 shows the approximate fraction of the corresponding paper dedicated to discussing interconnect topology improvements in each generation of Stratix.

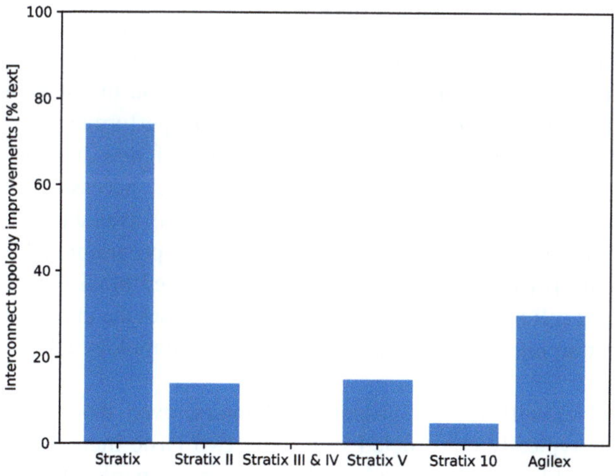

Fig. 2.18 Approximate fraction of text dedicated to discussing modifications of the programmable interconnect topology in papers presenting several generations of Altera/Intel FPGAs [Lew03, Lew05, Lew09, Lew13, Lew16, Chr20]

We can see that after the serious changes reported for the original Stratix, subsequent Stratix papers dedicate little attention to the question of improving interconnect topology. Needless to say, this does not mean that interconnect topology remained unchanged throughout this period: Each time the logic block was redesigned, some of the previous decisions about the lengths of wires in the routing channels potentially became suboptimal, requiring reconsideration, and in turn triggering a change in switch-pattern design; additional logic block inputs and outputs had to be accessed from the routing channels; and increased device capacity also called for increasing the capacity of the channels [Sch03]. Yet, these modifications were not significant enough to become the focus of the authors' attention. In fact, they mostly amounted to a quick optimization around a known local optimum provided by the previous generation of the architecture, as the description of modifications applied to Stratix V clearly states [Lew13]:

Consequently rather than completely re-architect the routing, we explored minor variations that could keep pace with the increase in routing demand as well as obtain performance improvement.

So how did it happen that evolution of the very essence of FPGAs was deliberately confined to minor variations for almost two decades? Stephen Trimberger's recollection provides an answer to this question (Sect. 2.4). First of all, scaling worked well during this period and availability of an increasing number of metal layers with different pitches provided a way to combat rising resistance by simply optimizing layer use [Lew13]. Second, major modifications to the interconnect topology were not as likely to bring such high density and performance benefits as hardening functionality did [Kuo06, Tri15]. Third, even if changes in the interconnect could produce significant gains, they came with a great risk: An FPGA with a suboptimal hardened multiplier could still perform very well on designs that do not use multiplication and having some hardened multiplier is certainly better for the applications that need them than having none at all; any suboptimalities could be rectified in the next generation of the architecture. On the other hand, having poor programmable interconnect would hamper the performance of all circuits implemented on that FPGA. In summary, addition of new hardened blocks or redesign of old ones is very forgiving to suboptimalities necessitated by the need to transition to the latest technology node as soon as it is available, while dramatic changes to interconnect topology are not. On top of this, programmable interconnect layout is usually performed manually, whereas hardened blocks are implemented using a standard cell flow [Yaz19], which further increases the urge to limit changes to interconnect architecture as much as possible, if technology adoption timelines are to be met.

Finally, as we will see in subsequent chapters, even though programmable interconnect is typically constructed from only two basic ingredients—wires and multiplexers—unlike highly complex blocks such as hardened multipliers and memories, optimizing it is inherently combinatorial and reduces to graph construction; this is entirely different from designing memory blocks, deciding LUT sizes, multiplier port widths, etc., which are all problems that can be effectively parameterized with only a handful of parameters taking on values from small ranges.

Hence, even if programmable interconnect design may be conceptually simpler than hardening different functionality, it is computationally much more intractable in practice.

To make this discrepancy even larger, years of ASIC development have led to designing fixed-functionality circuits being supported by various abstractions, design methodologies, optimization algorithms, and in many cases fully automated design flows. On the other hand, the aforementioned lack of a need to perform major changes and even outright pressure not to innovate has led to no analogous solutions existing for the problem of designing programmable interconnect. At best, existing design automation in that area enables solving precisely the only problem that was relevant in the past two decades: local optimization around a known local optimum.

2.9.2.1 Agilex: Juggling Resistance, Capacitance, and Routability at the Topology Level

As we have discussed previously, the biggest obstacles to reducing interconnect delay in early FPGAs were the stored-select multiplexers. However, technology scaling that inevitably makes transistors faster and wires slower resulted in hierarchical architectures with very long wires, such as APEX20K, being long left behind, despite their initial superiority. In the two decades that followed, technological improvements such as changing materials and increased availability of metal layers made it relatively easy to keep the effects of rising wire resistance in check. Around the 7 nm node, however, these effects became so pronounced [Che14] that a new major redesign of the programmable interconnect architecture was necessary.

As we can see from the plot of Fig. 2.18, the fraction of the text dedicated to the modifications of programmable interconnect topology in the paper presenting the latest Agilex family of Altera/Intel FPGAs increased to 30% [Chr20]—the highest fraction since the original Stratix. These modifications were numerous, as battling resistance required halving the multiplexer sizes to reduce capacitive loading on wires that they take their inputs from and to enable swapping pass transistors for transmission gates to increase performance [Chr20]; drastic reduction in the length of wires traced at lower metal layers, with implications on both intracluster routing and switches between channel wires; increasing multiplexer input sharing to reduce via count. All this on top of the usual local optimization of lengths of channel wires and switch-patterns that connect them, and while retaining and further improving the interconnect pipeline registers of Stratix 10, that can isolate the delays of individual wires, but cannot reduce them [Chr20].

2.10 Where Does This Book Come into Play?

How does one make all multiplexers in the programmable interconnect architecture half as large as they used to be, while ensuring that all circuits that the FPGA was previously able to route are still routable? This is just one of the questions that the architects of Agilex had to answer. From the recollection of Stephen Trimberger, we learned that a crucial ingredient in quickly designing a new FPGA architecture—in time for the next technology node—has been the experience and intuition of the design team. However, when a single technology node brings so much change that it requires reducing the number of inputs to all multiplexers in a state-of-the-art architecture by half, one could argue that the existing intuition developed from years of prior experience is no longer sufficient for rapid design. Unfortunately, with technological advances increasingly relying on one-off *scaling boosters* [Tok22, Mor23], it is likely that developing such intuition that could translate from one technology to the next will in the future be very difficult.

In Part II of this book, we present several novel techniques for automating the design of certain aspects of programmable interconnect architectures that have not been satisfactorily automated before. Although they do not constitute a complete automated design flow capable of producing fully customized programmable interconnect architectures optimized for the given fabrication technology and target applications, we believe that they represent an important step toward achieving this goal. Our hope is that using these techniques, reconfigurable architectures will be able to overcome the current technology scaling challenges and continue to deliver high performance in future nodes, as well as in all other settings discussed in Chap. 1 where hardware reconfigurability is starting to appear.

Before presenting these techniques and results of their application, in the next chapter, we first give some additional background information on island-style FPGAs and CAD algorithms used to map circuits onto them. We also define the problem of programmable interconnect design in the most abstract sense, which is useful for providing context for the proposed algorithms.

Readers who enjoyed the historical perspective of FPGA evolution and in particular the invaluable recollections of Stephen Trimberger [Com17] may also wish to refer to similar stories of Bill Carter[2]—the principle designer of XC2064 [Com15] and Bernie Vonderschmit—one of Xilinx's three cofounders and its first CEO, as well as the inventor of the fabless business model [Sta95].

References

[Ahm00] Ahmed, E., & Rose, J. (2000). The effect of LUT and cluster size on deep-submicron FPGA performance and density. In *Proceedings of the ACM/SIGDA international symposium on field programmable gate arrays* (pp. 3–12).

[Ahm09] Ahmed, T., Kundarewich, P. D., & Anderson, J. H. (2009). Packing Techniques for Virtex-5 FPGAs. *ACM Trans. Reconfigurable Technol. Syst.*, *2*, 3. ISSN: 1936-7406.

[Alo16] Alonso, G. (2016). *Data Processing on the fast lane (Keynote of the 2016 International Conference on Field-Programmable Logic and Applications (FPL))*. https://fpl2016.epfl.ch/slides/Alonso%20--%20Data%20Processing%20on%20the%20Fast%20Lane.pdf

[2] The interview was conducted by Dr. Trimberger that provides an even more interesting context.

[Alt90] Altera Corporation (1990). *Data Book.*

[Alt01] Altera Corporation (2001). *FLEX 10K Embedded Programmable Logic Device Family.* ver. 4.1. 101 Innovation Drive, San Jose, CA 95134.

[Alt04] Altera Corporation (2004). *APEX 20K Programmable Logic Device Family.* ver. 5.1. 101 Innovation Drive, San Jose, CA 95134.

[Ans23] Ansell, T. (2023). Google investment in open source custom hardware development including no-cost shuttle program. In *Proceedings of the 2023 international symposium on physical design.* Virtual Event.

[Bau20] Bauer, H., Burkacky, O., Kenevan, P., Lingemann, S., Pototzky, K., & Wiseman, B. (2020). *Semiconductor design and manufacturing: Achieving leading-edge capabilities.* Tech. rep. McKinsey.

[Bea08] Beauchamp, M. J., Hauck, S., Underwood, K. D., & Hemmert, K. S. (2008). Architectural modifications to enhance the floating-point performance of FPGAs. *IEEE Trans. Very Large Scale Integr. Syst. 16*(2), pp. 177–87. ISSN: 1063-8210.

[Bec17] Becker, T., Burovskiy, P., Nestorov, A. M., Palikareva, H., Reggiani, E., & Gaydadjiev, G. (2017). From exaflop to exaflow. In *Design, Automation & Test in Europe Conference & Exhibition (DATE), 2017* (pp. 404–9).

[Bet20] Betz, V. (2020). Parallel CAD for FPGAs: A personal retrospective and thoughts for the future. In *Proceedings of the 30th International Conference on Field-Programmable Logic and Applications (FPL).* Virtual Event.

[Bet99a] Betz, V. & Rose, J. (1999). FPGA Routing Architecture: Segmentation and Buffering to Optimize Speed and Density. In *Proceedings of the 1999 ACM/SIGDA Seventh International Symposium on Field Programmable Gate Arrays.* Monterey, CA, USA (pp. 59–68).

[Bet99] Betz, V., Rose, J., & Marquardt, A. (1999). *Architecture and CAD for Deep-Submicron FPGAs.* Kluwer Academic Publishers. ISBN: 0-7923-8460-1.

[Bou21] Boutros, A. & Betz, V. (2021). FPGA architecture: Principles and progression. *IEEE Circuits and Systems Magazine, 21*(2), 4–29. ISSN: 1558-0830.

[Bou18] Boutros, A., Yazdanshenas, S., & Betz, V. (2018). You cannot improve what you do not measure: FPGA vs. ASIC efficiency gaps for convolutional neural network inference. *ACM Transactions on Reconfigurable Technology and Systems, 11*, 3. ISSN: 1936-7406.

[Bry86] Bryant, R. E. (1986). Graph-based algorithms for boolean function manipulation. *IEEE Transactions on Computers, C-35*(8), 677–91. ISSN: 1557-9956.

[Cad21] Cadence Design Systems (2021) *Cadence Unveils Next-Generation Palladium Z2 and Protium X2 Systems to Dramatically Accelerate Pre Silicon Hardware Debug and Software Validation.* https://www.cadence.com/en_US/home/company/newsroom/press-releases/pr/2021/cadence-unveils-next-generation-palladium-z2-and-protium-x2-syst.html. Accessed on 04.04.2023

[Cha15] Chandrakar, S., Gaitonde, D., & Bauer, T. (2015). Enhancements in UltraScale CLB Architecture. In *Proceedings of the 2015 ACM/SIGDA International Symposium on Field-Programmable Gate Arrays* (pp. 108–16).

[Cha01] Chang, Y.-W., Lin, J.-M., & Wong, M.D.F. (2001). Matching-based algorithm for FPGA channel segmentation design. *IEEE Transactions on Computer-Aided Design of Integrated Circuits and Systems, 20*(6), 784–791.

[Che20] Chen, C., Xiang, X., Liu, C., Shang, Y., Guo, R., Liu, D., et al. (2020). Xuantie-910: A Commercial Multi-Core 12-Stage Pipeline Out-of-Order 64-bit High Performance RISC-V Processor with Vector Extension. In: *2020 ACM/IEEE 47th Annual International Symposium on Computer Architecture (ISCA)* (pp. 52–64). Valencia, Spain.

[Che14] Chen, J. H.-C., Standaert, T., Alptekin, E., Spooner, T., & Paruchuri, V. (2014). Interconnect performance and scaling strategy at 7 nm node. In: *IEEE International Interconnect Technology Conference* (pp. 93–96).

[Chi13] Chiasson, C. (2013). *Optimization and modeling of FPGA circuitry in advanced process technology.* MA thesis. University of Toronto.

[Chr20] Chromczak, J., Wheeler, M., Chiasson, C., How, D., Langhammer, M., Vanderhoek, T., et al. (2020). Architectural Enhancements in Intel®Agilex™FPGAs. In *Proceedings of the 2020 ACM/SIGDA International Symposium on Field-Programmable Gate Arrays* (pp. 140–49).

[Com15] Computer History Museum (2015). *Oral history of bill carter*. https://www.computerhistory.org/collections/catalog/102737926

[Com17] Computer History Museum (2017). *Oral history of steve trimberger*. https://archive.computerhistory.org/resources/access/text/2018/07/102740229-05-01-acc.pdf

[Con94] Cong, J. & Ding, Y. (1994). FlowMap: An optimal technology mapping algorithm for delay optimization in lookup-table based FPGA designs. *IEEE Transactions on Computer-Aided Design of Integrated Circuits and Systems, 13*(1), 1–12. ISSN: 1937-4151.

[Cut21] Ian Cutress (2021). *Intel's Process Roadmap to 2025: with 4 nm, 3 nm, 20A and 18A?!* AnandTech. https://www.anandtech.com/show/16823/intel-accelerated-offensive-process-roadmap-updates-to-10nm-7nm-4nm-3nm-20a-18a-packaging-foundry-emib-foveros. Accessed on 29.03.2023.

[DeH99] DeHon, A. (1999). Balancing Interconnect and Computation in a Reconfigurable Computing Array (or, Why You Don't Really Want 100% LUT Utilization). In *Proceedings of the 1999 ACM/SIGDA Seventh International Symposium on Field Programmable Gate Arrays* (pp. 69–78).

[Den22] Denton, M. & Schmit H. (2022). Direct spatial implementation of sparse matrix multipliers for reservoir computing. In *2022 IEEE International Symposium on High-Performance Computer Architecture (HPCA)* (pp. 1–11).

[Des04] Design & Reuse (2004). *Xilinx Ships 90-Nanometer Devices In Volume Production—Addresses Record Demand For It's Low Cost Spartan-3 FPGAs*. https://www.design-reuse.com/news/8546/xilinx-90-nanometer-devices-volume-record-demand-it-cost-spartan-3-fpgas.html. Accessed on 20.03.2023.

[Des05] Design & Reuse (2005). *Xilinx Virtex-II Pro World's Most Popular 130 nm FPGA*. https://www.design-reuse.com/news/10557/xilinx-virtex-ii-pro-popular-130nm-fpga.html. Accessed on 20.03.2023.

[Des06] Design & Reuse (2006). *Xilinx Unveils 65 nm Virtex-5 Family—Industry's Highest Performance Platform FPGAs*. https://www.design-reuse.com/news/13342/xilinx-65nm-virtex-5-highest-performance-platform-fpgas.html. Accessed on 20.03.2023.

[Des07] Design & Reuse (2007). *Altera Ships First Member of High-End Stratix III FPGA Family*. https://www.design-reuse.com/news/16566/altera-member-end-stratix-iii-fpga-family.html. Accessed on 29.03.2023.

[Des08] Design & Reuse (2008). *Altera Starts Shipping Its 40-nm Stratix IV FPGA Family*. https://www.design-reuse.com/news/19725/40-nm-stratix-iv-fpga.html. Accessed on 29.03.2023.

[Des09] Design & Reuse (2009). *Xilinx Starts Shipments of Virtex-6 FPGAs*. https://www.design-reuse.com/news/20394/xilinx-virtex-6.html. Accessed on 20.03.2023.

[Des10] Design & Reuse (2010). *Xilinx 7 Series FPGAs Slash Power Consumption by 50% and Reach 2 Million Logic Cells on Industry's First Scalable Architecture*. https://www.design-reuse.com/news/23745/xilinx-virtex-7-kintex-7-artix-7.html. Accessed on 20.03.2023.

[Des14a] Design & Reuse (2014). *Altera and TSMC Collaborate to Bring Advanced Packaging Technology to Arria 10 FPGAs and SoCs*. https://www.design-reuse.com/news/34406/altera-tsmc-packaging-arria-10-fpga.html. Accessed on 20.03.2023.

[Des14] Design & Reuse (2014). *Xilinx Tapes-out First Virtex UltraScale All Programmable Device as Part of Industry's Only High-End 20 nm Family*. https://www.design-reuse.com/news/33730/xilinx-virtex-ultrascale-tape-out-20nm-family.html. Accessed on 20.03.2023.

[Des15] Design & Reuse (2015). *Altera Reveals Stratix 10 Innovations Enabling the Industry's Fastest and Highest Capacity FPGAs and SoCs*. https://www.design-reuse.com/news/37587/altera-stratix-10-innovations.html. Accessed on 20.06.2023.

[Des15a] Design & Reuse (2015). *Xilinx Announces Publicly Available Tools and Documentation for 16 nm UltraScale+ Devices*. https://www.design-reuse.com/news/38857/xilinx-16nm-ultrascale-tools.html. Accessed on 20.03.2023.

[Des18] Design & Reuse (2018). *Xilinx Unveils Versal: The First in a New Category of Platforms Delivering Rapid Innovation with Software Programmability and Scalable AI Inference*. https://www.design-reuse.com/news/44859/xilinx-versal.html. Accessed on 20.03.2023.

[Des19] Design & Reuse (2019). *Intel Ships First 10 nm Agilex FPGAs*. https://www.design-reuse.com/news/46667/intel-10nm-agilex-fpga.html. Accessed on 20.03.2023.

[ElG81] El Gamal, A. (1981). Two-dimensional stochastic model for interconnections in master slice integrated circuits. *IEEE Transactions on Circuits and Systems, 28*(2), 127–38.

[Fer84] Ferranti Semiconductor (1984). *Quick Reference: The ULA*.

[Fer77] Ferranti Semiconductors (1977). *The 225 Cell Uncommitted Array Family*. ULA 2000 Series. Chadderton, UK.

[Gai19] Gaide, B., Gaitonde, D., Ravishankar, C., & Bauer, T. (2019). Xilinx Adaptive Compute Acceleration Platform: Versal Architecture. In *Proceedings of the 2019 ACM/SIGDA International Symposium on Field-Programmable Gate Arrays* (pp. 84–93).

[Gan16] Ganusov, I. & Devlin, B. (2016). Time-borrowing platform in the Xilinx UltraScale+ family of FPGAs and MPSoCs. In *2016 26th International Conference on Field Programmable Logic and Applications (FPL)* (pp. 1–9).

[Gor12] Gort, M. & Anderson, J. H. (2012). Analytical placement for heterogeneous FPGAs. In *Proceedings of the 22nd International Conference on Field Programmable Logic and Applications (FPL)* (pp. 143–50).

[Gor13] Gort, M. & Anderson, J. H. (2013). Combined Architecture/Algorithm Approach to Fast FPGA Routing. *IEEE Transactions on Very Large Scale Integration (VLSI) Systems, 21*(6), 1067–79. ISSN: 1557-9999.

[Hal96] Hall, E. C. (1996). *Journey to the Moon: The History of the Apollo Guidance Computer*. American Institute of Aeronautics and Astronautics. ISBN: 978-1-56347-185-8.

[Hau07] S. Hauck & A. DeHon (Eds.) (2007). *Reconfigurable Computing: The Theory and Practice of FPGA-Based Computation*. Morgan Kaufmann. ISBN: 9780123705228.

[Hen19] Hennessy, J. L. & Patterson, D. A. (2019). A new golden age for computer architecture. *Communications of the ACM, 62*(2), 48–60. ISSN: 0001-0782.

[Hil93] Hill, D., Britton, B., Oswald, B., Woo, N.-S., Singh, S., Chen, C.-T., et al. (1993). ORCA: A new architecture for high-performance FPGAs. In *Proceedings of the 2nd International Workshop on Field-Programmable Logic and Applications* (pp. 52–60).

[Hut01] Hutton, M., Adibsamii, K., & Leaver, A. (2001). Timing-driven placement for hierarchical programmable logic devices. In *Proceedings of the 2001 ACM/SIGDA Ninth International Symposium on Field Programmable Gate Arrays* (pp. 3–11).

[Hut02] Hutton, M., Chan, V., Kazarian, P., Maruri, V., Ngai, T., Park, J., et al. (2002). Interconnect enhancements for a high-speed PLD architecture. In: *Proceedings of the 2002 ACM/SIGDA Tenth International Symposium on Field-Programmable Gate Arrays* (pp. 3–10).

[Hut04] Hutton, M. D., Schleicher, J., Lewis, D. M., Pedersen, B., Yuan, R., Kaptanoglu, S., et al. (2004) Improving FPGA performance and area using an adaptive logic module. In: *Proceedings of the 14th International Conference on Field Programmable Logic and Application* (pp. 135–44).

[Int23] Intel Corporation (2023). *Intel®Stratix®Series FPGAs and SoCs*. https://www.intel.com/content/www/us/en/products/details/fpga/stratix.html. Accessed on 29.03.2023.

[Jam06] Jamieson, P. & Rose, J. (2006) Enhancing the area-efficiency of FPGAs with hard circuits using shadow clusters. In *2006 IEEE International Conference on Field Programmable Technology* (pp. 1–8).

[Jar22] Jaros, A. (2022). *The Future of Embedded FPGAs—eFPGA: The Proof is in the Tape Out*. Circuit Cellar. https://circuitcellar.com/insights/the-future-of-embedded-fpgas-efpga-the-proof-is-in-the-tape-out/. Accessed on 30.03.2023.

[Jou21] Jouppi, N. P., Yoon, D. H., Ashcraft, M., Gottscho, M., Jablin, T. B., Kurian, G., et al. (2021) Ten Lessons from Three Generations Shaped Google's TPUv4i. In *Proceedings of the 48th Annual International Symposium on Computer Architecture*. Virtual Event (pp. 1–14).

[Koc13] Koch, D. (2013). *Partial Reconfiguration on FPGAs*. Lecture Notes in Electrical Engineering (Vol. 153). Springer. ISBN: 978-1-4614-1225-0.

[Kuo06] Kuon, I. & Rose, J. (2006). Measuring the Gap between FPGAs and ASICs. In: *Proceedings of the 2006 ACM/SIGDA 14th International Symposium on Field Programmable Gate Arrays* (pp. 21–30).

[Kuo04] Kuon, I. C. (2004). *Automated FPGA design, verification and layout*. MA thesis. University of Toronto.

[Lei20] Leiserson, C. E., Thompson, N. C., Emer, J. S., Kuszmaul, B. C., Lampson, B. W., Sanchez, D., et al. (2020). There's plenty of room at the Top: What will drive computer performance after Moore's law? *Science, 368*(6495), 1–7.

[Lem04] Lemieux, G., Lee, E., Tom, M., & Yu, A. (2004). Directional and single-driver wires in FPGA interconnect. In *Proceedings of the 2004 IEEE International Conference on Field-Programmable Technology* (pp. 41–48).

[Lew05] Lewis, D., Ahmed, E., Baeckler, G., Betz, V., Bourgeault, M., Cashman, D., et al. (2005). The Stratix II logic and routing architecture. In *Proceedings of the 2005 ACM/SIGDA 13th International Symposium on Field-Programmable Gate Arrays* (pp. 14–20).

[Lew09] Lewis, D., Ahmed, E., Cashman, D., Vanderhoek, T., Lane, C., Lee, A., et al. (2009). Architectural enhancements in Stratix-III™and Stratix-IV™. In *Proceedings of the ACM/SIGDA International Symposium on Field Programmable Gate Arrays* (pp. 33–42).

[Lew03] Lewis, D., Betz, V., Jefferson, D., Lee, A., Lane, C., Leventis, P., et al. (2003) The Stratix routing and logic architecture. In *Proceedings of the 2003 ACM/SIGDA Eleventh International Symposium on Field Programmable Gate Arrays* (pp. 12–20).

[Lew13] Lewis, D., Cashman, D., Chan, M., Chromczak, J., Lai, G., Lee, A., et al. (2013) Architectural enhancements in Stratix V™. In *Proceedings of the ACM/SIGDA International Symposium on Field Programmable Gate Arrays* (pp. 147–56).

[Lew16] Lewis, D., Chiu, G., Chromczak, J., Galloway, D., Gamsa, B., Manohararajah, V., et al. (2016) The Stratix™10 Highly Pipelined FPGA Architecture. In *Proceedings of the 2016 ACM/SIGDA International Symposium on Field-Programmable Gate Arrays* (pp. 159–68).

[Lew12] Lewis, D. & Chromczak, J. (2012). Process technology implications for FPGAs. In *Proceedings of the 2012 International Electron Devices Meeting* (pp. 25.2.1–4).

[Li19] Li, W., Lin, Y., & Pan, D. Z. (2019). elfPlace: Electrostatics-based Placement for Large-Scale Heterogeneous FPGAs. In *2019 IEEE/ACM International Conference on Computer-Aided Design (IC- CAD)* (pp. 1–8).

[Luu14] Luu, J., Goeders, J., Wainberg, M., Somerville, A., Yu, T., Nasartschuk, K., et al. (2014). VTR 7.0: Next Generation Architecture and CAD System for FPGAs. *ACM Transactions on Reconfigurable Technology and Systems, 7*, 2. ISSN: 1936-7406.

[Mar92] Marple, D. (1992). An MPGA-Like FPGA. *IEEE Design Test, 9*(4), 51–60. ISSN: 0740-7475.

[Men+20] Mencer, O., Allison, D., Blatt, E., Cummings, M., Flynn, M. J., Harris, J., et al. (2020). The history, status, and future of FPGAs. *Communications of the ACM, 63*(10), 36–39. ISSN: 0001-0782.

[Mor23] Moroz, V. (2023). Gate-all-around technology is coming: What's Next After GAA? In *Proceedings of the 2023 International Symposium on Physical Design*. Virtual Event.

[Mor71] Morris, R. L. & Miller, J. R. (Eds.) (1971). *Designing with TTL Integrated Circuits*. Texas Instruments Electronics Series. McGraw-Hill Book Company.

[Nik21] Nikolić, S., Catthoor, F., Tőkei, Z., & Ienne, P. (2021). Global is the new local: FPGA Architecture at 5 nm and Beyond. In *The 2021 ACM/SIGDA International Symposium on Field-Programmable Gate Arrays* (pp. 34–44). Virtual Event.

[Par12] Parandeh-Afshar, H., Benbihi, H., Novo, D., & Ienne, P. (2012). Rethinking FPGAs: Elude the Flexibility Excess of LUTs with and-Inverter Cones. In *Proceedings of the ACM/SIGDA International Symposium on Field Programmable Gate Arrays* (pp. 119–28).

[Par13] Parandeh-Afshar, H., Zgheib, G., Novo, D., Purnaprajna, M., & Ienne, P. (2013). Shadow And-Inverter Cones. In *2013 23rd International Conference on Field programmable Logic and Applications* (pp. 1–4).

[Put23] Putnam, A. (2023). *If you build it, will they come? (Cornell ECE 6950 Guest Lecture).* https://www.youtube.com/watch?v$=$TuS6bAZ5hps

[Raj22] Rajarathnam, R. S., Alawieh, M. B., Jiang, Z., Iyer, M., & Pan, D. Z. (2022). DREAMPlaceFPGA: An open-source analytical placer for large scale heterogeneous FPGAs Using Deep-Learning Toolkit. In *2022 27th Asia and South Pacific Design Automation Conference (ASP-DAC)* (pp. 300–306).

[Ram80] Ramsay, F. R. (1980). Automation of design for uncommitted logic array. In *Proceedings of the 17th Design Automation Conference* (pp. 100–107).

[Ros93] Rose, J., El Gamal, A., & Sangiovanni-Vincentelli, A. (1993). Architecture of field-programmable gate arrays. *Proceedings of the IEEE, 81*(7), 1013–29. ISSN: 1558-2256.

[Ros89] Rose, J., Francis, R. J., Chow, P., & Lewis, D. (1989). The effect of logic block complexity on area of programmable gate arrays. In *1989 Proceedings of the IEEE Custom Integrated Circuits Conference* (pp. 5.3/1–5.3/5).

[Ros18] Rose, J. (2018). A user programmable reconfigurable logic array. In *FPGA and Reconfigurable Computing Hall-of-Fame Endorsement.*

[Sai17] Saint, C. & Saint, J. L. (2017). *Fabricating ICs.* Encyclopedia Britannica Online, https://www.britannica.com/technology/integrated-circuit/Fabricating-ICs. Accessed on 20.03.2023.

[San23] Sangiovanni-Vincentelli, A., Liang, Z., Zhou, Z., & Zhang, J. (2023). Automated design of chiplets. In *Proceedings of the 2023 International Symposium on Physical Design* (pp. 1–8). Virtual Event.

[Sch21] Schiavone, P. D., Rossi, D., Di Mauro, A., Gürkaynak, F. K., Saxe, T., Wang, M., et al. (2021). Arnold: An eFPGA-Augmented RISC-V SoC for Flexible and Low-Power IoT End Nodes. *IEEE Transactions on Very Large Scale Integration Systems, 29*(4), 677–90. ISSN: 1557-9999.

[Sch03] Schmit, H. (2003). Extra-dimensional Island-Style FPGAs. In *Proceedings of the 13th International Conference on Field Programmable Logic and Applications* (pp. 406–15).

[Sch22a] Schmit, H. & Denton, M. (2022). Multi-Input Serial Adders for FPGA-like Computational Fabric. In *Proceedings of the 2022 ACM/SIGDA International Symposium on Field-Programmable Gate Arrays* (pp. 35–41). Virtual Event.

[Sha49] Shannon, C. E. (1949). The synthesis of two-terminal switching circuits. *The Bell System Technical Journal, 28*(1), 59–98. ISSN: 0005-8580.

[Sin02] Singh, D. P. & Brown, S. D. (2002). Constrained clock shifting for field programmable gate arrays. In *Proceedings of the 2002 ACM/SIGDA Tenth International Symposium on Field-Programmable Gate Arrays* (pp. 121–26).

[Sin92] Singh, S., Rose, J., Chow, P., & Lewis, D. (1992). The effect of logic block architecture on FPGA performance. *IEEE Journal of Solid-State Circuits, 27*(3), 281–87.

[Sta95] Stanford Libraries (1995). *Interview with Bernard Vonderschmitt.* https://exhibits.stanford.edu/silicongenesis/catalog/pb421gq4809. Accessed: 27.09.2024.

[Tho21] Thompson, N. C. & Spanuth, S. (2021). The decline of computers as a general purpose technology. *Communications of the ACM, 64*(3), 64–72. ISSN: 0001-0782.

[Tok22] Tőkei, Z. (2022). Logic scaling options for the next 10 years: From FinFet to CFET, from dual damascene to semi damascene. In *Proceedings of the 2022 ACM/SIGDA International Symposium on Field-Programmable Gate Arrays.* Virtual Event.

[Tri94] Trimberger, S. (1994). *Field-programmable gate array technology.* Springer Science+Business Media New York.

[Tri15] Trimberger, S. M. (2015). Three Ages of FPGAs: A retrospective on the first thirty years of FPGA technology. *Proceedings of the IEEE, 103*(3), 318–31. ISSN: 1558-2256.

[Tri97] Trimberger, S., Duong, K., & Conn, B. (1997). Architecture issues and solutions for a high-capacity FPGA. In *Proceedings of the 1997 ACM Fifth International Symposium on Field-Programmable Gate Arrays* (pp. 3–9).

[TSM23] TSMC (2023). *Logic Technology*. https://www.tsmc.com/english/dedicatedFoundry/technology/logic. Accessed on 29.03.2023.

[Xil11] Xilinx Inc (2011). *Virtex-II Pro and Virtex-II Pro X Platform FPGAs: Complete Data Sheet*. DS083 (v5.0).

[Xil93] Xilinx Inc (1993). *XC2000 Logic Cell Array Families*.

[Xil98] Xilinx Inc (1998). *The Programmable LogicData Book*.

[Xil01] Xilinx Inc (2001). *Virtex-II Platform FPGA Handbook*.

[Yaz19] Yazdanshenas, S. & Betz, V. (2019). COFFE 2: Automatic modelling and optimization of complex and heterogeneous FPGA architectures. *TRETS, 12*(11), 3:1–27.

Chapter 3
Background

Contents

In the previous chapter, we introduced two fundamental building blocks of modern SRAM-based FPGAs: the LUT and the stored-select multiplexer. Furthermore, we described the development of island-style FPGAs from their MPGA origins to the present day. In this chapter, we will give further information about island-style FPGAs and CAD algorithms used to map user circuits onto them, which is necessary for understanding the subsequent chapters.

But first, let us briefly introduce the problem of programmable interconnect architecture design in its most abstract form, putting aside for now all the practical issues and usual solution approaches that are specific to island-style FPGAs.

© The Author(s), under exclusive license to Springer Nature Switzerland AG 2025

S. Nikolić, *Modern Programmable Interconnect Design*,
https://doi.org/10.1007/978-3-031-80629-2_3

3.1 The Problem of Programmable Interconnect Design

To be able to precisely define the problem of programmable interconnect design, we first need to formalize the notion of a fixed-functionality user circuit. Without loss of generality, we assume that it has already been expressed in terms of K-LUTs, after the process called *technology mapping* [Con94]. Furthermore, to simplify the definitions, we assume that flip-flops that exist in the user circuit have already been paired with appropriate LUTs (with some perhaps being configured as multiplexers), to conform to the BLE structure of Fig. 2.8. Hence, we will use the term "LUT" to designate both LUTs with registered and nonregistered outputs.

Definition 3.1 (Fixed-Functionality Circuit) A fixed-functionality circuit is a graph $G = (V, E)$ where each $v \in V$ is either an LUT, a *primary input*, or a *primary output* of the entire circuit. Every LUT-representing node has an in-degree $\leq K$, where K is the maximum LUT size used during technology mapping, every primary-input-representing node has a zero in-degree, and every primary-output-representing node has an in-degree of one and a zero out-degree.

Two examples of fixed-functionality circuits are shown in Fig. 3.1.

Definition 3.2 (Reconfigurable Circuit) An (n, K) reconfigurable circuit is a graph $H = (V_H, E_H)$, where each node is either a K-LUT, a primary input, a primary output, or a stored-select multiplexer. The number of nodes representing K-LUTs equals n. Each LUT-representing node has an in-degree of exactly K, while the in-degree of each multiplexer-representing node determines the corresponding multiplexer's size. Degrees of primary inputs and outputs are the same as in fixed-functionality circuits.

An example of a reconfigurable circuit is shown in Fig. 3.2. Because the in-degree of an LUT is set to exactly K, each of its inputs will be driven by exactly one other LUT, primary input, or multiplexer. Which physical input is driven by which direct predecessor is of secondary importance for now, since LUT inputs are all equivalent

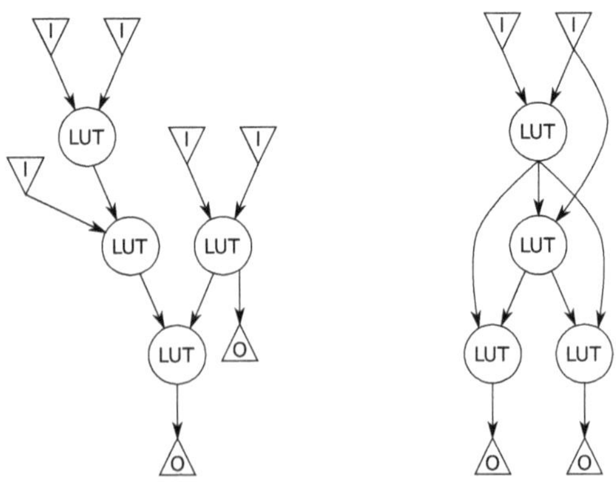

Fig. 3.1 Two examples of fixed-functionality circuits

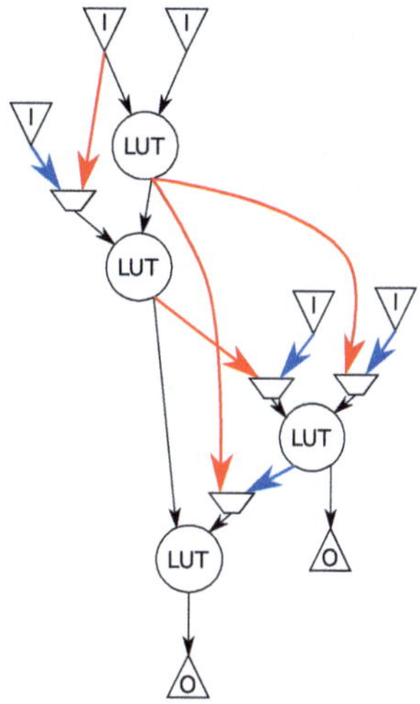

Fig. 3.2 An example reconfigurable circuit that can implement both fixed-functionality circuits of Fig. 3.1. Blue driver selection corresponds to a configuration implementing the left circuit, while the red one corresponds to a configuration implementing the right circuit

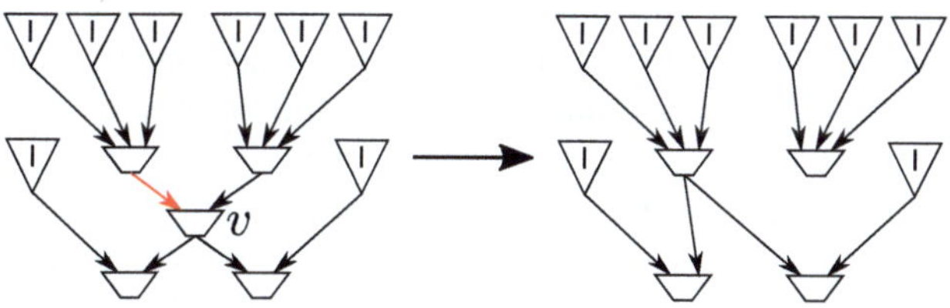

Fig. 3.3 Illustration of the driver selection operation

in terms of logic and only differ in terms of speed [Ahm09]. The same holds for multiplexers, though their inputs typically also have very similar delays [Lew16]. Since K-LUTs can implement any Boolean function with $\leq K$ variables, including a $K : 1$ multiplexer, strictly speaking, stored-select multiplexers are not necessary for implementing programmable interconnect [Kuc19]. Nevertheless, since silicon area of a $K : 1$ multiplexer rises linearly with K, whereas LUT area rises exponentially, it is very unlikely that architectures relying solely on LUTs for programmable connectivity will ever be competitive.

Let us now formalize the process of adapting the edge set of the reconfigurable circuit according to the needs of the given fixed-functionality user circuit.

Definition 3.3 (Driver Selection) Driver selection operation on a multiplexer- or LUT-representing node v is performed by selecting one of its direct predecessors, u, adding an edge from u to each direct successor of v, and removing v from the graph.

Driver selection operation is illustrated in Fig. 3.3.

Definition 3.4 (Configuration) Given a fixed-functionality circuit $G = (V, E)$ and a reconfigurable circuit $H = (V_H, E_H)$, we say that a mapping $f : V \mapsto V_H$

and a sequence of driver selections σ constitute a *configuration implementing G on H* iff after applying σ on H, f becomes an isomorphism from G to a subgraph of H.

Driver selections on the reconfigurable circuit of Fig. 3.2 corresponding to configurations that implement the left (right) fixed-functionality circuit of Fig. 3.1 are shown in blue (red).

We can now define the problem of programmable interconnect architecture design.

Definition 3.5 (Programmable Interconnect Architecture Design Problem) Given a set of fixed-functionality circuits Γ, find the lowest cost H, such that for every $G \in \Gamma$, there exists a configuration (f, σ) implementing G on H.

The missing part in the definition above is the definition of cost of a reconfigurable circuit. Likely the simplest reasonable way to define it is to count the number of edges, $|E_H|$. Since the number of LUTs in a (n, K) reconfigurable circuit is fixed, as are in-degrees of all LUT- and primary-output-representing nodes, minimizing $|E_H|$ minimizes the total number of multiplexer inputs, hence finding a balance between multiplexer sizes and their count. Traditionally, each multiplexer input has been called a *switch*, and minimizing the number of switches in the programmable interconnect architecture has been the focus of considerable research activity in 1990s and early 2000s [Lem04a]. The advantage of using edge count as the minimization objective is that perhaps the problem could be cast into that of finding a *minimum common supergraph* [Bun00] of all graphs in Γ, which is a far more general problem than that of programmable interconnect design. However, this metric is only a very rough abstraction of the actual goals in cost minimization: The programmable interconnect architecture must be efficient to lay out in silicon. For example, it should have a small area, be routable in a minimal number of metal layers, take the least amount of effort for the layout engineers to implement... Additionally, implementation of user circuits on the reconfigurable architecture usually has requirements in terms of achievable clock frequency and sometimes even runtime taken by CAD tools to find the configuration. All of these goals and constraints are difficult to capture in a closed-form minimization objective, leading the most common approaches to solving the programmable architecture design problem to rely on actual CAD tools to assess the quality of the proposed architectures. We shall use that approach in this book as well, although we will return to the appeal of direct objectives in Chap. 11.

Another problem with this definition of the programmable interconnect design problem is that it is very dependent on the choice of Γ. In principle, this would completely cease to be an issue only if Γ contained all circuits that may ever be mapped on the architecture under design. Since it is very difficult to predict a priori which exact circuits will be used on the FPGA, as this depends on both the user designing them and the synthesis tools, it may be desirable to guarantee that the programmable interconnect architecture will support all circuits satisfying certain constraints. Nevertheless, it is very difficult to even isolate relevant constraints that will not excessively limit the space for architecture optimization. Moreover, even if one could formally prove that it is possible to map a certain circuit on the given interconnect architecture, if there is no efficient algorithm to execute this mapping, the proof is of little practical use. This is another reason why most approaches to programmable interconnect design rely on using actual CAD tools in the design process, applying them to a finite set of circuits, both real and synthetic, with a hope that this set will be large enough to provide sufficient likelihood that circuits outside of it will also be implementable [Ros89, Tri97, Lew03, Cha15, Chr20]. Let us now see one very straightforward way to guarantee that *all* circuits are implementable.

3.2 Simplicity of the Complete Graph

The simplest way to guarantee that a programmable interconnect architecture H can implement any edge set is to make it a complete bipartite graph as follows: For each LUT, introduce K multiplexers that will drive its inputs. Then, for each LUT or primary input u, and each multiplexer v, add (u, v) to E_H. This is shown in Fig. 3.4.

Due to logic equivalence of LUT inputs, it is not necessary to be able to connect the output of every LUT (or every primary input) to all LUT inputs in order to guarantee that all pairs of LUTs (or primary-input-LUT pairs) can be independently connected together. Connectivity patterns minimizing the total number of switches $|E_H|$, while retaining this full connectivity, have been independently discovered by DeHon [DeH95] and Ye [Ye10]. Nevertheless, because a change in value of an input closer to the root of the LUT's multiplexer tree (e.g., input A in Fig. 2.7) has to propagate through fewer levels of pass transistors to become visible at the output than a change in value of an input closer to the truth table memory (e.g., input C in Fig. 2.7), it may still be useful to retain richer connectivity than the minimum allowed by input permutability. In this way, it can be ensured that the timing-critical signals—which are different in every circuit and hence not a priori known—can reach the faster inputs.

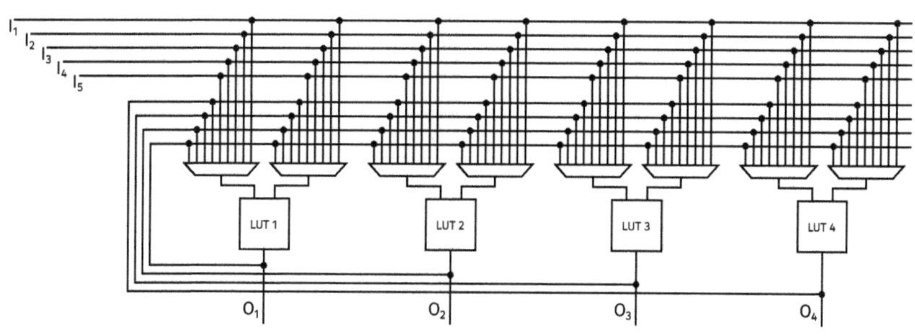

Fig. 3.4 Programmable connections implementing a complete graph on four 2-LUTs, with five primary inputs. Output of each LUT is also made a primary output, for simplicity

3.3 A Collection of Cliques

The complete graph routes every signal from its source to its destination using only a single level of multiplexing. In terms of the number of multiplexer hops, this is as fast as it gets in programmable interconnect, unless some LUT input is permanently fixed to a particular driver. The problem is, however, that as the size of the reconfigurable circuit increases, so does the size of each multiplexer. At some point, even though the number of hops remains one for all connections, the delay of this hop becomes exceedingly high. One idea that could help alleviate this problem could be to decompose the larger reconfigurable circuit into cliques of size N, where this N is determined as the number of LUTs beyond which a single-level multiplexer structure starts to become inefficient due to multiplexer size increase. One could then connect these cliques together using some sparser and more efficient structure.

Note the similarity between the complete graph of Fig. 3.4 and the logic cluster of the APEX 20K FPGA, previously shown in Fig. 2.12. In fact, cluster-based FPGAs, which are the predominant class today, employ exactly this approach of breaking the graph into a collection of sparsely connected cliques. A standard block diagram of a logic cluster—abbreviated from now on as *CLB*, inherited from Xilinx terminology [Hau07]—is shown in Fig. 3.5. Because the inputs to the cluster are no longer the primary inputs of the entire circuit, they also have to be programmably driven. Hence, the LUT input multiplexing structure is now made up of two levels: one to bring in external inputs to the cluster and the other to further dispatch them to the inputs of individual LUTs. The first level is typically called a *connection block* and the second a *(local) crossbar* [Bou21].

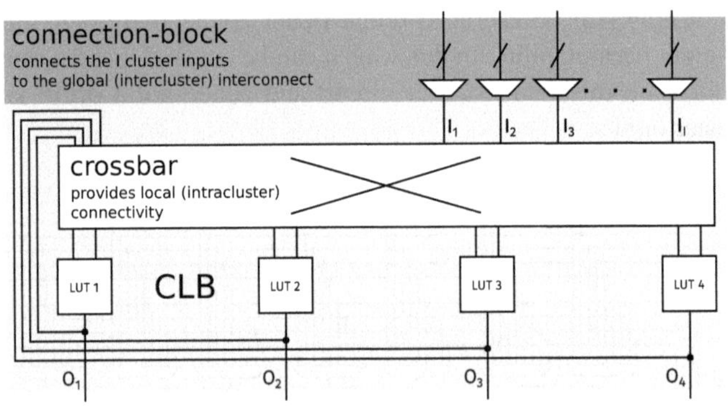

Fig. 3.5 Schematic of a logic cluster (CLB) composed of four 2-LUTs and a *connection block* that provides its inputs from the external routing

3.4 Rent's Rule

Although splitting H into a collection of sparsely connected cliques, rather than implementing it as a complete graph, was a necessity dictated by physical constraints, justifications for doing so are also firmly grounded in a fundamental empirical observation about the nature of connectivity of fixed-functionality circuits called the *Rent's rule* [Chr00]. Rent's rule was first observed in electronic circuits but has since been shown to even hold in nature [Red09]. It also forms the basis for a large part of our current theoretical understanding of programmable interconnect [DeH99, Pis03, Sch03]. For this reason we briefly review it in this section.

Let us assume that we have a method for bipartitioning a circuit G such that the node sets of the two halves, V^A and V^B, are roughly of equal size and that the number of nodes in one half that connect to at least one node in the other half, $\{u \in V^A : (\exists v \in V^B)((u, v) \in E)\} \cup \{u \in V^B : (\exists v \in V^A)((u, v) \in E)\}$, is minimized. Let us now apply this bipartitioning recursively on each half, quarter, etc. Taking one thus obtained subset of nodes at any level of recursion, V^R, we define its external connectivity as

$$S(V^R) = \{u \in V^R : (\exists v \in V \setminus V^R)((u, v) \in E\} \cup \{u \in V \setminus V^R : (\exists v \in V^R) \\ ((u, v) \in E)\}. \tag{3.1}$$

Rent's rule relates, on average, the size of $S(V^R)$ to the size of V^R [Chr00]:

$$\left|S(V^R)\right| = k \times \left|V^R\right|^p, \quad p \in (0, 1). \tag{3.2}$$

Here k is the average size of external connectivity of individual nodes of G, which is at most $K + 1$, for G mapped onto K-LUTs. More important is the power law with exponent p called the *Rent's exponent.* The exact value of p differs from circuit to circuit but is typically between 0.5 and 0.7 [Chr00]. What Rent's rule states is that a circuit can be partitioned into subsets of nodes such that connectivity tends to stay increasingly local to a subset of nodes, as the size of this subset increases. This can be observed in Fig. 3.6 plotting for varying size of V^R the ratio of the expected size of external connectivity of V^R computed using Rent's exponent of 0.7 and its theoretical maximum, if no drivers are generated within the subset or shared among its nodes. From the plot of Fig. 3.6, we can observe that when clusters of size ten are used, the amount of connectivity external to the clusters is reduced by almost 50%, even for the relatively high Rent's exponent of 0.7. Interestingly enough, APEX 20K contained exactly ten 4-LUTs in its cluster (Fig. 2.12), which may have resulted from a similar observation. Note that this does not depend on K, so the same observation would hold for newer Stratix architectures based on a cluster of ten 8-input ALMs [Lew05], although LUT fracturability

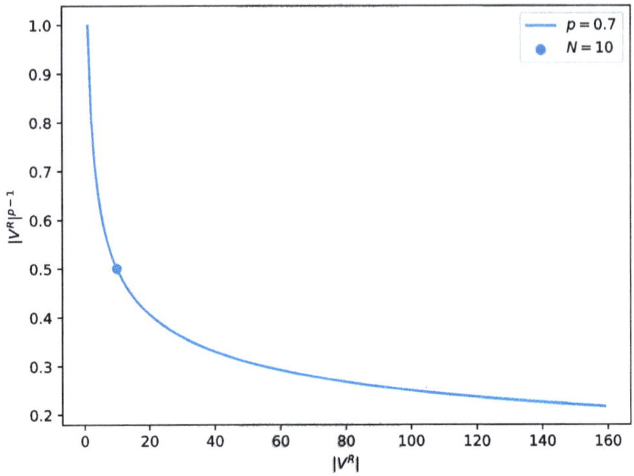

Fig. 3.6 Localization of connectivity. The plot shows the ratio of the expected size of external connectivity of V^R, computed using Rent's rule ($k\left|V^R\right|^{p=0.7}$) and its theoretical maximum, assuming that all nodes have k incident edges ($k\left|V^R\right|$). For $\left|V^R\right| = 1$, there is no sharing of drivers among the elements of the set, nor are any drivers generated locally. However, already for $\left|V^R\right| = 10$, irrespective of k, almost half of the drivers are shared or generated locally

complicates the analysis in that case. Of course, locality of connections can be further exploited to recursively create hierarchical clusters such as the APEX 20K Mega LAB (Fig. 2.12). Plot of Fig. 3.6 suggests that using a 160 4-LUT Mega LAB would reduce the amount of external connectivity by about 78%, provided that the sparse supercluster interconnect can implement all internal connections. However, the returns on recursively extending the cluster are diminishing, and most modern architectures have dropped this approach, although the new 32 6-LUT cluster of Xilinx/AMD Versal appears to bear some resemblance [Gai19].

3.4.1 *How Many Inputs Does a Cluster Need?*

Rent's rule provides some guidance in choosing the cluster size based on the number of connections that remain external as the size increases. It is even more useful for determining the number of inputs that a cluster of any given size needs, since this can be computed from the formula directly. For example, a cluster of ten 4-LUTs should contain $I = \lceil 4 \times 10^{0.7} \rceil = 21$ inputs to be able to implement an average group of ten 4-LUTs from a circuit with Rent's exponent equal to 0.7. Note should be taken, however, that the average number of inputs to LUTs of a circuit mapped onto K-LUTs is generally lower than K. For example, Hutton et al. observed that a typical circuit mapped onto 6-LUTs will have an average LUT size of about 4.67 [Hut04]. Moreover, Pistorius and Hutton determined that the average Rent's

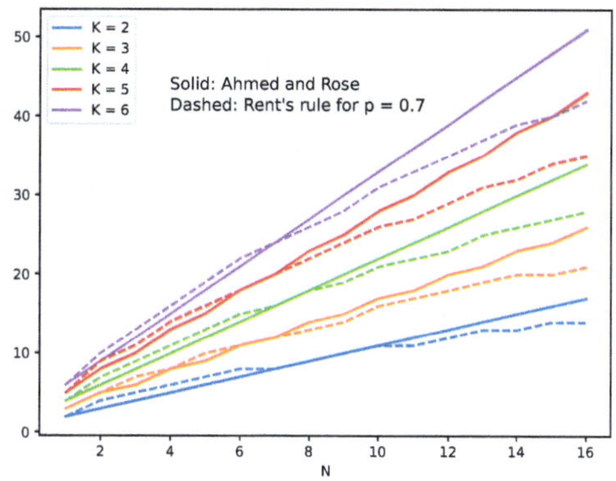

Fig. 3.7 Comparison of Rent's rule for $p = 0.7$ and the formula for determining the number of cluster inputs that guarantees dense packing (Eq. 3.4 [Ahm00]). Equation 3.4 can be considered a linearization of Rent's rule around $N = 8$

exponent of typical industrial circuits mapped on FPGAs is 0.6, meaning that 0.7 indeed corresponds to demanding circuits [Pis03]. Hence, a cluster with 21 inputs is likely to be able to implement groups of LUTs with above-average input demand too.

In fact, Betz and Rose demonstrated experimentally that allowing 22 inputs to a cluster of ten 4-LUTs enables an average cluster utilization of 98% [Bet98]. They also provided a formula for determining the number of inputs that are required to satisfy this utilization rate, which is valid for $N \in [1, 16]$ [Bet98]:

$$I = 2N + 2. \tag{3.3}$$

This formula was later generalized by Ahmed and Rose for any $K \in [2, 7]$:

$$I = \frac{K}{2}(N + 1). \tag{3.4}$$

As the plot of Fig. 3.7 shows, the Betz and Rose formula can be considered a linearization of Rent's rule for $p = 0.7$ around $N = 8$, which was the middle of the cluster size range that they explored. Same can be said about the generalization of Ahmed and Rose.

While leveraging Rent's rule to reduce the number of inputs to the cluster can result in a net area reduction, not all FPGA architectures employ this technique. A notable example is the 7-Series FPGA family of Xilinx [Pet21]. Retaining the maximum possible cluster input bandwidth makes it easier to perform flat placement, which is crucial for achieving high performance [Li19a] (see Sect. 3.9.3).

3.4.2 *Sparse Crossbars*

We have already mentioned the result independently derived by DeHon and Ye that permutability of LUT inputs allows crossbars to be sparsified while still guaranteeing that all combinations of cluster inputs can reach all LUTs independently [DeH95, Ye10]. Lemieux and Lewis demonstrated experimentally that the level of sparsification can be significantly increased, without any apparent loss of routability, but with a significant reduction in silicon area needed to implement the crossbar [Lem01]. In particular, they concluded that removing 50% of all switches in a fully populated crossbar almost always still leads to a routable design. Early commercial architectures adopting crossbar depopulation were also based on 50%-sparse crossbars [Lew03], whereas newer generations often rely on even sparser ones [Shr23]. Lemieux and Lewis also observed that giving up on some of the reduction in the number of cluster inputs that Rent's rule would enable allows sparser crossbars to remain routable, leading to a net silicon area reduction. This may be another reason why very sparse commercial architectures do not attempt to reduce the cluster input bandwidth [Shr23].

In their previous work, Lemieux et al. also observed that a highly routable sparse crossbar approximates a *superconcentrator* [Lem00]. Let $G_C = (V_C, E_C)$ be a *Directed Acyclic Graph* (DAG) [DeM94] where the nodes in $I \subset V_C$ are designated as *inputs* and the nodes in $O \subset V_C$ as *outputs*. Furthermore, let $|I| = |O|$, $I \cap O = \varnothing$, and in-degree of any $u \in I$ (out-degree of any $u \in O$) be zero. G_C is a superconcentrator iff for every $r \in [1, |I|]$ and every $A \subseteq I$ and $B \subseteq O$ for which $|A| = |B| = r$, there exist r node-disjoint paths between A and B [Alo84].

The problem with superconcentrators is that it is even difficult to detect them: Deciding whether a given DAG is a superconcentrator is generally a co-NP-complete problem [Blu81]; the only exception are depth-one DAGs (with the longest path length equal to one), for which it is in P [Blu81]. When perceived in isolation, a cluster crossbar is a depth-one DAG, however. By relying on Hall's theorem that can be used for deciding whether a depth-one DAG is a superconcentrator or not [Blu81], Lemieux et al. developed an efficient heuristic algorithm for constructing highly routable sparse crossbars [Lem00]. However, as mentioned before, connection block and crossbar together form a multilevel multiplexing structure and the combined graph representing it is no longer of depth one. Optimizing the two levels independently may result in lost opportunity for sparsification or reduced routability. On the other hand, optimizing them together using a superconcentrator approximation is likely significantly more difficult—which has been acknowledged by Lemieux and Lewis as well [Lem01]—since even detecting a superconcentrator becomes difficult [Blu81].

Superconcentrators provide a sound way of obtaining an upper bound on routability of crossbars with given density. Nevertheless, layout restrictions typically prevent their implementation in practice [Lew03]. A practical sparse crossbar that was used in the original Stratix is shown in Fig. 3.8. Forcing all LUTs to have identical input connectivity means that their outputs could be swapped without a need to reroute the inputs [Lew03]. Moreover, input sharing reduces the amount of metal and the number of vias needed to lay out the crossbar, simultaneously reducing the capacitive load on the drivers of the crossbar inputs [Chr20].

An alternative measure of routability, defined as the logarithm of the number of distinct configurations that a multilevel multiplexing structure can support, and

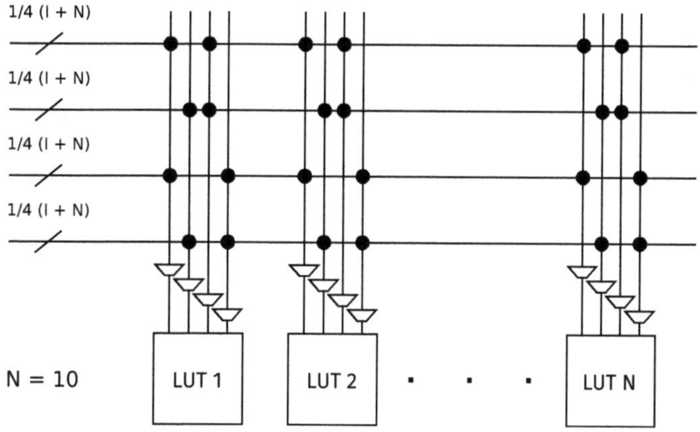

Fig. 3.8 50%-sparse crossbar used in Stratix [Lew03]. Since all LUTs have the same input connectivity pattern, they can be shuffled within the cluster without rerouting their inputs. This increases overall routability compared to other sparsification approaches [Lew03]

called *entropy*, has also been successfully used for analysis and joint optimization of such structures [DeH95, Fen08, Gre11]. This measure is related to the number of different input-to-output mappings that the structure can implement and it can be determined either in closed form, as has been done for some classes of multilevel multiplexer structures by DeHon [DeH95] and Kaptanoglu and Feng [Fen08], or using a more general Monte Carlo approach, as has been done by Lemieux et al. for evaluation of routability of sparse crossbars [Lem00].

3.5 Wire Sharing

In previous sections, we have determined that a programmable interconnect architecture can be efficiently constructed as a collection of sparsely connected cliques. We have also provided some information about how the size of the clique can be determined and how many inputs it needs to effectively connect to other cliques. As of now, however, we have not mentioned anything about connecting together different clusters, other than that this connectivity must be sparse.

Let us start by first considering a pair of clusters. One idea could be to simply sparsify the connection blocks as explained in Sect. 3.4.2. This approach is illustrated in Fig. 3.9a. As we can see from the figure, it is highly wasteful in terms of wires needed to implement the circuit in silicon, because all outputs of cluster *A* are brought to the inputs of cluster *B* before the sparsification occurs.

Recall that Rent's rule dictates that, on average, a portion of signals generated within the cluster will only be used locally and does not have to leave it. In particular, for a cluster of size ten, about half of the signals will remain local. Hence, it should suffice that only five instead of ten wires are brought from cluster *A* to cluster B. If

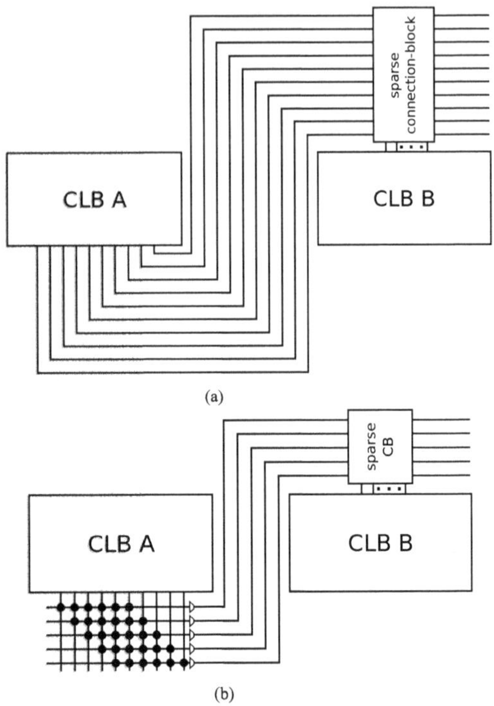

Fig. 3.9 Illustration of wire sharing through configuration-domain multiplexing. Rent's rule stipulates that only a portion of locally generated signals have to leave the cluster to reach another one. Leveraging this can greatly reduce the number of physical wires needed to implement the reconfigurable circuit. The remaining wires can be multiplexed across different configurations to select drivers appropriate in different circuits. (**a**) All wires. (**b**) Multiplexed wires. This connectivity pattern is due to DeHon [DeH95]

in some particular instance more than five signals need to pass from *A* to *B*, *A* can always be broken into multiple clusters instead.

What Rent's rule does not tell us, however, is exactly which five LUTs will need to pass their outputs to the other cluster. This will generally depend on the exact circuit being implemented. The problem is resolved by driving the five wires themselves through stored-select multiplexers, as illustrated in Fig. 3.9b. What this essentially achieves is multiplexing the same long metal wires over multiple configurations of the reconfigurable circuit. Time-domain-multiplexing of resources during the execution of a single configuration and not across configurations [Tri97, Fra08], as well as packet-switched routing provided by hardened *Networks on Chip* (NoCs) [Abd14], will be discussed in Chap. 9.

3.5.1 Tree-Based Hierarchical FPGAs

Rent's rule also stipulates that some of the signals passing between clusters *A* and *B* will stay local to that pair of clusters. Hence, only a portion of signals leaving *A and B* needs to be distributed further, to another pair. To achieve this, we can again apply multiplexing, as illustrated in Fig. 3.10 using the graph model of Sect. 3.1.

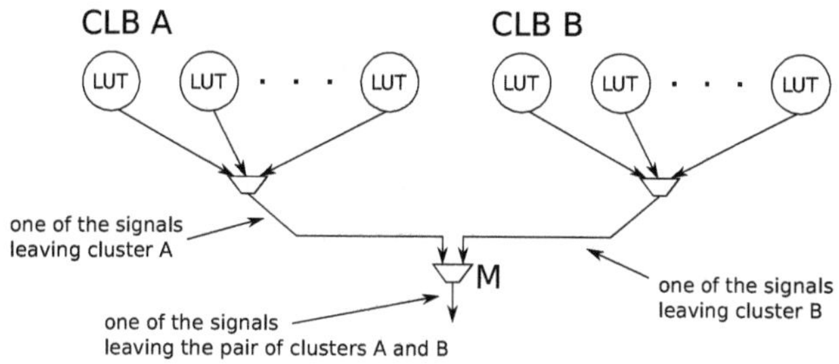

Fig. 3.10 Tree-based multiplexing. Rent's rule indicates that only a fraction of signals will have to leave the pair of clusters A and B, in order to connect to another pair. We can achieve this reduction by adding a level of multiplexers (M in the figure) to choose among the signals leaving CLB A and those leaving CLB B. Such recursive construction leads to a *tree-based* hierarchical FPGA [DeH04]

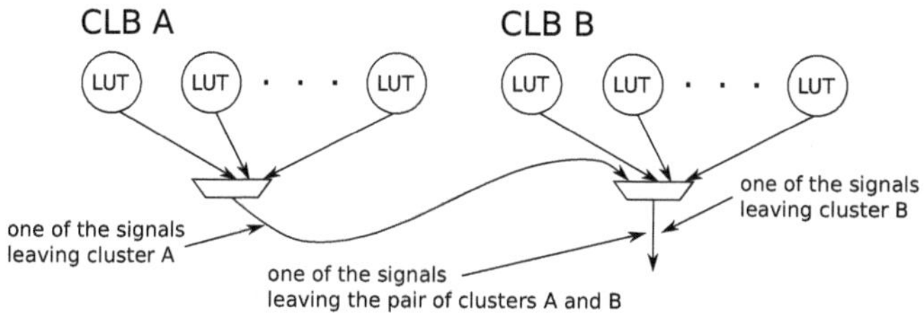

Fig. 3.11 Segmented-channel routing. Combining signals leaving the two clusters can also be achieved by cascading the appropriate multiplexers. This corresponds to a segmented-channel architecture described in Chap. 2

Recursive multiplexing in this manner leads to a *tree-based* hierarchical FPGA [DeH04]. In this type of a hierarchical FPGA, wires generally do not span half of the entire chip, and large devices contain many more levels of hierarchy than APEX 20K. However, wires implementing connections at higher levels of hierarchy are still very long and hence have to be routed at thick metal layers [DeH04] that are already needed for clock and power distribution [Fie15]. Finally, as we have already mentioned, porting ASIC CAD algorithms to hierarchical architectures is difficult, and tree-based ones are no exception to this.

3.5.2 Multiplexer Cascades

Programmably selecting between outputs leaving two clusters can also be achieved by cascading their output multiplexers as shown in Fig. 3.11. Note that the obtained graph is now composed of two identical subgraphs, with an edge between them. If we interpret this edge as an abstraction of a channel wire, for channel wires in an island-style FPGA also serve to pass signals between multiplexers in the vicinity of different clusters, we can see that this construct corresponds to segmented-channel routing of Sect. 2.2. The fact that the two subgraphs are now identical makes segmented-channel routing particularly appealing as the entire FPGA can be composed of identical tiles. The same cannot be said about the graph of Fig. 3.10 that contains an additional multiplexer, M. The tileablity property of constructed programmable interconnect architectures can be ensured by restricting the solutions of the design problem of Sect. 3.1 to a special class of graphs that we define next.

3.6 Periodic Graphs: A Unified Way of Representing Tiled Architectures

Tileable architectures naturally correspond to the notion of *periodic graphs* that we borrow from Höfting and Wanke [Hof94].

Definition 3.6 (Static Graph [Hof94]) A 2D static graph is an edge-weighted directed multi-graph $S = (V_S, E_S \subseteq V_S \times V_S \times \mathbb{Z}^2)$.

Multiple edges distinguished by their weight vectors can exist between any two nodes, as can loops, making the graph a multi-graph. An example of a static graph representing a simple FPGA architecture is shown in Fig. 3.12a.

Definition 3.7 (Periodic Graph [Hof94]) A 2D periodic graph P corresponding to a 2D static graph $S = (V_S, E_S)$ is the graph obtained by replicating V_S at the points of a 2D integer lattice, and for each $e = (u, v, \mathbf{w}) \in E_S$ and each $\mathbf{p} \in \mathbb{Z}^2$, adding an edge between the replica of u whose position vector is $\mathbf{p}$ and the replica of v whose position vector is $\mathbf{p} + \mathbf{w}$.

As we will see in Chap. 7, static graph representation is particularly useful for designing fixed-connectivity patterns of wires between individual LUTs [Nik20]. Nevertheless, they can represent any tileable architecture and have recently been used for analysis of commercial FPGAs as well [Gre23]. The periodic graph corresponding to the static graph of Fig. 3.12a is shown in Fig. 3.12b. It was deliberately drawn in the form usually used to represent island-style FPGA architectures [Bet99]. For example, multiplexer m_3 is represented as a driver of a channel wire going two tiles left. The static graph, however, merely specifies the connectivity between multiplexers. In particular, it states that each tile contains a multiplexer m_3 that is connected to multiplexers m_1, m_3, m_4, and m_6, two tiles to the left (due to the $(-2, 0)$ weight of the corresponding edges) as well as to multiplexer m_1 one tile to the left (due to the $(-1, 0)$ weight of the remaining edge). That this hyperedge is typically realized using a length-2 horizontal channel wire going left, with one intermediate tap, is essentially an implementational detail. Note also that there is no need for the weights of edges of the static graph connecting multiplexer-representing nodes to have exactly one nonzero component. Both components can be nonzero, to describe noncardinal wires that exist in some FPGA architectures, such as the

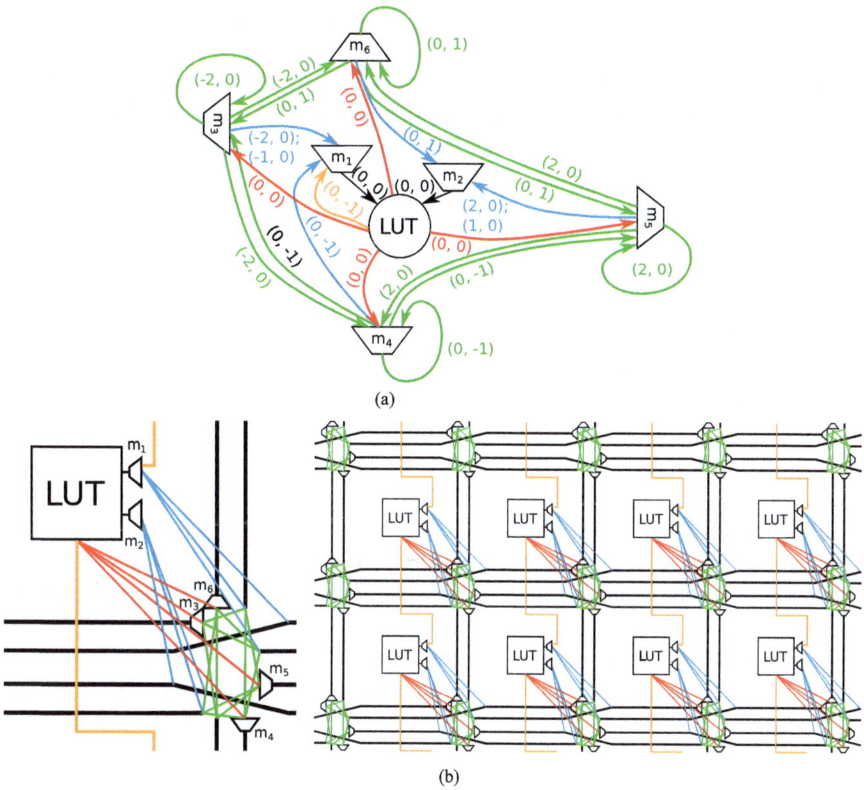

Fig. 3.12 Representation of tileable FPGA architectures using periodic graphs. (**a**) Static graph. (**b**) Periodic graph with a connectivity detail

Xilinx 7-Series [Pet21], or both can be zero, to describe multiplexer networks that are local to the tile, which also commonly occurs in commercial FPGAs [Pet21].

3.7 Academic Terminology of Island-Style FPGAs

Nevertheless, due to their historical origins in MPGAs that we discussed in Chap. 2, FPGAs are usually not thought of as networks of multiplexers, edges of which are implemented as wires, but rather as channels of prefabricated wires that have to be connected together through multiplexers and to which logic blocks must also be able to connect through multiplexers [Bou21]. This gave rise to a particular standard decomposition of the interconnect architecture into several interconnect blocks, many of which we have already seen. In particular, multiplexers that are direct predecessors of LUTs are usually said to belong to the *crossbar* [Bou21] (Sect. 3.3), while the edges with heads in these multiplexers are said to determine the *crossbar connectivity pattern*. Multiplexers that are direct predecessors of crossbar multiplexers are usually said to belong to the *connection block* [Bou21] (Sect. 3.3), although this structure does not always exist; that is the case in Fig. 3.12. We note that some commercial architectures such as those from Xilinx/AMD and Actel/Microchip jointly call the connection block and the crossbar an *Input Interconnect Block* (IIB) [Fen08, Shr23].

Analogously to the crossbar connectivity pattern, edges that have their heads in the multiplexers of the connection block are typically said to determine the *connection-block connectivity pattern* (blue in Fig. 3.12). Multiplexers that drive channel wires are typically said to belong to the *switch-block* [Bou21]. As we have already noted, static graph representation does not intrinsically carry the notion of channel wires as such, so to make this distinction when it is needed, it is necessary to appropriately label the corresponding multiplexers. Historically, when FPGAs commonly employed tri-stated buffers, the equivalent of edges between LUTs and switch-block multiplexers was considered to determine the *output connection-block connectivity pattern* [Bet99a] (shown in red in Fig. 3.12). On the other hand, edges between switch-block multiplexers are said to determine the *switch-block connectivity pattern*, which we will refer to simply as *switch-pattern* in the remainder of the book, unless stated otherwise. Switch-pattern is shown in green in Fig. 3.12.

Common abbreviations of connection- and switch-block are CB and SB, respectively [Bou21]. For easier comparison with commercial architectures, we note that in Altera/Intel terminology, closest analoga to CB, crossbar, and SB multiplexers are, respectively, *LAB Input Multiplexers* (LIMs), *Logic Element Input Multiplexers* (LEIMs), and *Driver Input Multiplexers* (DIMs) [Chr20].

When output of an LUT directly drives an input of another LUT (potentially through a decoupling multiplexer), without passing through any switch-block, connection block, or crossbar, it is said to provide a *direct connection* to that LUT [Nik20]. Such connections have been used in commercial FPGAs in a limited manner from XC2064 [Xil93] all the way to Versal [Gai19]. One example of a direct connection between LUTs is indicated in orange in Fig. 3.12.

3.8 Ideal and (Currently) Realistic Design Goals

Ideally, one would design the entire interconnect architecture—that is, the entire static graph describing it—at once. Otherwise, assuming orthogonality between decisions made in parts of the decomposed problem may lead to wrong conclusions [Yan02]. Nevertheless, as even adequately exploring the design space of individual subproblems has in many cases been a challenge, the decomposition is typically still retained. That is, channel segmentation (determining how many wires of which length the horizontal and vertical routing channels will be composed of), connection block, crossbar, and switch-block connectivity patterns are usually designed in isolation, with other aspects of the programmable interconnect architecture fixed.

3.9 FPGA CAD Flow

The purpose of FPGA CAD tools is to map a fixed-functionality user circuit onto a given FPGA architecture. In this section, we review the main stages of a typical FPGA CAD flow, introducing the problems that each stage is tasked with solving, and briefly discussing the main algorithms that are used to this end. Further details about some individual stages of the flow will be given in the appropriate subsequent chapters. For a more comprehensive recent overview of the entire topic, including *High-Level Synthesis* (HLS), which has here been entirely left out, interested readers may refer to Tu et al. [Tu24]. Valuable insight from the original authors of several fundamental FPGA CAD algorithms can be obtained from Hauck and DeHon [Hau07].

3.9.1 Synthesis

The first step of a typical FPGA CAD flow is actually identical to that of a typical ASIC CAD flow: A textual description of the user circuit, written in a *Hardware Description Language* (HDL) such as Verilog or VHDL, is first transformed into a technology-agnostic graph representation. One of the most popular representations in use today is the so-called *And-Inverter Graph* (AIG), in which each node represents a 2-input AND gate and polarity of edges can be optionally inverted [Mis06]. A number of technology-independent optimizations are then applied to this circuit graph in order to improve some metric of its quality. Common metrics are graph's size expressed in terms of the number of its gates, as well as *depth*—the length of the longest path between any two registers. A classical reference for various synthesis algorithms can be found in the textbook of De Micheli [DeM94], while a discussion of state-of-the-art SAT-based synthesis algorithms can be found in the thesis of Haaswijk [Haa19]. Likely the most popular open-source tool implementing various synthesis algorithms is *ABC* [Mis23]. It is integrated with a robust Verilog front end in an open-source tool called *Yosys* [Wol23].

3.9.2 Technology Mapping

Once the technology-independent circuit graph has been sufficiently optimized, it has to be rewritten in terms of the available gates in a step called *technology mapping* [DeM94]. This is where the FPGA and ASIC flows first diverge. Namely, when designing a standard cell-based ASIC, the product of technology mapping is a circuit graph in which nodes correspond solely to logic gates that exist in the standard cell library. Since the number of these gates is usually small, classical technology mapping algorithms have relied on matching different structural representations of each gate (different graphs composed of the gates used in the technology-independent subject graph—AND-2 and NAND-2 in case of AIG—which compute the given standard cell function) in the technology-independent subject graph [DeM94]. The main problem with applying the same approach to LUT mapping is that each K-LUT can compute 2^{2^K} Boolean functions, each with multiple structural representations. To combat this explosion, early technology mapping algorithms such as *Chortle-crf* [Fra91] used various heuristics, until Cong and Ding developed the *FlowMap* algorithm, which relies on enumerating input-count-feasible cuts in a topological order, and proved that it can perform depth-optimal technology mapping onto LUTs in polynomial time [Con94]. To the best of our knowledge, FlowMap remains at the core of all subsequent FPGA technology mapping algorithms.

One interesting example of such an algorithm is *WireMap* [Jan09]. Besides reducing LUT area under depth-optimality constraint, it also attempts to reduce the amount of connectivity between the LUTs. Considering connectivity at every stage of the CAD flow is of great importance for improving efficiency of programmable interconnect architectures but has unfortunately not been sufficiently practiced.

As far as the author is aware, depth-optimal technology mapping is the only major problem that occurs in transforming a fixed-functionality user circuit into its FPGA implementation for which an optimal polynomial-time algorithm is known.

3.9.3 Placement

Once the circuit graph has been represented in terms of LUTs, each of its nodes has to be assigned a physical LUT on the FPGA grid—that is, it has to be assigned coordinates. This is done in a step called *placement* [Bet99]. Sometimes, before placing nodes of the circuit graph, they are grouped in such a way that each group can be implemented by a single logic cluster. If this step exists, it is typically called *packing* or *clustering* [Bet99]. Afterward, instead of placing individual LUTs, entire clusters are placed, which can produce a great benefit for placement runtime, since fewer objects need to be assigned a location [Bet99]. However, it has long been known that grouping LUTs a priori, without knowing where they are going to be located, can result in inferior performance of the implemented circuit, since connections between nodes that end up in different clusters may be unnecessarily elongated [Che07]. As a remedy, early placement algorithms proposed breaking up clusters late in the placement process, to release nodes belonging to timing-critical signals and reposition them in different, more appropriate clusters [Che04, Bet20].

Many modern placement algorithms alleviate this problem by skipping packing altogether and placing LUTs in a flat manner, each circuit graph node being a placeable object and each physical LUT being a candidate location [Li19a]. What makes this order of magnitude increase in the number of placeable objects possible is that these modern algorithms moved away from the classical simulated annealing-based approaches that do not scale well with problem size [Bet20]. They instead adopt analytical techniques that do not only scale significantly better [Li19] but are also much better parallelizable, even allowing effective acceleration on GPUs [Raj22].

To enable flat placement, intracluster routing architecture of commercial FPGAs that are targeted by these algorithms is deliberately designed such as to impose as few placement constraints on the movable LUTs as possible [Pet21]. Often, the only constraints are the limited number of control signals available for flip-flops placed in the same cluster [Yan16]. If this were not the case, testing whether LUTs can be placed in the same cluster during the placement process itself could be prohibitively costly in terms of runtime, since it would require ensuring that all signals can be brought to their destination inside the cluster [Luu14a]. On the other hand, legalization of placement using a postprocessing step could result in too significant displacements of different nodes, nullifying much of the optimization that was previously obtained.

General FPGA CAD flows such as VPR, intended to support any FPGA architecture, even if it does not possess the aforementioned characteristics of those intended for flat placement, still perform packing before placement [Mur20]. Due to flexibility of the simulated annealing heuristic, VPR also still relies on it, despite its inferior scalability [Mur20]. Note that this is an issue only for leading-edge FPGAs with very high capacity, which have been the focus of a majority of recent FPGA CAD research [Mur15]. Nevertheless, as we have seen in Chap. 2, reconfigurable fabrics comprising just a few thousands of LUTs are increasingly present as blocks in larger SoCs, or as standalone affordable edge devices. For them, simulated annealing is still quite relevant—especially if the algorithm can be tuned in such a way as to maximize the resource-poor fabric utilization and performance at the expense of increased runtime. Recent research has shown that even this runtime can be reduced by using reinforcement learning to guide the selection of special *directed* moves [Elg2022].

The rationale behind all placement heuristics is to make the edges of the circuit graph as short as possible by placing their endpoints near each other, since the shorter the edge is, the faster the connection through programmable interconnect that implements it will be. The most common optimization objective is some approximation of total wirelength [Bet99]. Of course, not all signals of the circuit are likely to end up on the critical path and hence, in optimizing the lengths of various edges, those corresponding to more timing-critical signals are typically given higher importance; that is, higher weight in the total wirelength sum [Mar00].

Minimizing total wirelength has another, perhaps much more significant effect: It is strongly correlated with the ability to *route* (see the next subsection) the placed circuit with a limited number of prefabricated wires [Pis03]. Modern placement algorithms also include advanced techniques for estimating and minimizing routing congestion more directly, sometimes based on machine learning models [Maa18].

3.9.4 Routing

The final stage of an FPGA CAD flow is called *routing* [Bet99]. In routing, each net of the circuit is implemented as a tree of appropriate routing resources in the programmable interconnect, such that trees belonging to different nets are disjoint; otherwise, any overlaps would create a short circuit. Typically, the optimization goal is to minimize the critical path delay, by assigning greater importance to timing-critical signals, such that they make smaller (ideally no) detours from the shortest possible paths from the net's source to its sink terminals, which have been fixed during placement [McM95]. Once placement is complete, it is usually assumed fixed during routing and is not revisited, unless routing fails due to excessive congestion (the number of signals that need to pass through a certain region of the chip is such that they cannot be distributed among the limited number of FPGA's resources to produce a legal, overlap-free routing). This is precisely the reason why increasingly more effort is being put into accurately predicting the behavior of the router during placement [Maa18], so that the number of place and route iterations can be limited. One common exception to keeping placement fixed in commercial routers is that they can reposition LUTs within their respective clusters, which is made possible by specific construction of the crossbar [Lew03]. There have also been historical attempts to simultaneously place and route a circuit [Nag98] in order to maximize the quality of the implementation, but, due to high computational effort, this has recently been limited mostly to relatively small regions instead of the entire circuit [Fra18].

A majority of modern routers [Kap12] are based on the *PathFinder* algorithm [McM95], which uses *congestion negotiation* principles to iteratively produce a legal routing. We will review PathFinder in greater detail in Sect. 5.3.

3.9.5 Bridging the Gaps Between the Stages

The traditional split between the different CAD stages came out of necessity to manage the high computational complexity of the overall transformation of the specification of the user circuit into an actual physical implementation. On several occasions, it has been demonstrated that combining different stages together and solving their problems simultaneously yield better results [Nag98, Fra18]. Yet, the computational cost of doing this remains largely prohibitive, so the two most popular approaches to bridging the gap between different stages have been: (1) to locally revisit decisions of the previous stage once more information is available from the subsequent ones [Che04, Sin05] and (2) to try to predict in advance the behavior of the subsequent stages and guide the optimization in the present one accordingly [Che07, Maa18]. The second approach typically uses either advanced machine learning models [Maa18] or fast approximate surrogate algorithms such as global routing, which determines only the channels through which each signal should pass, such that channel capacity is never exceeded, without taking into consideration the actual distribution of signals among the individual routing resources [Bet99]. Machine learning in particular has recently attracted significant attention from major ASIC EDA tool vendors [Han22, Kha22], and it is probably just a matter of time until major FPGA vendors will start incorporating it in their CAD flows as well.

There has recently been a drive toward making FPGAs more appealing to software developers [Eve16]. For this effort to be successful, it is not sufficient to raise the level of abstraction of hardware description [Jos22]; compilation time of the FPGA CAD flow must be drastically reduced as well, in order to make it comparable to what software developers are used to [Xia22]. This is one side of the story, intended to provide a solution to flattening out of general-purpose CPU performance. However, note should be taken that there is still significant room for improvement of software performance, simply through writing better software or developing better compilers [Ans23]. The other side of the story is that a large number of hardware designers who have been engaged in developing ASIC designs now have to shift to FPGA development, due to simple economics of prohibitive ASIC design costs [Bau20]. Even for them, fast compilation is a great asset in the prototyping phase, but contrary to software engineers, they are used to very long ASIC compilation times and are ready to wait significantly longer to achieve better performance [Kah21]. Hence, for final implementation, especially in small programmable fabrics of embedded devices or complex SoCs, algorithms with high computational complexity, possibly solving multiple problems simultaneously and/or solving some of the subproblems optimally, are likely to become more appealing than before. Since FPGA architecture design is ultimately a codesign problem with that of designing CAD algorithms, this is something that should be kept in mind.

With this, we conclude the presentation of the necessary background material and are ready to move on to the main contributions of the book.

References

[Abd14] Abdelfattah, M. S. & Betz, V. (2014). The case for embedded networks on chip on field-programmable gate arrays. *IEEE Micro, 34*(1), 80–89. ISSN: 1937-4143.

[Ahm00] Ahmed, E. & Rose, J. (2000). The effect of LUT and cluster size on deep-submicron FPGA performance and density. In *Proceedings of the ACM/SIGDA International Symposium on Field Programmable Gate Arrays* (pp. 3–12).

[Ahm09] Ahmed, T., Kundarewich, P. D., & Anderson, J. H. (2009). Packing Techniques for Virtex-5 FPGAs. *ACM Transactions on Reconfigurable Technology and Systems*, *2*(3). ISSN: 1936-7406.

[Alo84] Alon, N. & Milman, V. D. (1984). Eigenvalues, Expanders and Super-concentrators. In *Proceedings of the 25th Annual Symposium on Foundations of Computer Science* (pp. 320–22).

[Ans23] Ansell, T. (2023). Google investment in open source custom hardware development including no-cost shuttle program. In *Proceedings of the 2023 International Symposium on Physical Design*. Virtual Event.

[Bau20] Bauer, H., Burkacky, O., Kenevan, P., Lingemann, S., Pototzky, K., & Wiseman, B. (2020). *Semiconductor design and manufacturing: Achieving leading-edge capabilities*. Tech. rep. McKinsey.

[Bet20] Betz, V. (2020). Parallel CAD for FPGAs: A personal retrospective and thoughts for the future. In *Proceedings of the 30th International Conference on Field-Programmable Logic and Applications (FPL)*. Virtual Event.

[Bet98] Betz, V. & Rose, J. (1998). How much logic should go in an FPGA logic block? *IEEE Design Test Computer, 15*(1), 10–15.

[Bet99a] Betz, V. & Rose, J. (1999). FPGA routing architecture: Segmentation and buffering to optimize speed and density. In *Proceedings of the 1999 ACM/SIGDA Seventh International Symposium on Field Programmable Gate Arrays* (pp. 59–68).

[Bet99] Betz, V., Rose, J., & Marquardt, A. (1999). *Architecture and CAD for Deep-Submicron FPGAs*. Kluwer Academic Publishers. ISBN: 0-7923-8460-1.

[Blu81] Blum, M., Karp, R. M., Vornberger, O., Papadimitriu, C. H., & Yannakakis, M. (1981). The complexity of testing whether a graph is a supercon-centrator. *Information Processing Letters, 13*(4), 164–67. ISSN: 0020-0190.

[Bou21] Boutros, A. & Betz, V. (2021). FPGA architecture: Principles and progression. *IEEE Circuits and Systems Magazine, 21*(2), 4–29. ISSN: 1558-0830.

[Bun00] Bunke, H., Jiang, X., & Kandel, A. (2000). On the minimum common supergraph of two graphs. *Computing, 65*(1), 13–25. ISSN: 0010-485X.

[Cha15] Chandrakar, S., Gaitonde, D., & Bauer, T. (2015). Enhancements in UltraScale CLB Architecture. In *Proceedings of the 2015 ACM/SIGDA International Symposium on Field-Arrays, P. G.* (pp. 108–16).

[Che07] Chen, D. T., Vorwerk, K., & Kennings, A. (2007). Improving Timing-Driven FPGA Packing with Physical Information. In *2007 International Conference on Logic, F. P. and Applications* (pp. 117–23).

[Che04] Chen, G. & Cong, J. (2004). Driven, S. T. Clustering and Placement for FPGAs. In *Proceedings of the 14th International Conference on Logic, F. P. and Applications* (pp. 158–67).

[Chr00] Christie, P., & Stroobandt, D. (2000). The interpretation and application of Rent's rule. *IEEE Transactions on Scale, V. L. Integration (VLSI) Systems, 8*(6), 639–48. ISSN: 1557-9999.

[Chr20] Chromczak, J., Wheeler, M., Chiasson, C., How, D., Langhammer, M., Vanderhoek, T., et al. (2020). Architectural Enhancements in Intel®Agilex™FPGAs. In *Proceedings of the 2020 ACM/SIGDA International Symposium on Field-Arrays, P. G.* (pp. 140–49).

[Con94] Cong, J. & Ding, Y. (1994). FlowMap: An optimal technology mapping algorithm for delay optimization in lookup-table based FPGA designs. *IEEE Transactions on Computer-Aided Design of Integrated Circuits and Systems, 13*(1), 1–12. ISSN: 1937-4151.

[DeM94] De Micheli, G. (1994). *Synthesis and Optimization of Digital Circuits*. McGraw-Hill. ISBN: 0070163332.

[DeH04] DeHon, A. & Rubin, R. (2004). Design of FPGA interconnect for multilevel metallization. *IEEE Transactions on Very Large Scale Integration (VLSI) Systems, 12*(10), 1038–50. ISSN: 1557-9999.

[DeH95] DeHon, A. (1995). *Entropy, counting, and programmable interconnect*. Transit Note 128. University of CA.

[DeH99] DeHon, A. (1999). Balancing interconnect and computation in a reconfigurable computing array (or, Why You Don't Really Want 100% LUT Utilization). In *Proceedings of the 1999 ACM/SIGDA Seventh International Symposium on Field Programmable Gate Arrays* (pp. 69–78).

[Elg2022] Elgammal, M. A., Murray, K. E., & Betz, V. (2022). RLPlace: Using reinforcement learning and smart perturbations to optimize FPGA placement. *Transactions on Computer-Aided Design of Integrated Circuits and Systems, 41*(8), 2532–45. ISSN: 0278-0070.

[Eve16] Evensen, T. (2016). A software developer's journey into a deeply Heterogeneous World. In *Proceedings of the 26th International Conference on Field Programmable Logic and Applications*.

[Fen08] Feng, W. & Kaptanoglu, S. (2008). Designing efficient input interconnect blocks for LUT clusters using counting and entropy. In *ACM Transactions on Reconfigurable Technology and Systems, 1*(1). ISSN: 1936-7406.

[Fie15] Fiévet, N., Raghavan, P., Baert, R., Robert, F., Mercha, A., Verkest, D., et al. (2015) Impact of device and interconnect process variability on clock distribution. In *2015 International Conference on IC Design & Technology (ICICDT)* (pp. 1–4).

[Fra18] Fraisse, H. & Gaitonde, D. (2018). A SAT-based timing driven place and route flow for critical soft IP. In *2018 28th International Conference on Field Programmable Logic and Applications (FPL)* (pp. 8–87).

[Fra91] Francis, R., Rose, J., & Vranešić, Z. (1991). Chortle-Crf: Fast technology mapping for lookup table-based FPGAs. In *Proceedings of the 28th ACM/IEEE Design Automation Conference* (pp. 227–33).

[Fra08] Francis, R., Moore, S., & Mullins, R. (2008). A network of time-division multiplexed wiring for FPGAs. In *Proceedings of the Second ACM/IEEE International Symposium on Networks-on-Chip*. Newcastle upon Tyne (pp. 35–44).

[Gai19] Gaide, B., Gaitonde, D., Ravishankar, C., & Bauer, T. (2019). Xilinx Adaptive Compute Acceleration Platform: Versal Architecture. In *Proceedings of the 2019 ACM/SIGDA international symposium on field-programmable gate arrays* (pp. 84–93).

[Gre11] Greene, J., Kaptanoglu, S., Feng, W., Hecht, V., Landry, J., Li, F., et al. (2011). A 65 nm Flash-Based FPGA fabric optimized for low cost and power. In *Proceedings of the 19th ACM/SIGDA International Symposium on Field Programmable Gate Arrays* (pp. 87–96).

[Gre23] Greene, J. W. (2023). FPGA mux usage and routability estimates without explicit routing. In *Proceedings of the 2023 ACM/SIGDA International Symposium on Field Programmable Gate Arrays* (pp. 141–47).

[Haa19] Haaswijk, W. J. (2019). SAT-Based exact synthesis for multi-level logic networks. PhD thesis. École Polytechnique Fédérale de Lausanne.

[Han22] Hanchate, N. (2022). Improving chip design performance and productivity using machine learning. In *Proceedings of the 2022 International Symposium on Physical Design*. Virtual Event.

[Hau07] Hauck, S. & DeHon, A. (Eds.) (2007). *Reconfigurable computing: The theory and practice of FPGA-based computation*. Morgan Kaufmann. ISBN: 9780123705228.

[Hof94] Höfting, F., & Wanke, E. (1994). Polynomial time analysis of toroidal periodic graphs. In *Proceedings of the 21st International Colloquium on Automata, Languages and Programming* (pp. 544–55).

[Hut04] Hutton, M. D., Schleicher, J., Lewis, D. M., Pedersen, B., Yuan, R., Kaptanoglu, S., et al. (2004). Improving FPGA performance and area using an adaptive logic module. In *Proceedings of the 14th International Conference on Field Programmable Logic and Application* (pp. 135–44).

[Jan09] Jang, S., Chan, B., Chung, K., & Mishchenko, A. (2009). WireMap: FPGA technology mapping for improved routability and enhanced LUT Merging. *ACM Transactions on Reconfigurable Technology and Systems, 2*(2). ISSN: 1936-7406.

[Jos22] Josipović, L., Guerrieri, A., & Ienne, P. (2022). From C/C++ code to high-performance dataflow circuits. *IEEE Transactions on Computer-Aided Design of Integrated Circuits and Systems, 41*(7), 2142–2155.

[Kah21] Kahng, A. B. (2021). Advancing placement. In *Proceedings of the 2021 International Symposium on Physical Design*. Virtual Event (pp. 15–22).

[Kap12] Kaptanoglu, S. (2012). Pathfinder: A negotiation-based performance-driven router for FPGAs. In *FPGA and Reconfigurable Computing Hall-of-Fame Endorsement*.

[Kha22] Khandelwal, V. (2022). Machine-Learning Enabled PPA Closure for Next-Generation Designs. In *Proceedings of the 2022 International Symposium on Physical Design*. Virtual Event.

[Kuc19] Kucherenko, A., Nikolić, S., & Ienne, P. (2019). On feasibility of FPGAs without dedicated programmable interconnect structure. In *Proceedings of the 2019 ACM/SIGDA International Symposium on Field-Programmable Gate Arrays*.

[Lem04a] Lemieux, G., & Lewis, D. (2004). *Design of interconnection networks for programmable logic*. Kluwer Academic Publishers. ISBN: 1402077009.

[Lem00] Lemieux, G., Leventis, P., & Lewis, D. (2000). Generating highly-routable sparse crossbars for PLDs. In *Proceedings of the 2000 ACM/SIGDA Eighth International Symposium on Field Programmable Gate Arrays* (pp. 155–64).

[Lem01] Lemieux, G. & Lewis, D. (2001). Using Sparse Crossbars within LUT Clusters. In *Proceedings of the 2001 ACM/SIGDA Ninth International Symposium on Field Programmable Gate Arrays* (pp. 59–68).

[Lew05] Lewis, D., Ahmed, E., Baeckler, G., Betz, V., Bourgeault, M., Cashman, D., et al. (2005). The Stratix II logic and routing architecture. In *Proceedings of the 2005 ACM/SIGDA 13th International Symposium on Field-Programmable Gate Arrays* (pp. 14–20).

[Lew03] Lewis, D., Betz, V., Jefferson, D., Lee, A., Lane, C., Leventis, P., et al. (2003). The stratix routing and logic architecture. In *Proceedings of the 2003 ACM/SIGDA Eleventh International Symposium on Field Programmable Gate Arrays* (pp. 12–20).

[Lew16] Lewis, D., Chiu, G., Chromczak, J., Galloway, D., Gamsa, B., Manohararajah, V., et al. (2016) The Stratix™10 Highly Pipelined FPGA Architecture. In *Proceedings of the 2016 ACM/SIGDA International Symposium on Field-Programmable Gate Arrays* (pp. 159–68).

[Li19] Li, W., Lin, Y., & Pan, D. Z. (2019). elfPlace: Electrostatics-based placement for large-scale heterogeneous FPGAs. In *2019 IEEE/ACM International Conference on Computer-Aided Design (ICCAD)* (pp. 1–8).

[Li19a] Li, W. & Pan, D. Z. (2019). A new paradigm for FPGA placement without explicit packing. *IEEE Transactions on Computer-Aided Design of Integrated Circuits and Systems, 38*(11), 2113–26. ISSN: 1937-4151.

[Luu14a] Luu, J. (2014). *Architecture-aware packing and CAD infrastructure for field-programmable gate arrays*. Ph.D. Thesis. University of Toronto.

[Maa18] Maarouf, D., Alhyari, A., Abuowaimer, Z., Martin, T., Gunter, A., Grewal, G., et al. (2018). Machine-learning based congestion estimation for modern FPGAs. In *2018 28th International Conference on Field Programmable Logic and Applications (FPL)* (pp. 427–34).

[Mar00] Marquardt, A., Betz, V., & Rose, J. (2000). Timing-driven placement for FPGAs. In *Proceedings of the 2000 ACM/SIGDA Eighth International Symposium on Field Programmable Gate Arrays* (pp. 203–13).

[McM95] McMurchie, L. & Ebeling, C. (1995). PathFinder: A negotiation-based performance-driven router for FPGAs. In *Proceedings of the 1995 ACM Third International Symposium on Field-Programmable Gate Arrays* (pp. 111–17).

[Mis06] Mishchenko, A., Chatterjee, S., & Brayton, R. (2006). DAG-aware AIG rewriting: a fresh look at combinational logic synthesis. In *2006 43rd ACM/IEEE Design Automation Conference* (pp. 532–535).

[Mis23] Mishchenko, A. (2023). *ABC: A System for Sequential Synthesis and Verification*. Berkeley Logic Synthesis and Verification Group. https://people.eecs.berkeley.edu/~alanmi/abc/. Accessed on 05.05.2023.

[Mur20] Murray, K. E., Petelin, O., Zhong, S., Wang, J. M., Eldafrawy, M., Legault, J.-P., et al. (2020). VTR 8: High-performance CAD and Customizable FPGA Architecture Modelling. *ACM Transactions on Reconfigurable Technology and Systems (TRETS), 13*(2), 9:1–60.

[Mur15] Murray, K. E., Whitty, S., Liu, S., Luu, J., & Betz, V. (2015). Timing-Driven Titan: Enabling large benchmarks and exploring the gap between academic and commercial CAD. *ACM Transactions on Reconfigurable Technology and Systems, 8*(2). ISSN: 1936-7406.

[Nag98] Nag, S. K. & Rutenbar, R. A. (1998). Performance-driven simultaneous placement and routing for FPGAs. *IEEE Transactions on Computer-Aided Design of Integrated Circuits and Systems, 17*(6), 499–518.

[Nik20] Nikolić, S., Zgheib, G., & Ienne, P. (2020). Straight to the Point: Intra-and Intercluster LUT connections to mitigate the delay of programmable routing. In *FPGA '20: The 2020 ACM/SIGDA International Symposium on Field-Programmable Gate Arrays, Seaside, CA, USA, February 23–25, 2020* (pp. 150–60).

[Pet21] Petersen, M. B., Nikolić, S., & Stojilović, M. (2021). NetCracker: A peek into the routing architecture of Xilinx 7-Series FPGAs. In *Proceedings of the 2021 ACM/SIGDA International Symposium on Field Programmable Gate Arrays*. Virtual Event (pp. 11–22).

[Pis03] Pistorius, J. & Hutton, M. (2003). Placement rent exponent calculation methods, temporal behaviour and FPGA architecture evaluation. In *Proceedings of the 2003 International Workshop on System-Level Interconnect Prediction* (pp. 31–38).

[Raj22] Rajarathnam, R. S., Alawieh, M. B., Jiang, Z., Iyer, M., & Pan, D. Z. (2022). DREAM-PlaceFPGA: An open-source analytical placer for large scale heterogeneous FPGAs using deep-learning toolkit. In *2022 27th Asia and South Pacific Design Automation Conference (ASP-DAC)* (pp. 300–306).

[Red09] Reda, S. (2009). Using circuit structural analysis techniques for net-works in systems biology. In *Proceedings of the 11th International Workshop on System Level Interconnect Prediction* (pp. 37–44).

[Ros89] Rose, J., Francis, R. J., Chow, P., & Lewis, D. (1989). The effect of logic block complexity on area of programmable gate arrays. In *1989 Proceedings of the IEEE Custom Integrated Circuits Conference* (pp. 5.3/1–5.3/5).

[Sch03] Schmit, H. (2003). Extra-dimensional Island-Style FPGAs. In *Proceedings of the 13th International Conference on Field Programmable Logic and Applications* (pp. 406–15).

[Shr23] Shrivastava, S., Nikolić, S., Ravishankar, C., Gaitonde, D., & Stojilović, M. (2023). IIBLAST: Speeding up commercial FPGA routing by decoupling and mitigating the Intra-CLB Bottleneck. In *Proceedings of the 2023 IEEE/ACM International Conference on Computer Aided Design (ICCAD)*.

[Sin05] Singh, D. P., Manohararajah, V., & Brown, S. D. (2005). Two-stage physical synthesis for FPGAs. In *Proceedings of the IEEE 2005 Custom Integrated Circuits Conference, 2005* (pp. 171–78).

[Tri97] Trimberger, S., Duong, K., & Conn, B. (1997). Architecture issues and solutions for a high-capacity FPGA. In *Proceedings of the 1997 ACM Fifth International Symposium on Field-Programmable Gate Arrays* (pp. 3–9).

[Tu24] Tu, K., Tang, X., Yu, C., Josipović, L., & Chu, Z. (Eds.) (2024). *FPGA EDA: Design Principles and Implementation*. Springer. ISBN: 9789819977543.

[Wol23] Wolf, C. (2023). *Yosys Open SYnthesis Suite*. https://yosyshq.net/yosys/. Accessed on 05.05.2023.

[Xia22] Xiao, Y., Micallef, E., Butt, A., Hofmann, M., Alston, M., Goldsmith, M., et al. (2022). PLD: Fast FPGA compilation to make reconfigurable acceleration compatible with modern incremental refinement software development. In *Proceedings of the 27th ACM International Conference on Architectural Support for Programming Languages and Operating Systems* (pp. 933–45).

[Xil93] Xilinx Inc (1993). *XC2000 Logic Cell Array Families*.

[Yan02] Yan, A., Cheng, R., & Wilton, S. J. E. (2002). On the Sensitivity of FPGA architectural conclusions to experimental assumptions, tools, and techniques. In *Proceedings of the 2002 ACM/SIGDA Tenth International Symposium on Field-Programmable Gate Arrays* (pp. 147–56).

[Yan16] Yang, S., Gayasen, A., Mulpuri, C., Reddy, S., & Aggarwal, R. (2016). Routability-Driven FPGA placement contest. In *Proceedings of the 2016 on International Symposium on Physical Design* (pp. 139–43).

[Ye10] Ye, A. G. (2010). Using the minimum set of input combinations to minimize the area of local routing networks in logic clusters containing logically equivalent I/Os in FPGAs. *IEEE Transactions on Very Large Scale Integration (VLSI) Systems, 18*(1), 95–107. ISSN: 1557-9999.

Part II
Designing Programmable Interconnect for Scaled Technologies

This part of the book contains its main contribution: demonstrating that automating the design of programmable interconnect is both possible and useful in the era of highly resistive, scaled fabrication technologies. Needless to say, before we can automate an optimization process, we have to be able to measure its outcome. For this reason, in Chap. 4, we introduce a new physical modeling framework for programmable interconnect designed in advanced technology nodes, down to 3nm. After using this framework to explain some of the architectural changes in the most recent Agilex family of Intel/Altera FPGAs and confirming that reduction of capacitive load on highly resistive wires by altering the switch-patterns without reducing routability of the architecture is of utmost importance, we proceed to introduce a new algorithm for automatic switch-pattern design. This algorithm leverages the dominant heuristic of *congestion negotiation* used in all common FPGA routers, and it is described in detail in Chap. 5. In order to produce high performance, but regular switch-patterns that can be efficiently fabricated and used by the existing CAD tools, in Chap. 6, we couple the previously introduced algorithm with an *Integer Linear Programming* (ILP) approach, to limit the search space to architectures satisfying arbitrary constraints. Finally, in Chaps. 7 and 8, we pose the question of whether programmability can selectively be removed from portions of FPGA interconnect, relying on fast and cheap direct connections between LUTs instead. We first propose an algorithm for automated design of useful patterns of hardened interconnect and apply it to finding architectures that answer the above question affirmatively. We then introduce a new dedicated placement algorithm that significantly increases the benefits of the additional direct connections.

Chapter 4
Modeling Programmable Routing in Advanced Technologies

Contents

© The Author(s), under exclusive license to Springer Nature Switzerland AG 2025

S. Nikolić, *Modern Programmable Interconnect Design*,
https://doi.org/10.1007/978-3-031-80629-2_4

In Chap. 2, we mentioned that the FPGAs implemented in 7 nm technologies brought about the most significant changes to the programmable interconnect architecture in almost two decades [Gai19, Chr20]. That something was bound to change should not come as a surprise, since experts have already anticipated a dramatic rise in wire resistance as technology scales beyond this point [Che14]. Why does resistance increase? Intuition about this is given already by high-school physics: Resistance of a wire is $\rho\frac{L}{W\times H}$, where L, W, and H are, respectively, the wire's length, width, and height. If transistor dimensions shrink, so must the width of the wires. However, the height must shrink as well, since there is a limit to the aspect ratio beyond which the structure would collapse [Lin23]. Even though high-school physics allows us to anticipate that problems with high resistance will arise, it does not allow us to predict when precisely they will become so pronounced as to call for a drastic change in architectural design. For that, more complex resistance models are required, taking into account phenomena such as line-edge-roughness and the characteristics of the barrier [Che14, Tok16]. In this chapter, we develop a framework for physical modeling of programmable interconnect architectures at advanced FinFET nodes, going down to 3 nm. The goal is to be able to analyze and explain the design choices of the latest commercial architectures, as well as to enable architectural exploration in advanced technologies that can anticipate future developments. In doing so, we adapt state-of-the-art wire and via resistance models, previously developed at IMEC, to fit the level of abstraction required for rapid FPGA architecture exploration.

For widths and heights of different wires, we use dimensions representative of publicly available information about existing commercial metal stacks, along with predictive data from leading research institutions. While these dimensions are negatively impacted by technology scaling, the length of wires receives a positive impact: As devices shrink, so do the distances that wires have to span. To precisely determine the lengths of various wires that comprise a programmable interconnect architecture, detailed layout models are required. Yet, these models need to be simple enough for fast evaluation required by architectural exploration.

Unfortunately, most existing tools for physical modeling of FPGAs, such as *COFFE* [Chi13a] and *FPRESSO* [Zgh16], have been designed for use in older planar technologies where the effects of high resistance were not nearly as problematic as they are today. Moreover, they were intended to be compatible with technology-agnostic area models of VPR, based on counting minimum width transistor equivalents [Bet99]. For this reason, they utilized very rough approximations of actual layouts, assuming for instance that any block can be laid out with an arbitrary aspect ratio [Yaz19], including a perfect square [Chi13a, Zgh16]. In order to rectify this, we develop new scalable layout models, drawing both from prior academic work, and especially from the publicly available data about commercial architectures.

Precise tile dimensions and wire pitch information allow us to analyze all architectures in terms of the available routing channel track capacity—a fundamental question that is rarely addressed in academic work. However, in this chapter we primarily use the developed framework to explain some of the changes made in the interconnect architecture of Intel Agilex FPGAs [Chr20]. In particular, we show that technology scaling beyond 7 nm has such a profound effect that even some long-standing rules of thumb, such as the choice of cluster size for optimizing performance, no longer hold.

Besides reexamining the cluster sizing problem, we also return to that of channel segmentation, confirming the observation of previous authors that scaled technologies favor shorter channel wires [Lin10]. In doing so, quick evaluation of the proposed physical model proves to be an important feature, allowing us to evaluate a large number of architectures in a reasonable amount of time. This will become even more important in the next two chapters, where evaluating the model upon changing even a single switch allows us to successfully solve the problem of automated switch-pattern design, which we identify here as crucial.

This chapter is largely based on our paper entitled "Global Is the New Local: FPGA Architecture at 5 nm and Beyond," previously published at the 2021 ACM/SIGDA International Symposium on Field-Programmable Gate Arrays [Nik21]. The paper was prepared in collaboration with Paolo Ienne from EPFL, and Francky Catthoor and Zsolt Tőkei from IMEC, who provided us with advanced resistance models and invaluable expert knowledge of interconnect fabrication and modeling.

4.1 A Nanometer Asteroid Strike

In Chap. 2, we saw that the very first FPGA—XC2064—had only one 4-LUT in its logic block [Xil93]. It did not take a long time, however, for FPGAs like FLEX 10K to adopt larger logic blocks, comprising eight LUTs [Alt01]. In APEX 20K, this further increased to ten, which is the number retained in the subsequent Stratix family of FPGAs, over its six generations [Lew03, Lew16]. In Sect. 3.3, we introduced the notion of a logic cluster, as a model for this kind of block in which multiple LUTs are grouped together and share some local interconnect. One of the reasons why this has been useful is that the local interconnect provides a fast way to connect LUTs within the cluster; we even argued that this is the fastest way possible, since the signals going through the local interconnect pass through only one level of multiplexing.

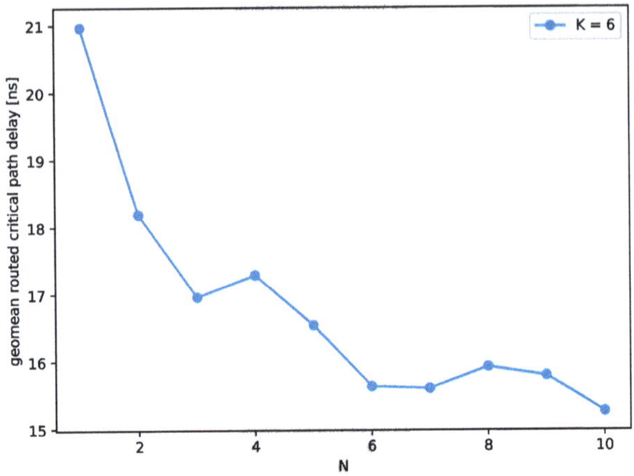

Fig. 4.1 Routed critical path delays for clusters comprising different numbers of 6-LUTs in 180 nm technology [Ahm01]

When transistor delay dominated that of metal, this was certainly true. Figure 4.1 shows the average routed critical path delay of circuits implemented on top of a 180 nm 6-LUT-based FPGA, for a varying number of LUTs in the cluster. It is adopted from a seminal work of Ahmed [Ahm00, Ahm01] that is still widely used to justify the choice of LUT and cluster sizes [Chi13, Zgh17]. The authors note that increasing the cluster size *N* beyond 3 leads to increasingly diminishing performance gains, suggesting that taking any *N* in the range between 3 and 10 would be a good choice [Ahm01]. Nevertheless, as demonstrated by the Stratix V [Lew13] and UltraScale+ [Gan16] time-borrowing extensions, the era when 10% gains could be easily dismissed [Bet98] is long gone. Hence, it should not be surprising that most high-end commercial architectures have relied on logic clusters of 8–10 LUTs [Cha15, Lew16].

How does the wire resistance surge beyond 7 nm impact the performance benefits of these large (compared to the single-LUT block of XC2064) clusters? Will they persist or will they be swept away by technological change in favor of smaller clusters, much like the Cretaceous–Paleogene extinction left no place for large dinosaurs?

4.1.1 Global Is the New Local?

To illustrate why one might have concerns about that, let us take a look at floorplan sketches of a cluster of ten LUTs in two different technologies. They are shown in Fig. 4.2. Annotated are the resistance and the intrinsic delay of a local (intracluster)

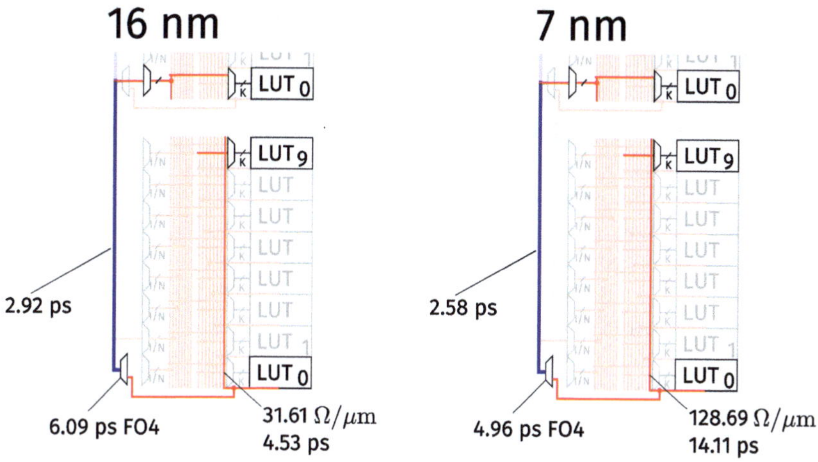

Fig. 4.2 Evolution of local (intra-) and global (intercluster) interconnects. The local lines in red use thin lower metal layers, while the global blue ones employ thick higher layers [Gai19]. The two technologies are approximately representative of commercial 16 and 7 nm technologies (full details are available in Sect. 4.3)

wire and a vertical global (intercluster/channel) wire spanning the height of the cluster. The former is naturally drawn in one of the fine-pitch metal layers, closer to the active devices [Gai19], whereas the latter, as customary for longer connections, uses a thicker metal layer from higher up in the stack. The numbers, accounting for reduction of all three dimensions of a wire, as well as logic delay scaling, are shown for nodes approximately representative of the commercial 16 and 7 nm. It seems plausible that reaching the top LUT from the bottom one in the same cluster will at some point take more time than reaching the bottom LUT in the cluster above it. Such a situation would make for an interesting challenge to a packing algorithm: It should try to spread timing critical connections across clusters instead of keeping them internal. To make things even worse, any signal would incur a higher delay penalty to enter the cluster than it took to reach it from a distant source through the global interconnect. All this suggests that clusters should be made smaller than they are today.

Of course, this is merely a back-of-the-envelope calculation and things are not this simple in reality: One should also consider the cost of the additional multiplexing, horizontal offset between the channel wire and the signal source and destination, the via stack required to reach the thicker metal, etc. Hence, in practice, it is not necessarily clear whether the described hypothetical situation will actually happen, and, if so, when. As we have mentioned, the purpose of this chapter is to develop a framework for modeling FPGA fabrics at advanced technology nodes, so as to be able to answer quantitatively this and similar questions.

4.1.2 Wait a Minute, Isn't Industry Going in the Opposite Direction?!

The benefits of having the ability to model programmable interconnect architectures in advanced FinFET nodes go far beyond enabling reassessment of the usual rules of thumb used for choosing cluster sizes. We shall be able to see that already in this chapter and especially in the following two. However, before beginning to present the details of the proposed model, it is of interest to compare our main motivating example of Fig. 4.2 with some recent commercial developments. This will provide us with a way to assess the validity of our conclusions.

As we have already mentioned, fabrication technology literature anticipated that at 7 nm there will be a turning point in how interconnect resistance impacts architecture design [Che14]. Fortunately, both major high-performance FPGA vendors (Xilinx/AMD and Altera/Intel) have already taped out 7 nm devices, and their design teams have presented the architectural modifications at length [Gai19, Chr20]. Changes to the programmable interconnect architecture were in both cases indeed considerably more drastic than in most prior generations since the start of the century. When listing some of the main modifications introduced in Agilex (see Sect. 2.9.2.1), we have not mentioned any reduction in cluster size, however. In fact, the authors clearly state that the cluster (LAB) still comprises ten 6-LUTs (ALMs) [Chr20]. Nevertheless, we believe that the change in this respect is merely hidden in terminology. Here is a quote from the authors of Agilex [Chr20]:

> *First, the LIM/LEIM network is now broken into four "lanes" that each service 2.5 ALMs, newly omitting much connectivity between the lanes. In previous architectures, all LIMs drove LEIMs on all 10 ALMs, but in Agilex, most LIMs drive LEIMs on ALMs in only one lane, allowing shorter wires for the remaining connectivity. Shorter LIM to LEIM wires can also move to a better place in the metal stack.*

We note once more that LIM is equivalent to a connection-block multiplexer in academic terminology that we use in this book, while LEIM corresponds to a crossbar multiplexer. Since Stratix 10, feedback connections do not go back to the crossbar multiplexers, but to those of the connection block instead, increasing the minimal number of multiplexer levels required for connecting two LUTs in the same cluster to two [Lew16]. How does breaking the local interconnect into four almost isolated pieces differ from reducing cluster size 4×? Most likely, other uses of the cluster, such as control signal routing, LUTRAM circuitry, and body bias selection [Lew09], are still implemented at the level of a ten-LUT group, which justifies still calling this group a cluster. However, in terms of interconnect topology, which has traditionally been the main reason for cluster-based organization of FPGAs—especially in academia, where control signals, LUTRAM, and body bias are routinely neglected—we believe that clusters of ten LUTs essentially ceased to exist.

What about the AMD/Xilinx Versal architecture? Not only do the authors not report any decrease in cluster size, but a 4× increase, from eight to 32 6-LUTs [Gai19]. We believe that in this case, the new cluster is in a way similar to the APEX 20K supercluster of Fig. 2.12. The main difference, however, is that the supercluster-level interconnect is reserved only for routing feedback connections between the LUTs of the supercluster, while the signals entering the supercluster from without do not have to go through it. This prevents the supercluster-level interconnect from becoming a performance bottleneck, allowing it to be slow. Even the most critical feedback connections can avoid the supercluster interconnect by routing through the channel wires as they would if the cluster size had not been increased. These signals are typically relatively few, however, and hence, the supercluster interconnect can provide benefit in terms of reducing the pressure on global routing, by catering to the majority of the remaining feedback signals [Gai19].

We will return to Versal in Sect. 4.1.2. Let us now start developing the proposed model. The first step is to be able to measure the length of various wires.

4.2 Area and Wirelength Modeling

In order to determine the lengths of different wires that occur in an FPGA—which has a decisive influence on delay of programmable interconnect in scaled technologies [Che14]—we need a precise model of dimensions and positions of various blocks that form the endpoints of these wires. The model that we developed for this purpose is introduced in this section.

4.2.1 Tile Floorplan

Island-style FPGAs are typically thought of as having routing channels that surround logic blocks, like in Fig. 2.3. In the days when metal layers were scarce, this was actually quite accurate, since channel wires could not be routed above the logic block simply because there, the metal was already taken by connections belonging to the logic block itself [Cho91]. In the meantime, this situation has changed, and in modern FPGAs there is no longer a need to reserve physical space in the 2D plane for the channel wires themselves: Instead, they are traced above the active devices, and the dimensions of the abutted tiles are entirely determined by the logic blocks and the routing multiplexers [Lew13]. We illustrate this change in Fig. 4.3.

Because the FPGA is constructed by abutting identical tiles,[1] in order to determine the lengths of wires, we only need to know the dimensions and positions within

[1] This book is concerned with answering fundamental questions about designing the reconfigurable fabric, and in particular its programmable interconnect part. Hence we limit ourselves to modeling only the logic clusters. Appropriately modeling the hard IPs such as DSP and memory blocks would require integration of entirely different techniques [Yaz19], which goes beyond our present scope. As we have mentioned before, since design and analysis of these hardened-functionality blocks can leverage a large portion of ASIC development tools and methodologies, we do not believe that postponing their treatment for future work significantly reduces the contribution presented here. Inability to model these blocks will have repercussions on which open-source benchmark circuits we can use in architectural evaluation, however. We will discuss this issue further in Sect. 4.7.7.

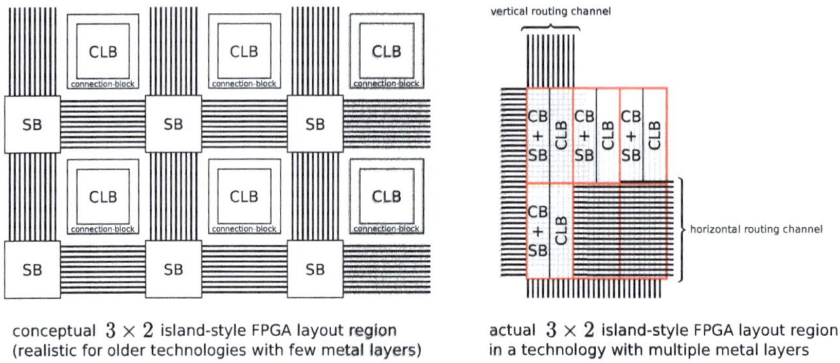

Fig. 4.3 Island-style FPGAs in technologies with multiple metal layers. In older technologies with few metal layers, channel wires had to be routed around the logic blocks (left). When more layers became available, tracing channel wires above the active area of the tile became possible (right) [Lew13]

one tile of logic elements and various multiplexers that are part of the programmable interconnect architecture. A simplified floorplan of a Stratix V tile, which we adopt in this work as well, is shown on the right of Fig. 4.4. As we have already seen in the discussion of the APEX architectures in Sect. 2.9.1.1, the logic cluster itself is composed of a single vertical stack of the given number of LUTs [Lew13]. Routing multiplexers belonging to the crossbar, the connection block, and the switch-block are stacked immediately to the left of the LUT stack, with no deliberate spacing between them [Lew13]. We note that COFFE 2 [Yaz19] uses a similar floorplan model but divides the stack into two columns. While this may be useful for reducing the average length of the local wires, we chose to retain the single column floorplan of the Stratix and the Agilex architectures [Lew13, Chr20], since it is our interest to be able to explain the developments that occurred between them.

When designing the interconnect architecture, we pay attention that the multiplexers can be evenly divided between the LUTs, similarly to Agilex [Chr20], and match the height of the multiplexers adjacent to one LUT to the height of the LUT itself [Lew13]. The floorplan of one concrete example architecture is shown on the left of Fig. 4.4.

4.2.2 LUT Dimensions

LUTs play a dominant role in determining the layout of the tile, since they take up around 50% of its area and since the routing multiplexers are pitch-matched to them [Lew13]. We based our layout assumptions on a layout due to Abusultan and Khatri [Abu14], shown in Fig. 4.5. It consists of a multiplexer tree with select inputs coming from the left and the output produced at the bottom, horizontally centered. We assume that two-gate-pitch SRAM cells [You15] are placed next to

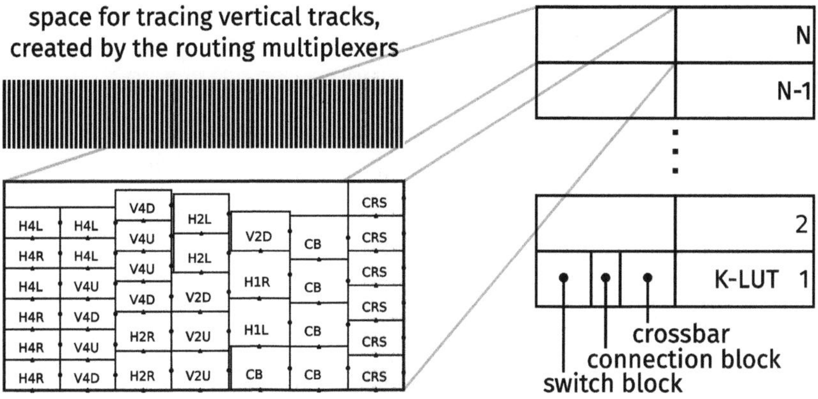

Fig. 4.4 A fully stacked floorplan of a tile similar to that occurring in the Stratix V architecture [Lew13], showing the positions of the LUTs and the various routing multiplexers that comprise it. A concrete example of a 3 nm architecture based on a cluster of four 6-LUTs is shown on the left. It illustrates a simple greedy multiplexer positioning algorithm, which stacks multiplexers sorted by decreasing input count, starting from those of the crossbar, right next to the LUT, then proceeding with those of the connection block, appending them to the left, and finally, finishing with those of the switch-block. Each time the height of one column exceeds the height of the adjacent LUT, a new column is appended to the left. This creates additional space for tracing vertical wires above the tile

each other, above the multiplexer tree. To provide the necessary stability [Lew12], the SRAM cells are assumed to be sized as *1:2:3*—i.e., that the NMOS transistors of the two inverters (M2 and M6 in Fig. 4.5) have three fins, the PMOS (M3 and M5 in Fig. 4.5) have one fin, and that the access transistors (M1 and M4 in Fig. 4.5) have two fins [Cla16]. We also assume that there is a one-fin spacing between the NMOS and the PMOS transistors, as allowed by the ASAP7 design rules [Cla16]. This means that the width of the LUT mask (truth table memory) amounts to $10 \cdot 2^K$ fin pitches, where K is the LUT input count. The LUT mask is considered to fully determine the width of the entire LUT, as it leaves ample space for increasing the size of the multiplexer tree transistors. The height of the LUT is determined as $2K$ gate pitches for the multiplexer tree, two gate pitches for the mask, and two more gate pitches for the mask buffers (not shown in Fig. 4.5).

Because the width of this layout increases exponentially with the increase in the number of LUT inputs, the distance that the input signals need to travel before reaching the most distant transistor of the multiplexer tree quickly becomes intolerable. A similar situation occurs for the output that must reach the routing multiplexers. For this reason, we consider a 4-LUT—the size explored by Abusultan and Khatri—to be the largest for which the resistance of the horizontal wires connecting the LUT to the routing multiplexers is acceptable. To create larger LUTs, we stack the required number of 4-LUTs on top of each other. This greatly reduces the distances that the LUT input and output signals need to cross, as can be seen in Fig. 4.6. It also creates more vertical space for the routing multiplexers.

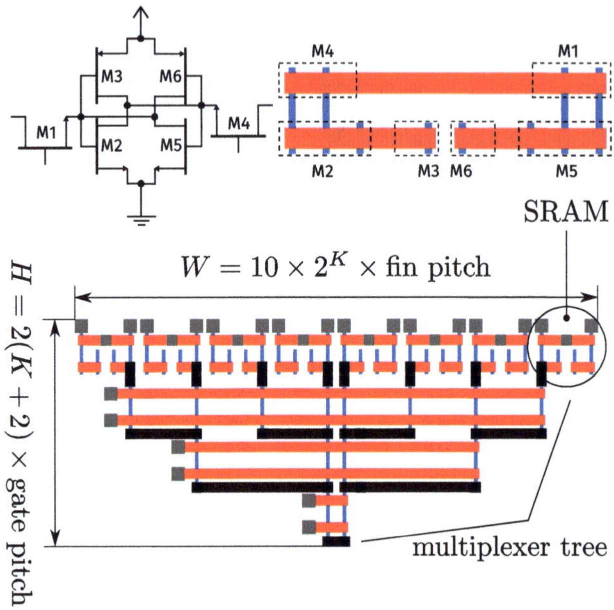

Fig. 4.5 A sketch of the assumed LUT layout, based on the one due to Abusultan and Khatri [Abu14]. The SRAM design is adopted from Young et al. [You15]. Fins are drawn in blue, gates in red, contacts in gray, while black designates metal. SRAM cell output buffers are not shown in the figure but are accounted for by the area model

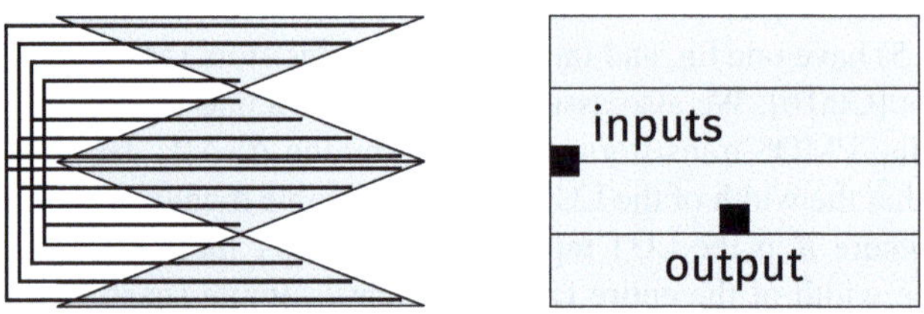

Fig. 4.6 A 6-LUT composed of four stacked 4-LUTs (represented by triangles), with their input connections. In a monolithic 6-LUT, these lengths would amount to 4× the width of the depicted 4-LUT. Assumed pin locations are shown on the right

We assume that the flip-flops and the register multiplexing circuitry can fit inside the empty space created by the triangular shape of the (4-)LUT. As an illustration backing this assumption, a flip-flop of the ASAP7 [Cla16] 7 nm standard cell library takes up approximately 116 square gate pitches, whereas the empty space left by a minimally sized 4-LUT equals approximately 600 square gate pitches, also according to the ASAP7 design rules.

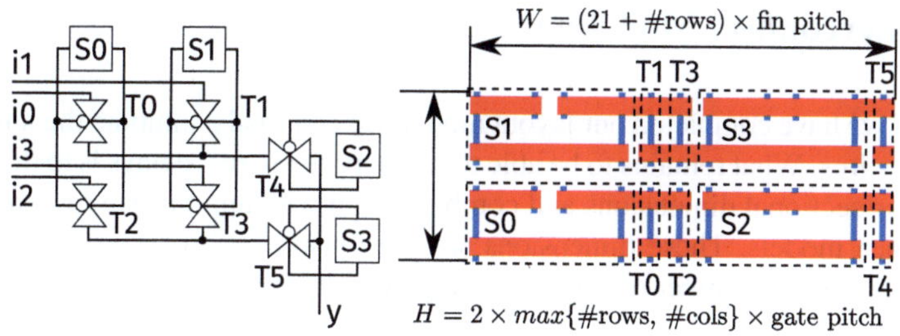

Fig. 4.7 A two-level multiplexer and its layout

4.2.3 Routing Multiplexers

We assume that all routing multiplexers are built of transmission gates, which follows the trends visible in Agilex [Chr20]. We also adopt the previous results of Chiasson [Chi13] that showed that already in planar technologies, it is sufficient that all transistors in the multiplexer transmission gates are minimally sized. A sketch of the layout of a 4:1 stored-select multiplexer is shown in Fig. 4.7. We assume that the 2-gate-pitch SRAM cells of Young et al. [You15] are stacked on top of each other, one for each column of the multiplexer. Immediately to the right, first-level transmission gates are aligned with the appropriate SRAM cells, followed by the SRAMs of the rows and the transmission gates of the second level, still to the right.

This results in a multiplexer height of twice the maximum between the number of rows and columns of gate pitches, and a width equal to 20 fin pitches for the SRAMs, one for the second-level transmission gates, and one for each multiplexer row. We note that the proposed layout sketch may be slightly optimistic for transmission gates as it may be underestimating the required well spacing.

The connection- and switch-block multiplexers require buffers at the output, whereas the crossbar multiplexers drive only one LUT input pin that has a buffer of its own, thus removing the need for additional buffering. We assume that, where required, buffers are placed below the appropriate multiplexer. The increase in the total multiplexer height is determined from the buffer's drive strength, after folding it so as to fit into the horizontal space used by the multiplexer itself.

In general, it is possible to optimize the aspect ratio of the individual multiplexers by adjusting their row and column counts, while optimizing multiplexer placement, so that their combined area is minimized (see Fig. 4.4). This goes beyond the scope of the present work and for the moment we rely instead on optimizing each multiplexer type individually, to minimize the number of SRAM cells used. Then we populate columns from the LUT left, starting from placing all crossbar multiplexers, then all connection-block multiplexers, and finally all switch-block multiplexers. Each time the LUT height is exceeded, a new column is appended to the left. An example result of application of this simple algorithm is shown on the left of Fig. 4.4.

4.3 Interconnect Modeling

Now that we have completed our layout model, we can easily compute the lengths of different wires. What remains to be able to measure the wires' delays is to determine their cross-sectional dimensions and combine them with appropriate resistance and capacitance models. These dimensions are later also used to assess feasibility of tracing the desired number of tracks over a given tile area.

4.3.1 Layers

Representative metal stacks for several of the technology nodes of interest are shown in Fig. 4.8. We assume that two pitch options are used to route all wires, referring to the tighter as Mx and to the more relaxed as My. In most cases these correspond to the tightest and the second tightest pitch in the interconnect

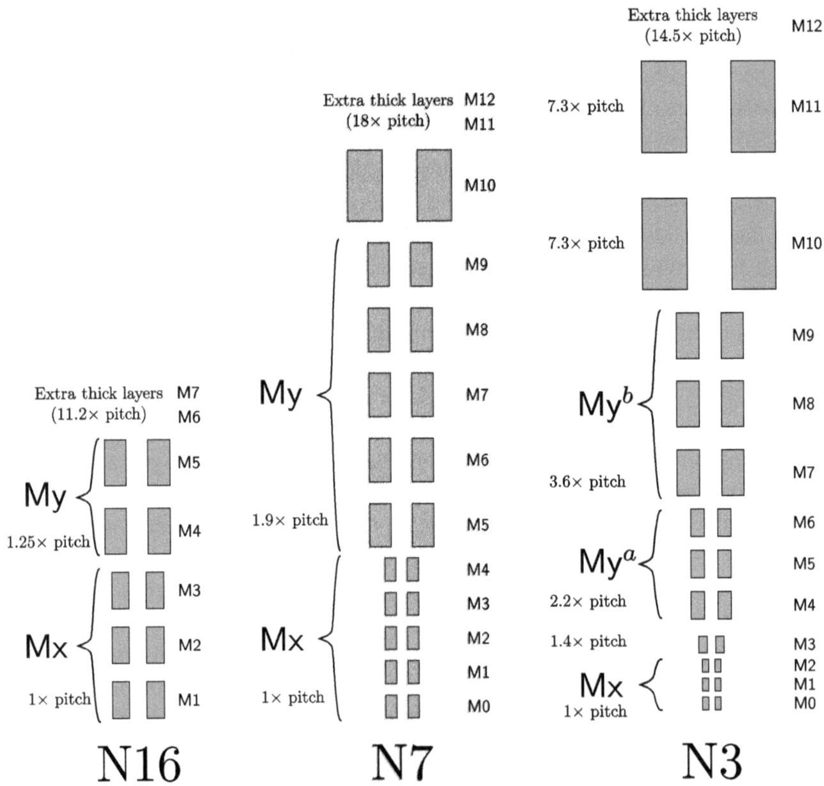

Fig. 4.8 Representative metal stacks for the 16 nm [Wu13], 7 nm [Wu16], and 3 nm [Pra19] nodes. All wires are drawn to scale

stack. The exception is the 3 nm node, where we also explore a possibility of promoting My one step further, as illustrated in the figure. In all cases, however, the layer group labeled as My is considered to be immediately above the layer group labeled as Mx, since the intermediate layers can be omitted. We assume that all connections within the individual blocks (LUTs, multiplexers, etc.) are routed at M2 or below, as customary for basic cells, whereas the intracluster connections connecting different LUTs together and passing external inputs to the crossbar multiplexers (*LAB lines* in the Stratix architectures [Lew13]), as well as all the intercluster wires are routed at M3 and above. Commercial FPGAs typically also make use of the extra thick Mz metal layers, for tracing the power and the clock signals, but also to implement very long bidirectional channel wires, sometimes spanning as many as 24 tiles [Tyh15, Lew16]. However, since these wires, only useful for very large designs, are peculiar in their bidirectionality and the way in which they can be accessed [Lew13], we leave modeling them for future work.

4.3.2 Cross-Sectional Wire Dimensions

All dimensions relevant to computing per-unit-length resistance and capacitance of wires, using the models of this section, are shown in Fig. 4.9a. Typically, the only one that is readily available in public data from the foundries is the pitch, and therefore we use it to derive all the other dimensions. Representative pitches for all the technology nodes of interest are shown in Table 4.1. As their names and the references in the footnotes suggest, the considered technology nodes are represented through parameters strongly inspired by corresponding commercial technologies (e.g., F16 resembles TSMC 16 nm). Speculative nodes (e.g., F3a/b representing hypothetical 3 nm nodes) are derived following the progression suggested in relevant literature and some manufacturability considerations. The F4 node is a speculative

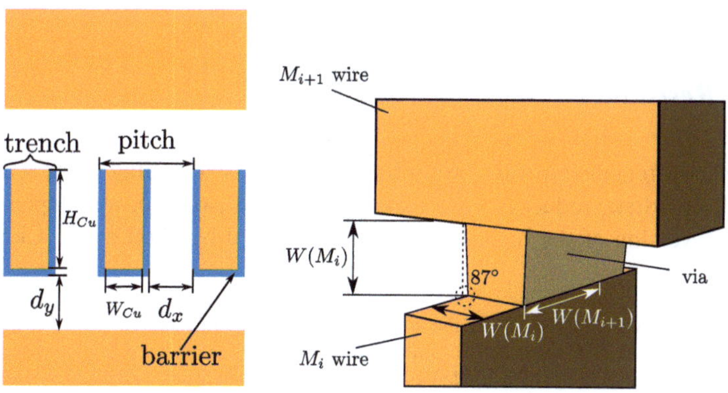

(a) Wire cross section. (b) Via.

Fig. 4.9 Relevant dimensions of a wire and a via. (**a**) Wire cross section. (**b**) Via

Table 4.1 Metal pitches for all the technology nodes of interest. F7, for instance, is our hypothetical 7 nm node

	F16	F7	F5	F4	F3a	F3b
Mx [nm]	64[a]	40[b]	38[c]	26[d]	22[e]	22[e]
My [nm]	80[a]	76[b]	72[f]	50[f]	48[g]	80[h]

[a] TSMC 16 nm [Wu13]
[b] TSMC 7 nm [Wu16]
[c] Fit to match the Mx RC increase from F7, amounting to about 16% [Yea19]
[d] Close to limits of single-patterned EUV and a reasonable intermediate point between F5 and F3.
[e] IMEC prediction [Pra19]
[f] Assuming 1.9× as in F7 [Wu16]
[g] IMEC prediction for the M4–M6 layers [Pra19]
[h] IMEC prediction for the M7–M9 layers [Pra19]

node that we suppose foundries might want to introduce as an intermediate step if transitioning directly to a 3 nm node would prove too dramatic a move.[2]

We assume that the trench width equals 1.1× half pitch, to mitigate the resistance increase and ease via contacting. Representative barrier thicknesses that are used to obtain the final copper dimensions of the Mx layers are 3 nm until F5 inclusive and 2 nm from F4 onward [Cio16]. For the My layers, we assume a constant barrier thickness of 4 nm across all nodes. Spacing between the layers (d_y in Fig. 4.9a) is set to equal the trench width, reflecting the typical via aspect ratio (height over width) being close to 1. Finally, the height of the wire (H_{Cu} in Fig. 4.9a) is determined through a sweep that seeks to minimize the $R'C'$ product (see below). The maximum allowed aspect ratio is set to 2 for all layers apart from Mx of F3b, for which we assumed possible an aspect ratio of 3 to mitigate the resistance surge at the expense of increased capacitance. With F3b we explore a different trade-off that could be made in a future advanced node. Taller wires are considered difficult to manufacture, so it is unlikely that it will be possible to further reduce resistance through aspect ratio optimization [Lin23].

4.3.3 *Resistance*

Resistance suffers the greatest impact from the aggressive scaling of the wire pitch, due to the quadratic reduction of the cross-sectional area. Here, we adopt a slightly simplified version of the resistivity model introduced by Ciofi et al. [Cio16] that

[2] When this work was first published, the latest available technology was the TSMC's 5 nm node. In the meantime, both a 4 nm intermediate and a 3 nm full node were introduced [TSM23]. Nevertheless, to the best of our knowledge, the only information that was made available about these nodes is that the tightest metal pitch at 3 nm is 23 nm [Wu22] (merely 1 nm more than what we have used) and that the gate pitch is 45 nm [Cha22] (which is slightly less scaled than the predictions that we used). Given the lack of further information and that different foundries will inevitably use different dimensions, we decided to retain the dimensions that we originally used.

is valid for all the technology nodes of interest to us. By assuming no tapering (i.e., wire sides are completely vertical, as in Fig. 4.9a), integration of equation (1) of Ciofi et al. simplifies substantially, and we obtain the following expression determining the resistance per unit length of a wire:

$$R' = \frac{1}{H_{Cu}W_{Cu}} \left(32.05 + 615 \left(\frac{\tanh(0.133W_{Cu})}{W_{Cu}} + \frac{\tanh(0.133H_{Cu})}{H_{Cu}} \right) \right). \tag{4.1}$$

Variables W_{Cu} and H_{Cu} correspond to the definition of Fig. 4.9a, while the constants have been empirically determined for a 7 nm technology node [Cio16], which is in the middle of the range that we intend to explore.

4.3.4 Capacitance

Capacitance is less impacted by the pitch scaling than resistance. Pitch reduction does decrease the distance to neighboring wires (d_x and d_y in Fig. 4.9a), thus increasing the coupling capacitance; yet, line width and height (W and H) decrease as well, balancing this out. Hence, for modeling capacitance, we use a less recent model due to Wong et al. [Won00], available at the PTM website [Nan11], without any modification. For all technology nodes, we assume a relative permittivity of 2.8 for the lower metal layers and 3.0 for the intermediate ones, which is representative of the current trends in industry. The obtained resistance and capacitance per unit length are reported in Table 4.2. In all cases, the $R'C'$ optimization resulted in the maximum allowed aspect ratio. As predicted, we can see a substantial rise in R' between consecutive nodes, due to the shrinking of cross-sectional area, whereas C' remains constant since the dimensions with opposing influence scale uniformly.

Table 4.2 Wire resistance and capacitance per micrometer length. Maximum aspect ratio for Mx of F3b was increased to 3, to reduce the resistance at the expense of increased capacitance

	F16	F7	F5	F4	F3a	F3b
Mx						
W_{Cu} [nm]	29.2	16.0	14.9	10.3	8.1	8.1
H_{Cu} [nm]	67.4	41.0	38.8	26.6	22.2	34.3
R' [Ω/μm]	31.6	128.7	151.6	392.9	666.4	396.7
C' [fF/μm]	0.22	0.22	0.22	0.22	0.22	0.28
My						
W_{Cu} [nm]	36.0	33.8	31.6	19.5	18.4	36.0
H_{Cu} [nm]	84.0	79.6	75.2	51.0	48.8	84.0
R' [Ω/μm]	18.7	21.6	25.1	75.7	86.4	18.7
C' [fF/μm]	0.24	0.24	0.24	0.24	0.24	0.24

Table 4.3 Resistance of vias. The reported values correspond to the resistance of a single via connecting two neighboring layers in the Mx group, or the buffer output at an Mx layer and a wire at an My layer, in case of stacked vias

	F16	F7	F5	F4	F3a	F3b
Mx-Mx [Ω]	10.9	34.8	39.9	58.9	92.9	92.9
Stacked vias (M2-M5)						
H [nm]	246.4	154.0	146.3	100.1	84.7	108.9
R [Ω]	19.2	30.5	34.7	69.8	88.0	44.9

4.3.5 *Vias*

To avoid the effects of high resistance increase at lower metal layers, more and more signals are routed at higher ones. That requires traversing long vertical distances, so it is important to accurately model via resistance, itself affected by technology scaling.

A via connecting layers M_i and M_{i+1} is shown in Fig. 4.9b. We assume a classical 87°-tapered via [Cio17]. We compute the width of the via at half the height and use it in place of H_{Cu} in Eq. (4.1). $W(M_{i+1})$ is used in place of W_{Cu}, to obtain the via resistance per unit length. Here we note that for connecting layers of different pitch, the shape of the via is typically different and cannot be accurately modeled with this approach. However, as this is a reasonably small penalty that needs to be paid only twice per connection, we chose to prioritize modeling simplicity over accuracy. As stated before, we assume a unit aspect ratio for vias, so the final resistance requires multiplication by $W(M_i)$. To account for the resistance of the top and the bottom barrier, we assume a constant resistivity of 1,200 Ωnm [Cio17], while the barrier thicknesses correspond to those of the layers that the via connects. Values of single via resistances obtained for all technology nodes of interest, using the pitch information from Table 4.1, are reported in Table 4.3. The table also shows the corresponding resistances of stacked vias connecting buffer outputs to the My wires. We assume that the buffer output pin is at M2 and that the target My layer is M5, i.e., that the via traverses the height of two Mx trenches and three Mx vias.

4.4 Device Modeling

For device modeling, we rely on the PTM [Nan11] and ASAP7 [Cla16] predictive models. We leave the 16 nm PTM models for F16 completely unchanged. For the nodes scaled further down, we update the fin dimensions and the gate lengths of ASAP7 SLVT devices at the fast-fast (FF) corner [Cla16], as indicated in Table 4.4. We leave the remaining parameters that have a less pronounced effect on the drive current unchanged. The same fin and gate pitches are used to convert the wirelengths computed by the scalable model of Sect. 4.2 to metric units. The choice of SLVT devices at the FF corner may underestimate the contribution of transistor delays to

Table 4.4 Representative device geometry and nominal supply voltages. All values are taken from Wu et al. [Wu20], apart from F4 which is an interpolation between F5 and F3, and F16 which comes from FreePDK15 [Bha15] and PTM [Nan11]

	F16	F7	F5	F4	F3
Gate pitch [nm]	64	56	48	44	41
Fin pitch [nm]	40	30	28	24	22
Gate length [nm]	20	18	16	15	14
Fin height [nm]	26	35	45	50	55
Fin width [nm]	12	6.5	6	5.5	5.5
Vdd [V]	0.85	0.75	0.7	0.65	0.65

Table 4.5 FO4 delays at nominal voltages and at 0.7 V. The delays at 0.7 V are useful to validate the relative speedup, shown in the last row. For this, note that F16 and F7 are two generations apart and that F4 models a possible half-node between F5 and F3. The values indicate that a reasonable speedup, roughly around 10%, is maintained between the consecutive nodes

	F16	F7	F5	F4	F3
At nominal Vdd [ps]	6.09	4.96	4.69	4.69	4.48
At 0.7 V [ps]	7.02	5.09	4.69	4.52	4.30
Δ		−27%	−8%	−4%	−5%

routed critical path delays, compared to the conclusions that could be made using slower devices. However, in our measurements, all transmission gates are powered by the same nominal voltage as other transistors, and we do not allow gate boosting that other authors have considered for further speed improvement [Chi13].

4.5 Delay Extraction Methodology

Many aspects related to delay modeling were described in the previous sections. Here we present the final steps needed to obtain the delays of all individual components.

4.5.1 Look-Up Tables

As our focus in this work is interconnect, we take the LUT delays reported for Stratix IV in the *K6_N10_mem32K_40 nm* architecture distributed with VTR 8 [Mur20] and scale them using the equations of Stillmaker and Baas [Sti17], from and to the closest nodes, until F7. From F7 onward, we assume the scaling of *fanout-of-4 inverter* (FO4) delays at nominal voltage values, reported in Table 4.5.

Table 4.6 Average input to output delays of a 6-LUT. The values are scaled from the *K6_N10_mem32K_40 nm* VTR architecture file [Mur20]

	F16	F7	F5	F4	F3
Average delay [ps]	94	68	64	64	61

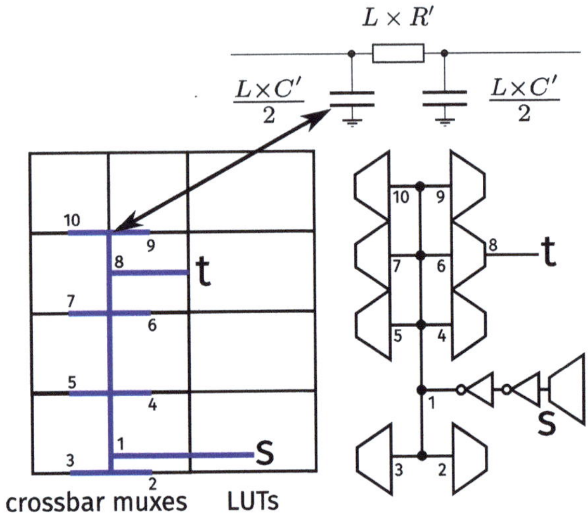

Fig. 4.10 Setup to measure local wire delay. Each section is modeled as a single RC Π-segment, shown at the top of the figure. L does not designate the section's inductance, but its length

In doing so, we somewhat underestimate the importance of wires inside the LUTs themselves, but we leave addressing this issue for future work. The resulting delays are reported in Table 4.6.

4.5.2 *Local Wires*

Long local connections traced at the highly resistive Mx layers are particularly sensitive to capacitive load of the various multiplexers that connect to them. For this reason, we fairly precisely model all the wires that participate in local signal distribution: We assume that one long vertical line distributes the signal and that shorter horizontal wires bring it to individual multiplexer inputs (Fig. 4.10). We also assume that two loading multiplexers are fully switched on, which corresponds to the typical fanout of a net in a real circuit [Hut97]; one of them is assumed to be in the middle of the line, while the other one is that at which the delay is being measured. The fraction of the loading multiplexers that have only the first-level transmission gates turned on is determined as an inverse of the average number of columns in these multiplexers, which is the probability that the one column SRAM that is high is controlling the transmission gate connected to the wire [Chi13a]. The horizontal position of the long vertical line is assumed to be at the center between the output of the LUT and the most distant-driven multiplexer. We sweep the buffer sizes to minimize delay, assuming that the maximum strength of the first inverter is 5, to avoid overloading the minimally sized transmission gates of the driving multiplexer, and that the second inverter can be at most 5× larger than the first one. A similar setup is used for measuring the global wire to LUT input delay, with the only difference being the position of the signal source, the connection-block delay being counted besides that of the crossbar, and the long vertical line positioned at half the distance between the connection-block output and the furthest driven crossbar multiplexer input. Each wire section in Fig. 4.10 is modeled as a single Π-segment.

4.5.3 Global Wires

For measuring global wire delay, we again determine the exact position of the loading multiplexers (see Fig. 4.4). We assume that the My wire brings the signal to the average of the loading multiplexer input coordinates, from which point it is further distributed using a simple rectilinear Mx tree, similar to the one modeling the routing of signals within the cluster. This is illustrated in Fig. 4.11. The long My wire is modeled using $N \times 2^{K-4}$ Π-segments (i.e., one per 4-LUT height in the cluster's LUT stack; see Fig. 4.6). This applies to both vertical and horizontal global wires.

We also assume that the vertical (horizontal) global wires are positioned at the center of the tile (LUT), horizontally (vertically), and account for the horizontal (vertical) spans needed to access them as well as to take the signal back to the target multiplexers. Because the driver sizes influence the multiplexer stacking, we predetermine them by simulations of a simplified load model, with the same overall buffering approach as the one used for local wires.

4.6 Extracted Local Wire Delays

In this section we present the resulting delays of local connections, extracted using the methodology that we have just described. The values support the interest to explore anew different cluster sizes, as previously indicated in Sect. 4.1.

4.6.1 The Low Performance of Low Metal Layers

Figure 4.12 shows for all considered technology nodes the delays from the output of the bottom LUT of an eight 6-LUT cluster (*K6N8*) to each of the other LUT inputs that it can access through a 50% sparse crossbar. We can see that for F16

Fig. 4.11 Setup to measure global wire delay

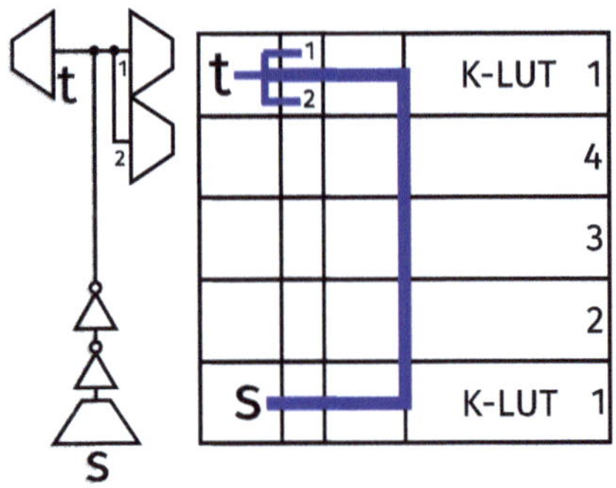

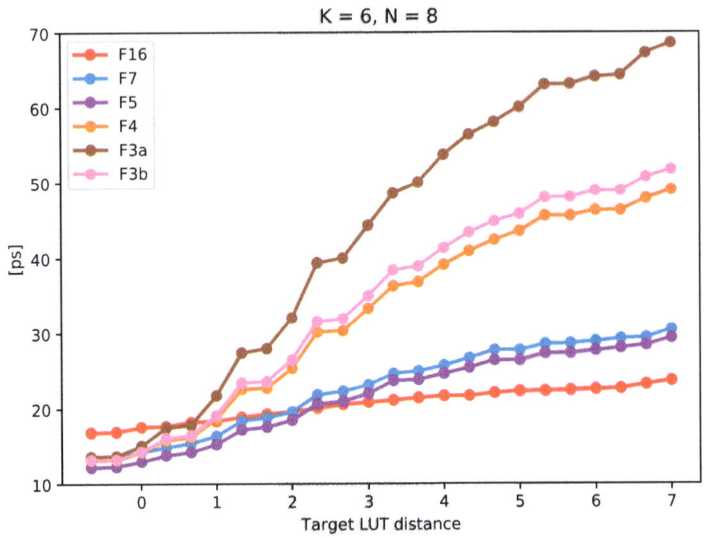

Fig. 4.12 Delays from the output of the bottom LUT in a K6N8 cluster to all LUT inputs reachable through a 50% sparse crossbar. Each point marks an input, and the last of the three inputs of a particular LUT is labeled by the corresponding number on the x-axis

it is business as usual: The delay increase with distance is reasonably modest. As the resistance rises in more advanced nodes, however, the delay increase rate grows rapidly—which is intuitive and somehow predictable. It is the magnitude of this increase that is interesting: Eventually, the delay of connections to the other end of the cluster becomes comparable with and even surpasses that of a 6-LUT (Table 4.6). Connections between immediately adjacent LUTs, dominated by the logic delay, are faster in newer technologies until F5, when device performance increase decelerates. At the other end, between faraway LUTs, F16 achieves the fastest connectivity—often by far. Finally, it is interesting to note that using an average delay as a single number representing local connection delays (which is often done for architectural research) could be justified for older technologies, as supported by the relatively flat curve of F16. For scaled technologies, however, it is imperative that CAD tools are aware of the delay disparity and that they can place the LUTs within the cluster accordingly. Ideally, this would happen during routing, as in Quartus [Lew03], since only then accurate information about signal criticality becomes available.

4.6.2 *Can You Repeat, Please?*

One way of mitigating the effect of delay increase due to higher resistance is repeater insertion. To see what an effect this could make, we consider two situations: (1) an optimistic setting where the repeaters can be inserted in the long vertical local wire itself and (2) a more realistic setting where the repeaters are located close to the LUT output, in the cavity created by two constituent 4-LUTs. In both cases, we vary the repeater number and size, assuming that they are located at equal space, aligned with LUT output heights, and that their size equals the size of the second inverter of the main driver. The results for F4 are plotted in Fig. 4.13. We can see that in-line buffer insertion does mitigate the delay to an extent but that it still remains substantial. Yet, as mentioned, it is not realistic to assume that each local wire can be buffered in line, because the repeaters would make it hard to maintain the dense spacing between the lines, even if there were sufficient space left by the crossbar multiplexers. The more realistic buffering scenario shows some benefit, but delays to faraway LUTs remain almost unchanged.

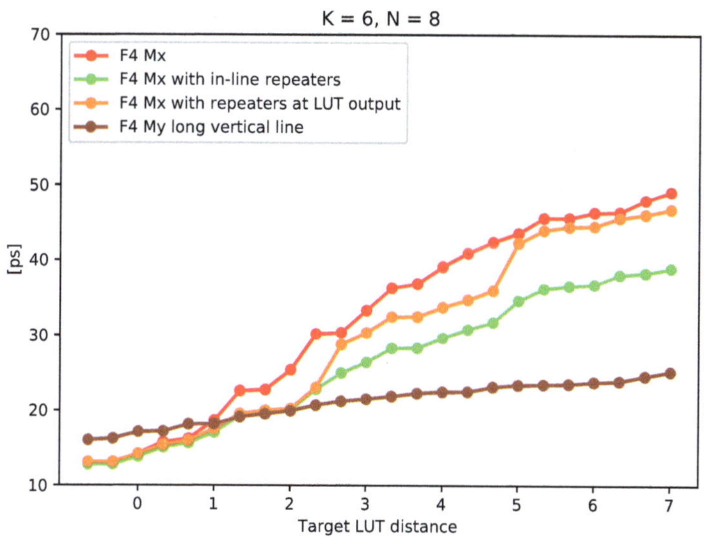

Fig. 4.13 Delays from the output of the bottom LUT in an F4 K6N8 cluster to all LUT inputs reachable through a 50% sparse crossbar, for different delay reduction options. The plot shows values for optimistic (in line) and realistic (at LUT output) repeater insertion; it also shows the effect of raising the long vertical line to My, leading to a much more dramatic delay reduction

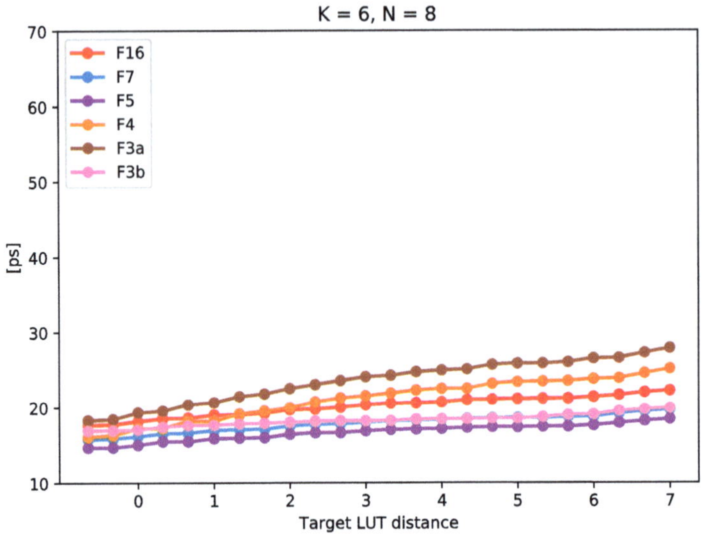

Fig. 4.14 Delays from the output of the bottom LUT in a K6N8 cluster to all LUT inputs reachable through a 50% sparse crossbar, with the long vertical wire raised to an My layer

4.6.3 The Rise of Thick Metal Wires

Together with different buffering options, Fig. 4.13 shows the delay resulting from raising the long vertical line to the My layer, with access wires remaining at Mx. The significant delay benefit is immediately apparent: The slope of the curve reduced dramatically, and the most distant LUT input can be reached within about a third of the representative LUT delay. We may note, however, that connections to about two LUTs away are still faster or roughly equal when performed on Mx, that is due to the lower wire capacitance of Mx. Figure 4.14 shows analogous plots for all explored technologies. We can see that the behavior is similar in other cases as well.

4.6.4 Thick Metal Wires Are Scarce

The previous section could suggest that lifting the long local connections to My would provide a solution for the delay increase with technology scaling. Yet, this may not be such a wise idea. Until now, we have focused on connections between LUTs in the same cluster, to provide motivation. However, the increase of the delay penalty that intercluster connections need to pay upon entering the cluster, while being dispatched from the connection-block output to the appropriate crossbar multiplexer over increasingly resistive local wires is likely even more important,

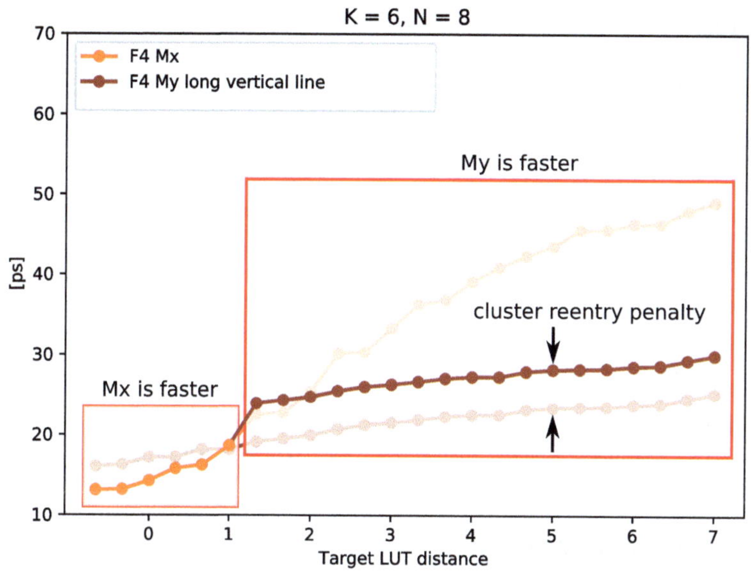

Fig. 4.15 Layer optimality for different distances

for intercluster connections occur more often on a typical critical path [Lew12]. Hence, to really see the benefit of raising local wires to the My layers, all of them would need to be raised, and not only those routing the LUT outputs. For a K6N8 cluster, that would mean occupying 40 tracks, while the tile width of such a cluster in F4 (including the routing multiplexers) can typically accommodate about 180 My tracks. This means that about 20% of the available routing space would be locked inside the cluster and unavailable to intercluster signals, potentially inducing a large impact on routability. This impact may be reduced by increasing the number of metal layers in newer technologies, but the recent trends have shown that it is desirable to keep as many connections at the lower layers as possible, likely to reduce fabrication cost, increase yield, and free up higher layers for new hardened additions to the reconfigurable SoCs [Gai19].

An alternative solution is again suggested by Fig. 4.13. By observing that communication within a two-LUT range is faster at Mx, we may suspect that a smaller cluster (e.g., $N = 2$) could be efficient in providing local communication, while communication with more distant LUTs can be achieved through global routing. This way, the performance gain from moving to an upper layer is reduced, but the tracks are not locked within the cluster. To provide a clearer illustration of the expected effect, we plot a combination of the corresponding sections of the Mx and the My curves in Fig. 4.15, the later shifted up by the delay of distributing signals within a distance of two LUTs, to simulate the cost of reentering an $N = 2$ cluster.

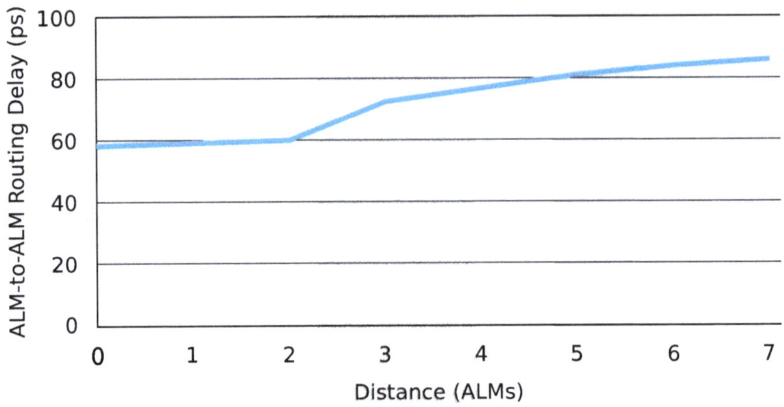

Fig. 4.16 Agilex feedback delays (Adopted from Chromczak et al. [Chr20])

4.6.5 *This Looks Familiar...*

As a reference, in Fig. 4.16 we also reproduce a plot from Chromczak et al. [Chr20], corresponding to that of our Fig. 4.12. We can immediately observe a similar shape of the curve, indicating that the approach taken by the designers of Agilex could indeed have been to utilize faster global routing for connections between LUTs at a distance of more than two. We note that the absolute values and slopes reported by Chromczak et al. are substantially higher than the ones we measured. A likely reason for this is that the actual Agilex logic element is significantly more complex than the 6-LUT followed by a flip-flop which we use: Each fracturable LUT has eight inputs requiring crossbar multiplexers, instead of one flip-flop, and it is accompanied by four, along with hardened arithmetic circuitry, complex output multiplexing, and local interconnect pipeline registers [Chr20]. At the cluster level, feedback connections pass through the connection block in addition to the crossbar, thus adding another level of multiplexing, equipped with bypassable pipeline registers that certainly must incur some delay [Lew16]. Additionally, LUTRAM circuitry, as well as control signal multiplexing, is likely spread across the cluster. This makes both multiplexing intrinsically slower and, more importantly, the local wires significantly longer, which could well be the cause why we observed a dramatic rise in local connection delays only at F4, whereas Agilex appears to have experienced it already at 7 nm.

Before we proceed with reevaluating the optimal cluster sizes across technologies, which the results presented until now suggest is of great interest, let us briefly return to the quote of Chromczak et al. that we used earlier in this chapter [Chr20]:

Shorter LIM to LEIM wires can also move to a better place in the metal stack.

This seems to suggest that the authors may have actually contemplated raising the long local wires (LAB lines [Lew16]) to an My layer, before deciding to break the large cluster into pieces.

4.7 To Minimize or Maximize the Channel Width? That Is the Question

We perform cluster size exploration in the classical manner of Ahmed and Rose [Ahm00]: For each $N \in [2, 4, 8, 16]$, we place and route benchmark circuits to obtain postrouting critical path delays, which we use to rank different architectures in terms of performance. However, rather than search for a minimal routable channel width, which was a great way of reducing FPGA area at the time when routing channels were physically surrounding the logic blocks, we seek the best way to fill the track space created by the active area, without making the tile dimensions metal bound. Indications that this is more appropriate than minimizing channel width for modern FPGAs in which channel wires are routed on top of logic blocks are given by Lewis et al., who tried to maximize the number of available tracks without increasing the number of multiplexer columns in the tile of Stratix V [Lew13]. In this section, we detail this process and list our other assumptions on the interconnect architecture. Later, we will also comment on other possibilities to use the available track space.

4.7.1 Crossbar

We compute the number of cluster inputs from the Rent's rule, as described in Sect. 3.4.1. However, we use a slightly higher exponent of 0.8, since it better corresponds to the Stratix architectures [Lew16]:

$$I = \left\lceil \frac{K \times N^{0.8}}{N} \right\rceil \times N. \tag{4.2}$$

The division and multiplication by N guarantee that the connection-block multiplexers can be evenly divided between LUTs, which eases layout modeling. Unlike the latest Altera/Intel architectures [Lew16], we assume the classical setting in which the feedback connections directly enter the crossbar, without passing through the connection block [Lew13]. The crossbar is assumed to be 50% sparse in delay measurements, to be representative of commercial architectures [Lew13]. In the final VPR experiments, its connectivity is modeled as fully populated, however. This is to remove one more possible source of routability impacting the delay results, even though Lemieux and Lewis have previously demonstrated that 50% sparse crossbars are almost always routable [Lem01] and Xilinx/AMD FPGAs utilize significantly sparser input multiplexing structures [Shr23].

4.7.2 Routing Channels: General Approach

Similarly to Agilex, we assume that an equal number of wires of each length and direction begin and end at the height of each LUT and that they drive only the switch-blocks at their end [Chr20]. We consider only unidirectional wires occurring in pairs of opposing direction.

4.7.3 Routing Channels: Maximum Wire Spans

Before exploring exact channel compositions, we determine the maximum lengths of wires for each cluster size in each technology. We do this by finding the longest wire that is still faster than two wires half its length connected in a sequence through a switch-block multiplexer. Wires longer than that would make little sense, since a faster connection could be achieved by composing two shorter wires that can also be used more flexibly than a single long one. In performing this computation, we assume a fixed tile width of $2\times$ the LUT width, which corresponds to the situation in Stratix V [Lew13]. In subsequent experiments, actual tile widths, accounting for the precise contributions of the various routing multiplexers, are used.

The comparison with halves naturally leads to consideration of wires of lengths that are powers of two. Although that does not encompass segmentations of some commercial architectures [Chr20, Pet21], it aligns well with prior work of Betz and Rose [Bet99a], so we do not consider the restriction a major loss of generality.

We note that Lee et al. have explored the effect of intermediate buffering of long channel wires, observing some performance improvement [Lee06]. However, their experiments were performed in 180 and 90 nm technologies, making it reasonable to altogether omit from consideration the negative impact of via resistance. In scaled technologies, this is no longer possible, and the delay penalty of descending the via stack in order to reach the buffer is much better amortized if a multiplexer is included as well, so that the metal track can be used more flexibly and with less waste. More recent academic research has also avoided intermediate buffering [Lin10], and, to simplify modeling, we avoid it too.

The resulting maximum spans are shown in Table 4.7. We observe that with increasing physical height of the tile, maximum logical distance that makes sense with respect to delay decreases. On the other hand, maximum logical horizontal distance stays constant for a given technology, since, with the floorplan assumptions of Fig. 4.4, tile width is not directly a function of N.

Table 4.7 Maximum wire spans for F16–F5 and F3b as a function of cluster size N. For F4 and F3a, all entries are halved, because the tighter My pitch reduces the distance that can be optimally traversed before buffering. V and H designate vertical and horizontal channel wires, respectively

N	2	4	8	16
V	16	8	4	2
H	8	8	8	8

4.7.3.1 A Note on Tile Aspect Ratios

We may note that the maximum spans of the horizontal and the vertical wires are equal for $N = 4$. This is because in that case the aspect ratio of the tile is closest to 1. For higher aspect ratios (higher N leading to taller tiles), horizontal wires can travel larger logical distances than the vertical ones, while for the aspect ratios smaller than 1 ($N < 4$), the opposite holds. Although academic research usually assumes that the aspect ratio of the tile can always be made 1 and hence also utilizes the same channel segmentation for both channels [Mur13], actual commercial FPGAs typically have a tile aspect ratio that is different from 1 [Lew03, Pet21]. This impacts both the lengths of wires in the different channels and, as we will shortly see, the channel track capacity. It is useful to remember that a wire of logical length L requires L tracks to be traced, since there exists a point in the channel where the instances of the wire starting in L consecutive tiles pass each other and must do so without colliding. Hence, the effects of increased track capacity and maximum wirelength in the channel parallel to the shorter side of the tile balance each other out. We also note that, for various fabrication and packaging reasons, it can be advantageous to make the entire die square [Lew03]. If this is the case, and tiles are not square, there will be a different number of them in each of the two dimensions. The dimension composed of a larger number of tiles will require more bandwidth [Bet96, Lew03], hence also profiting from the larger track space offered by the longer side of the tile. A detailed discussion of the practical consequences of different tile and device aspect ratios is provided by Lewis et al. [Lew03], while die shots of Altera/Intel Stratix II and Xilinx/AMD Artix 7 FPGAs, illustrating the described design decisions, are made available by Martjin Boer [Boe17] and John McMaster [McM23], respectively.

4.7.3.2 Impact of Resistance on Maximum Spans

Similarly, from Table 4.7, we can see that for technology nodes with higher My resistance (F4 and F3a), the maximum spans that can be efficiently realized further reduce. This is in line with the observations about the impact of technology scaling on channel segmentation that was previously made by Lin et al. [Lin10].

4.7.4 Routing Channels: Reference Composition

We determine the exact combination of wires of different lengths by enumerating all possibilities to (partially) fill the channel with wires longer than 1. We limit the exhaustive exploration by forcing the tile to be active-area bound. Then, we pad the remaining space in each combination by length-1 wires until it is the metal that determines the tile dimensions. After each additional wire set is inserted (two wires per LUT, in opposing directions), the multiplexer positions are recomputed (Fig. 4.4), to account for a possible increase of the tile width, due to the increased size of the multiplexers driven by the newly added wires, as well as due to the addition of the new multiplexers driving the inserted wires themselves. Since this process does not influence the capacity of the horizontal channel, as it is fully determined by the height of the stacked LUTs, but it may increase the capacity of the vertical channel, due to the tile width increase, horizontal channel is padded first.

In all cases, vertical wires are assumed to be traced in one My layer and horizontal wires in another. We note, however, that the designers of commercial FPGAs with a large number of metal layers available may choose to use them more liberally [Sax03]. An example of a vertical channel composition, corresponding to the floorplan of Fig. 4.4, is shown in Fig. 4.17, with the padded length-1 wires drawn in gray.

4.7.5 Routing Channels: Taps and Scaling

Out of the explored cluster sizes, $N = 8$ is the most representative of the state-of-the-art commercial FPGAs at 16 nm and before [Lew16, Pet21]. Our goal is to assess whether the rise of resistance at lower metal layers calls for a reduction in cluster size. In order to make a fair comparison of performance of different cluster sizes, the interconnect architecture should ideally be independently optimized for each. However, besides merely comparing different cluster sizes, we would also like to analyze the modifications introduced in the Agilex architecture. Given that the vertical channel segmentation reported for Agilex does not include significantly shorter wires than in Stratix 10 [Chr20] (at least not 4× shorter, which we anticipate the effective cluster size reduction to have been), we can assume that the lengths of channel wires are still expressed in terms of the previous reference cluster size—$N = 10$ for Intel and $N = 8$ for us. Hence, in each technology, we optimize the channel segmentation on the architecture with eight 6-LUT clusters by placing and routing a subset of benchmarks (see Sect. 4.7.7). Then, we scale the logical length of wires for other clusters so that the physical length is maintained. Because the short logical wires may disappear from architectures with smaller clusters, we introduce taps to maintain routability, as suggested in Fig. 4.17. This is conceptually consistent with Agilex wires maintaining the logical length of the large cluster,

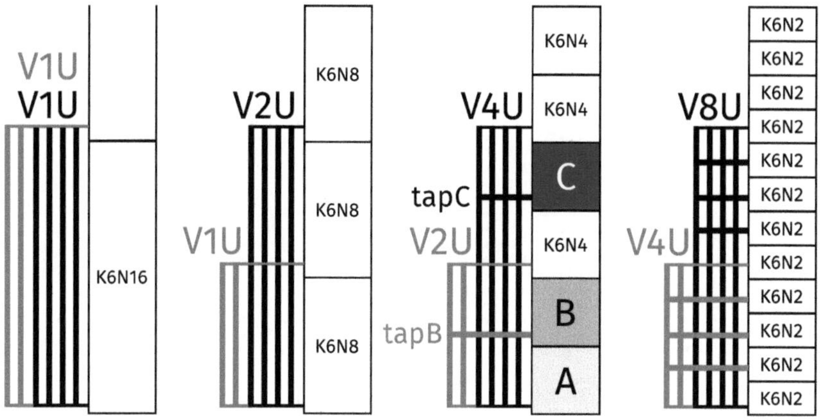

Fig. 4.17 Wirelength scaling and taps. Wirelengths are explored for the eight 6-LUT cluster ($K6N8$) to minimize delay. Combinations of wires longer than 1 (black in the figure) are enumerated in a brute-force manner, while to each particular combination, length-1 wires (gray in the figure) are added until the tile width becomes metal bound, including the active-area extension due to wire addition (Fig. 4.4). Smaller clusters inherit the solution adapted such as to maintain the physical length of the wires. Taps are added to offset less capable local interconnects. For instance, without *tapB*, the $K6N4$ cluster B would not be reachable from cluster A, while without *tapC*, C would not be reachable from it in one hop. The $K6N16$ cluster also inherits the solution, with length-1 wires replacing those of length <1, after scaling

but allowing multiple entry points into it [Chr20]. This method may be suboptimal for smaller clusters, but it is hence conservative: If a smaller cluster demonstrates an advantage, this advantage could only increase if its routing channels were individually optimized.

4.7.6 Switch-Block Patterns

In the most advanced technologies, the switch-pattern must also minimize the distances crossed at the lower metal layers. To address this, we allow connections only between wires ending and starting at the same LUT height (see Fig. 4.4). The exact pattern is formed in such a way that each wire drives at least one wire of each type (length and direction) and is in turn driven by at least one wire of each type, unless it is coming from the direction in which the driven wire is going (which would require it to make a *U-turn*). To further increase routability, length-1 wires are allowed to also drive length-1 wires in their own direction, starting at the height of the immediately adjacent LUTs. This again draws from concepts reported for Agilex [Chr20].

We allow the LUTs to drive all wires that start at their height. Similarly, all wires ending at a particular LUT height are evenly distributed among the connection-block

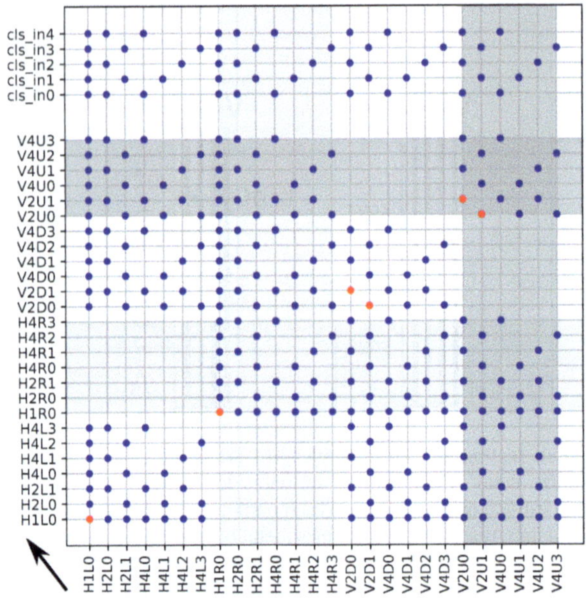

Fig. 4.18 Adjacency matrix of the switch-pattern of the 3 nm $K6N4$ architecture whose floorplan was shown in Fig. 4.4. All points correspond to connections between wires ending and starting at the same LUT height, apart from those marked in red, which are replicated also to the wires starting at the height of the adjacent LUTs. V and H stand for vertical and horizontal wires, respectively, while U, D, L, and R designate the up, down, left, and right directions, respectively. The first number in the name stands for the wire's logical length, while the second is its index within its type

multiplexers of the LUT. The pattern generation algorithm itself is available in the accompanying source code repository (see Sect. 4.10), while Fig. 4.18 gives as an illustration one concrete example, corresponding to the floorplan of Fig. 4.4.

4.7.7 *Experimental Setup*

We rely on the 20 largest MCNC [Yan91] benchmarks in this study. Although there have been concerns about their fitness for architectural exploration, it has been shown that, with respect to basic architectural parameters, they result in the same decisions as larger and newer benchmarks [Zgh17]. Besides, we are now essentially reassessing the same decisions that were made on these very benchmarks in older technologies [Ahm04]. We exclude *bigkey*, *dsip*, and *des* that underutilize the minimum-sized FPGA. The 10 smallest of the remaining 17 circuits are used for ranking channel segmentations, while all 17 are used for the main experiments.

To suppress the experimental noise possibly induced by the choice of the particular channel segmentation, we perform final experiments on the three best-ranked segmentations for which all circuits are routable for all considered cluster sizes. The FPGA grid dimensions in the number of tiles are computed so that the physical width and height of the grid are both approximately equal and minimum, given the requirements of the particular circuit.

We implement all circuits with VTR 8 [Mur20], taking the median routed delay of the three different placements. Then, for each circuit, the median delay obtained on the three chosen channel segmentations is taken as representative, and finally, a geometric mean of such representative delays is computed over all circuits, to represent the particular cluster size in the given technology. All circuits are routed using the *map* lookahead (A* heuristic) of VPR [Mur20].

4.8 The Asteroid Strikes

Results of the architectural study are reported here. We also discuss how they may relate to the recent trends visible in commercial architectures, as well as what could be their influence on the future outlook.

4.8.1 A Future of Small Clusters

Figure 4.19 shows the performance of all cluster sizes at all technology nodes. We can see that until F5 the cluster size ranking is largely maintained as in older technology nodes, with $N = 8$ being the best, or very nearly the best option for optimizing delay [Ahm04]. Yet, as interconnect resistance grows, we visualize the trend we were suspecting: smaller clusters emerge as the best solution. The turning point is, in our technology sequence, between F5 and F4, when $N = 4$ takes a clearer lead against $N = 8$, and $N = 2$ surpasses it as well; it is perhaps worth reiterating here that we do not claim this to be an absolute point, since different foundries evolve technology nodes differently and in ways that are impossible for us to know.

Needless to say, we are also aware that these results are comparable to the noise margin typical of such experiments [Rub11], possibly still worsened by the somewhat unconventional and not thoroughly explored routing channel and switch-pattern designs used in the study. Yet, we believe the observed trend, which follows both the theoretical expectations, as well as a plausible interpretation of the changes applied to a recent commercial architecture, is clear and inescapable.

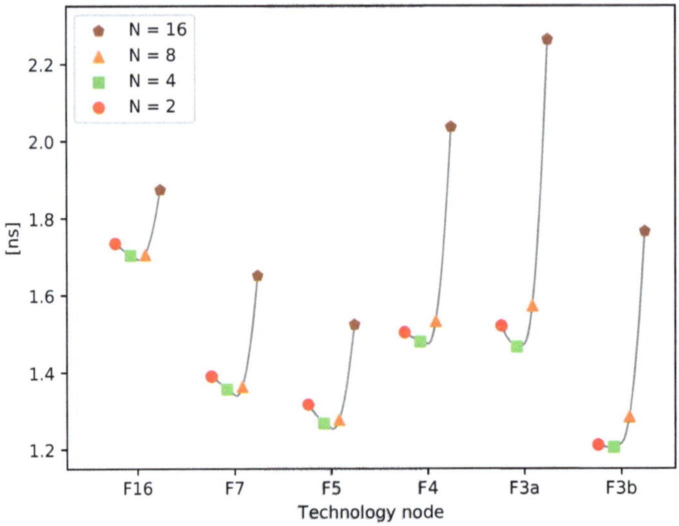

Fig. 4.19 Geometric mean delays of the MCNC benchmarks for different cluster sizes over all considered technology nodes

4.8.2 What Can the Large Clusters Tell Us?

It is worth observing the behavior of the largest, $N = 16$ cluster as well. Its delay is more sensitive to resistance increase than that of the others, both due to the larger load of the more numerous crossbar multiplexers, and because its height is approaching the range where the capacitive load of the long local wires themselves starts to become an important factor in determining their delay (Figure 11 of Ciofi et al. [Cio16]). As we have already noted, the physical dimensions of the Agilex tile are likely significantly larger than the ones of the architectures explored here; both for $N = 10$, each ALM having eight inputs that require local routing, and the higher overall complexity of the cluster. Hence, the effects observed by the designers of Agilex may have been closer to the impact of scaling on $K6N16$, which could have advanced the reduction of the effective cluster size to the 7 nm node. We must note here, however, that while the performance trend of $K6N16$ follows the theoretical expectations, it may be slightly disadvantaged compared to other clusters for two reasons: First, due to wirelength saturation (see Fig. 4.17), it may receive more vertical wires than other cluster sizes, which increases its tile width and thus negatively impacts the delay as well; second, for the smallest benchmarks, the grid height in the number of clusters that would make the physical grid square drops below that for which VPR can compute the router lookahead maps [Pet16] without adverse edge effects. Hence, some grids retain a higher aspect ratio than desired. Nevertheless, neither of these effects has influence on the delay of the local connections and should thus merely impact the speed of $K6N16$ relative to other clusters, but not its decelerating trend with respect to resistance increase across technologies.

4.8.3 So, Is Global the New Local?

Let us return to the question that started this paper: Does it ever become faster to route to the next cluster than to the furthest LUT in the same one? According to our measurements, yes, it does, but after all the cost of exiting and entering again the cluster, this happens only for F3a, and by an extremely small margin, contrary to the expectations of Fig. 4.13.[3] A human designer could reduce this additional cost, e.g., by positioning the global wires replacing the intracluster interconnect of the larger clusters closer to the LUTs. On the other hand, the stark rise in the delay of the local wires distributing intercluster signals to LUT inputs is obvious already with the present modeling[4] and demonstrates itself in the results of Fig. 4.19. Hence, likely the greatest benefit of the aforementioned decomposition of the large cluster in Agilex is in fact the reduction of the penalty that intercluster signals need to pay upon entering the target cluster.

4.8.4 Return of the Supercluster

In Sect. 4.1.2, we mentioned that Versal—also a 7 nm architecture—has increased the cluster size from 8 to 32 [Gai19]. A quick extrapolation of Fig. 4.19 to $N = 32$ immediately casts doubt that increasing the cluster size was motivated by a desire to increase performance. A plausible reason to effectuate such an increase could have been the need to free up track space at upper metal layers for integration of the new hardened fixed-functionality blocks that were added to Versal [Gai19]. This is a well-known constraint for implementing reconfigurable fabrics within a complex SoC [Koc21]. Nevertheless, suffering the performance deterioration anticipated by Fig. 4.19 would be unacceptable.

We hypothesize that the actual idea behind the cluster size increase was to provide an optional alternative for feedback connections within the new large clusters, without adding any additional constraints on how the signals can enter the constituent clusters, nor removing the possibility for constituent clusters to be

[3] It takes 21.4 ps for the output of the LUT to reach a $V4$ intercluster wire in $K6N2$, the delay of the $V4$ wire is 27.8 ps, and a further 18.7 ps is needed for the signal to reach the LUT input, starting from the global wire. In total, 67.9 ps, compared to the maximum delay of 68.6 ps inside $K6N8$.

[4] For instance, a typical value for $K6N8$ at F3a is 39.5 ps, compared to the above 18.7 ps in case of $K6N2$. At F16, the difference is 27.3 ps to 20.7. In the worst case, the impact is even higher, following the trends similar to those of Fig. 4.12.

connected through the faster global interconnect. We illustrate this in Fig. 4.20. When the additional supercluster-level interconnect is used only for non-timing-critical connections, it can be very slow, without impacting the performance of the implemented circuits. The key ingredient of this approach is that the additional supercluster-level interconnect architecture does not take away from the routing flexibility that existed before its addition, but only augments it. Of course, routing channel capacity is likely reduced after the feedback connections are consumed by the added structure, as otherwise the freeing up of the upper layers for the fixed-functionality blocks could not be achieved. However, there is a fundamental difference between the additional interconnect of Fig. 4.20 and the case of the APEX 20K Mega LAB where all connections were *forced* to pass through the supercluster-level interconnect, turning it into a performance bottleneck.

In principle, the purely additional supercluster interconnect structure can do no harm. However, it adds to the tile area and, as we have mentioned in Chap. 2, the effect that this area increase has on overall delay, through lengthening of wires, is profound. Of course, some of the area will be compensated by reduction in routing channel multiplexer number; however, this is not sufficient to offset the cost of the added supercluster interconnect multiplexers, since if it was, large clusters would have been detected as area optimal and they were not [Bet98, Ahm01]. Since the added interconnect can tolerate large delays, the problem could be solved by using area-efficient multilevel multiplexing structures, including Clos and Beneš networks [Wan14]. Enhancements to the placement algorithm could be leveraged to further sparsify supercluster interconnect by shuffling LUTs within the supercluster to align the feedback connections to what the interconnect can support [Nik20].

In Chap. 7, we will extensively rely on the "additional-only" interconnect enhancement approach, while in Chap. 8, we will explore the possibilities for utilizing a dedicated detailed placement algorithm to further enhance the effectiveness of these area-efficient enhancements. While we have no way to be sure that Versal really takes the approach of Fig. 4.20, Figures 4 and 8, as well as Section 3.2 of Gaide et al. [Gai19], strongly suggest that this may indeed be the case. At any rate, leveraging the fact that only a small number of connections in a typical circuit are timing-critical could be very useful for creating more efficient interconnect architectures in the future, despite the fact that the design of Agilex has deliberately tried to reduce the delay variation between routing resources that existed for this reason in the previous generations of Altera/Intel FPGAs [Chr20].

4.8.5 *The End of an Era*

There is perhaps a much more worrying effect to be noticed in Fig. 4.19: Not only newer technology nodes bring disarray to well-established architectural tenets, but, even with corrections to past habits, they do not appear to bring any speed advantages. On the contrary: The grim image of Fig. 4.19 confirms what we have postulated in Chap. 2: The era of simple scaling of programmable interconnect

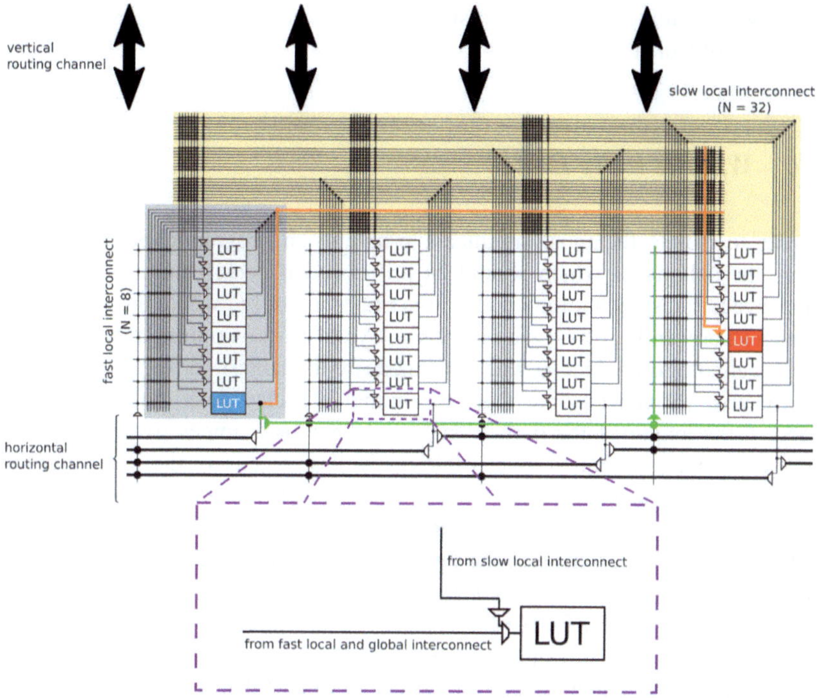

Fig. 4.20 Plausible organization of a Versal cluster comprising 32 LUTs. Non-timing-critical connections can be routed through the slow but metal-efficient additional supercluster-level interconnect shown in yellow (see the orange trace), while the rare timing-critical connections can continue to use the faster global interconnect. Decoupling multiplexers that allow the additional interconnect to be used only optionally ensure that it never becomes a performance bottleneck. Moreover, they do not have to be added to all LUT inputs; for instance, the fastest ones can be reserved exclusively for the signals entering the local 8-LUT cluster through the faster global interconnect. Commercial [Lew12] and latest academic [Moh23] routers will rotate the LUT pins so as to route the timing-critical signals through the fastest ones. On the other hand, most LUTs do not use all six inputs [Lew05], or at least not all of them are generated within the 32-LUT supercluster [Gai19]. Hence, disallowing access of some LUT pins to the supercluster-level local interconnect in order to avoid the additional delay of the decoupling multiplexers is unlikely to reduce the benefits of this newly introduced routing structure; on the contrary, it could make it smaller and faster

architecture without drastic redesign that Dr. Trimberger [Com17] recalled is over. We note that while our architectural exploration in this chapter has been somewhat limited and certainly allows for significant improvement, we have essentially performed the two main optimization steps that have been applied to prior generations of Stratix architectures (see Sect. 2.9). Namely, we: (1) optimized channel segmentation for each particular technology and (2) ensured that the switch-pattern design automatically adapts to changes in segmentation. Of course, the parametric switch-pattern may be suboptimal for any particular segmentation, as we will see in Chap. 5, but it at least offers balanced connectivity between different types of wires, unlike the classical disjoint pattern used, for example, as the baseline in the work of Lin et al. [Lin10].

4.8.6 What Is to Be Done?

In order to isolate the most significant problems in programmable interconnect design, the solving of which should be prioritized if performance is to continue to improve in latest technology nodes, it is important to see where the performance bottlenecks are. In Fig. 4.21, we plot the delays of various routing resources in F5 and F4 technologies, which is where we measure the largest performance decrease. As expected, we can observe that the amount of delay deterioration is directly related to the length of the particular wire. Hence, it is imperative to reduce the length of all wires as much as possible, as well as to ensure that the longer and more resistive wires are not excessively loaded. However, it is highly unlikely that a uniform area shrink could be effectuated without a major loss in routability. Instead, it is the multiplexers catering to the few timing-critical signals of the user circuits that should be carefully positioned in such a way that the wires connecting them are as short as possible. Furthermore, care should be taken that these wires are minimally loaded. The rest of the connectivity can be implemented by slower and longer wires, as we have already discussed in Sect. 4.8.4. The elusive nature of

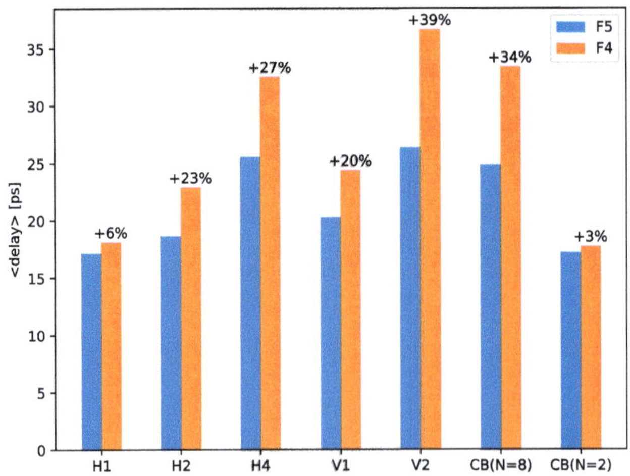

Fig. 4.21 Scaling of wire delays between F4 and F5. Horizontal (vertical) channel wires are indicated by *H* (*V*), followed by the span, while connection blocks are labeled with *CB*. Reported delays include driving multiplexer delays. Longer wires are more sensitive to per-unit-length resistance increase, as expected [Che14]. Similarly, the thinner local wires of the connection block suffer larger performance degradation for a comparable length (e.g., *CB(N=8)* is comparable to *V1*)

hardware reconfigurability comes to the forefront here too: Whereas in an ASIC critical paths are a priori known and can be adequately optimized in this manner, in an FPGA this is not the case and a balance must be struck between how good a physical implementation of a certain interconnect topology is and how well the CAD tools can actually utilize it when mapping user circuits. This will be a recurring topic in this book.

As expected, Fig. 4.21 also confirms that delays deteriorate more for connections implemented on lower and more resistive metal layers. Hence, it is imperative to reduce the distance that connections travel at these layers, for example, within the switch-block. This will be the topic of the next chapter.

4.8.7 Custom Technology Nodes for FPGAs

Another solution, somewhat complementary to architecture enhancements, could be to customize the interconnect stack to the very needs of FPGAs. This is addressed to an extent by our speculative F3b node. As evidenced by Table 4.8, the delay improvement of F3b over F4, due to a more relaxed My pitch and its lower resistance (see Tables 4.1 and 4.2), comes at the expense of almost no density increase and reducing the available track count above the tile by more than a third (Table 4.8). This means that while F3b offers some tangible speed benefits, perhaps insufficient to justify moving from a 5 nm to a 3 nm node, its utility cannot be properly assessed without taking the adverse impact on routability into consideration.

Design Technology Co-optimization (DTCO) is widely considered to be a key technique for allowing continued performance improvement in new technologies [Mor23]. In FPGA design, this would certainly include fine-tuning the pitches and aspect ratios of different layers in the metal stack, as has been done for Agilex [Chr20]. Nevertheless, modeling these possibilities goes beyond the scope of the present work. What also could have been done is to increase the spacing between channel wires at the expense of reduced track capacity, or by using

Table 4.8 Areas and channels in the various technologies. Area A_m is used by the channels and area A_a is the active area. Each column corresponds to the median-area architecture of the three that were chosen for $K6N8$ for the particular technology. All architectures are slightly metal-area bound

	F16	F7	F5	F4	F3a	F3b
A_m [μm^2]	393	239	203	154	136	149
Δ		−39%	−15%	−24%	−12%	−3%
A_a [μm^2]	374	230	186	144	124	123
Δ		−38%	−19%	−23%	−14%	−15%
A_a(routing)	58%	55%	56%	55%	55%	55%
H-tracks	320	288	272	352	336	208
V-tracks	192	144	144	176	176	112

more metal layers in each direction. This technique, previously explored by Betz and Rose, among others, reduces mutual capacitance and improves interconnect speed [Bet99a]. Multiple layers could, of course, also be used to reduce the mutual capacitance of local wires. By repeating the same connection at the layer above and connecting the two wires with multiple vias, even the effect of increased resistance could potentially be mitigated. This has been employed in QuickLogic FPGAs, for example, [Sax03]. Nevertheless, increasing the number of layers also increases the cost of the chip and complicates the process of physical modeling. Since our goal is to develop techniques for automated optimization of programmable interconnect architecture topology, we leave exploration of further DTCO possibilities for future work.

4.8.8 And What About Density?

Table 4.8 shows the evolution of tile area and channel width over the technology nodes. Compared to performance improvement, density scaling seems a lot more promising. In the most advanced nodes, observed area reduction meets the typical expectations (about 40% per node; remember that F4 is an intermediate node). We must note, however, that the reported area is influenced by the employed methodology intended to maximize the available wiring bandwidth above the tile (see Sect. 4.7.4). In reality, routability requirements of more complex circuits and the availability of multiple My layers must be taken into account as well. We leave this for future work.

Table 4.8 reports only the data for $N = 8$. For $N = 4$ and $N = 2$, the metal area and horizontal channel width are two and four times smaller, respectively, while the vertical channels maintain the width, due to the way in which they are composed (see Fig. 4.17). The only slight variation exists in the active area, due to the different crossbar and connection-block multiplexer size and count. The aforementioned wirelength saturation effect causes the vertical channels to be somewhat wider for $N = 16$, and its area a bit more than twice that reported in Table 4.8. Finally, it is interesting to note that active area of the programmable interconnect (including all multiplexers of the crossbar, the connection block, and the switch-block) is about 55% of the entire tile, which is in line with the values reported for commercial architectures [Lew13].

4.8.9 The Issue of Nonshrinking SRAM

One of the problems with programmable interconnect is the need to use SRAM cells to store the select inputs of routing multiplexers. As we can see from Fig. 4.7, stored-select multiplexer area is in fact dominated by the SRAM cells. For example, in a 16:1 multiplexer, 84% of the area not including the output buffer is consumed by

SRAM. If the output buffer is included, the fraction of the area consumed by SRAM typically falls down to 67%. We note once more that our transmission-gate layout model may be slightly optimistic, which could slightly overemphasize SRAM area. Nevertheless, already in 20 nm, Chromczak and Lewis report that SRAM consumes about 50% of the total routing area [Lew12] and in scaled technologies, optimal buffer size reduces, due to the large series resistance of the wire itself. In prior work, which neglects the issues of stability, this was not visible and SRAM was generally not considered a major concern [Chi13a]. In older technologies, in fact, effort was made even in commercial FPGAs to reduce the number of pass transistors that a signal had to pass while going through a multiplexer, all at the expense of increased SRAM cell count [Lew05]. Yet, when per-unit-length resistance of wires increases, it is imperative to reduce their length. This in turn calls for reducing active area and SRAM cells stand in the way.

As we have already mentioned, for the models developed here to be useful for architectural exploration, they had to be reasonably simple and scalable across technologies. Furthermore, in order to be representative of general trends and not peculiar to any particular foundry, we had to rely on only approximate values representing different dimensions. Finally, it is worth noting that the SRAM cell design we adopted from Young et al. [You15] was specifically tailored for use in FPGAs and is different from a typical 6T SRAM cell. Hence, it is only reasonable to expect that the SRAM cell area computed from the models developed in this work will differ from the values reported by the foundries. Yet, it is also important to compare the predicted trends with those recently made available by the foundries, in the period that followed the original publication of this work [Nik21].

Figure 4.22 compares the actual SRAM scaling reported by TSMC with that predicted by our models. As we can see, although the absolute values differ, the

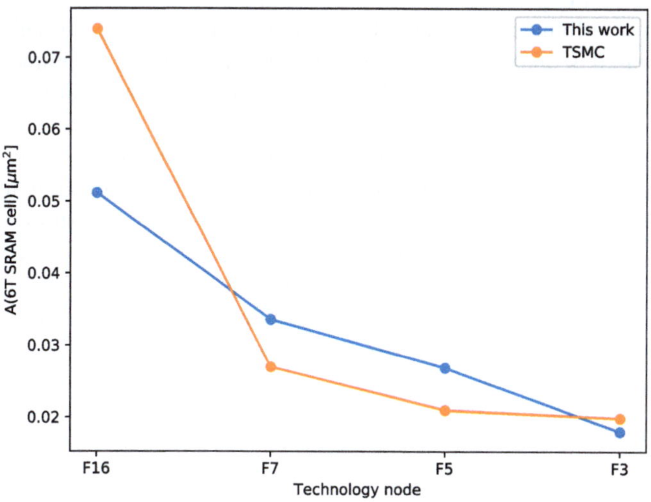

Fig. 4.22 SRAM area scaling as predicted by our model and as reported by TSMC [Sch22]

general trends of SRAM scaling predicted by our model are in line with those reported by TSMC until F5, inclusive. We note that the large discrepancy at F16 is an instance of foundry variability: Intel reports their 14 nm SRAM cell to take up 0.0588 μm^2, which is much closer to 0.0512 μm^2 computed from our model. What could be very worrying is that unlike the predictions of our model, scaling of TSMC's SRAM cell area between F5 and F3 is minute, and is even expected to entirely reduce to zero in future refinements of the 3 nm node [Sch22]. Although the overall area reduction that we measure is in line with what TSMC reports [Sch22], TSMC's gains are driven by reduction of logic and not SRAM area. If SRAM indeed stops scaling, this may create problems for the SRAM-heavy FPGAs, and in particular for their programmable interconnect in which, unlike in LUT masks, SRAM cells do not serve a dual purpose of creating user-accessible read-write memories. Perhaps multiplexer designs that minimize the number of transistors on signal paths should be abandoned in favor of architectures that increase configuration encoding efficiency, such as those explored by DeHon [DeH95]. Fully encoded multiplexers, as well as SRAM sharing between multiplexers (e.g., by using multi-bit buses [Ye06]) and even multiplexers and LUT masks, could be good candidates for alleviating this potentially very serious problem. Of course, a different area-switching-depth trade-off can be implemented for a small number of programmable connections intended to support timing-critical signals. All this, however, is out of scope of the present book, and we leave it for future work.

4.9 Low-Hanging Fruit Fell to the Ground

Significant attention has been given to the problem of optimizing channel segmentation over the past few decades, with approaches ranging from brute-force exploration of a limited set of possible compositions [Bet99a], through simulated annealing-based search [Lin10, Qia21], to matching-based techniques [Cha01]. However, we have shown through this study that the constraints of modern FPGA design—namely, that the number of wires of the given length and direction is divisible by cluster size [Chr20]—and the increased resistance that limits the maximum length of feasible wires limit the search space of channel segmentation to an extent when it becomes possible to essentially exhaustively explore it. Although we only considered lengths of wires that are powers of two, which does not encompass all lengths commonly observed in commercial architectures [Pet21, Chr20], removing this constraint would not render exhaustive exploration infeasible. Such exploration becomes easier with technology scaling, as indicated in Table 4.9, owing to the

Table 4.9 Total number of feasible channel segmentations in each technology

F16	F7	F5	F4	F3a	F3b
197	134	130	51	51	64

reduction in maximum feasible wirelength caused by resistance increase. Hence, we can conclude that channel segmentation optimization is no longer a complex problem requiring a complex solution—it can simply be brute-forced. Nevertheless, to make such exploration practical, physical models must be fast to evaluate. In particular, we explored 627 different channel segmentations in this study. If evaluating each of them took about ten hours as would have been the case with COFFE, for example [Chi13a], that would hardly be feasible. However, a typical evaluation of our model, together with the generation of the routing-resource graph and the architecture description file for VPR, takes only about 90 seconds.

This evaluation speed allows accurate exploration of other aspects of the interconnect architecture. For example, it enables optimizing the switch-block connectivity pattern to reduce the multiplexer area and capacitive load on channel wires, as well as to reduce the length of wires inside the switch-block itself, which are traced on highly resistive lower metal layers. This problem has a much larger solution space than that of designing channel segmentation and has not been satisfactorily solved in prior work. Due to its relevance for programmable interconnect performance in scaled technologies, it becomes our focus in the next chapter.

4.10 Conclusions and Future Work

In this chapter, we have presented a framework for fast physical modeling of FPGAs at advanced technology nodes. This framework allowed us to revisit a fundamental rule of thumb applied to FPGA architecture design over the course of the past twenty years—namely, that the cluster size optimal for performance should be around ten 6-LUTs. That in turn helped us explain the departure from this rule observed in the most recent commercial FPGAs. We also tried to predict how FPGA performance scaling may look like in the next few generations of silicon technologies and the outlook does not seem bright. Rather than see this as an end of programmable interconnect evolution, we merely interpret the result as empirical support for what we postulated in Chap. 2: A long-lasting era of simple scaling with only local optimization of the interconnect architecture has come to an end.

In subsequent chapters, we will focus on developing techniques for automating programmable interconnect architecture design far beyond what is typically done. In particular, through experiments performed in this chapter, we have identified that the key problem is that of designing efficient switch-patterns that minimize the area consumed by routing multiplexers, thus allowing wires to be shorter, along with decreasing capacitive load and distance traveled at the lower metal layers within the switch-block itself. This will be the topic of Chaps. 5 and 6. We have also seen that the price of entering a cluster remains very significant even in small clusters, due to the high resistance of local wires. As a possible solution, we mentioned allowing a small number of wires intended to support timing-critical signals to bypass the local interconnect altogether, minimizing the distance traveled at lower metal layers and avoiding significant capacitive loading exerted by many routing multiplexers on more general wires. We dedicate our attention to designing such specialized connectivity patterns in Chap. 7, noting that the approach is relevant even out of the technology scaling context, as it allows bypassing a number of transistors in routing multiplexers, which formed the performance bottleneck in older technologies. We have also mentioned that for these specialized connections to be effective in boosting the performance of user circuits, CAD tools must be aware of them and adjust the circuit mapping accordingly. Developing such tools is the focus of Chap. 8.

The present analysis was limited to architectures composed solely of LUTs. On the other hand, modern FPGAs have long been heterogeneous, incorporating columns of hardened DSP modules, for example, [Lew03]. Given that such blocks are often designed using standard cells [Yaz19], they will inevitably scale much better than the SRAM-dominated LUTs [She18, Sch22]. It would hence be very useful to extend the framework proposed in this chapter to support modeling hardened blocks as well. That would in turn also enable experiments with larger and more modern benchmarks circuits that can potentially reduce the impact of low density of LUTs on average length of wires implementing the circuit's connections, by relying on the hardened blocks instead. Extending the modeling framework to support standard cell-based blocks would even allow analysis of architectures that are based on logic elements different than LUTs, which too can profit from the better scaling of the standard cells. One such example could be the NAND-NORs introduced by Huang et al. [Hua17].

Finally, it is important to note that in Sect. 4.8.7, we observed that significant benefits can be obtained by codesigning the programmable interconnect architecture and the BEOL aspects of the technology node. While we illustrated this possibility by including in our experiments the hypothetical F3b technology node, any further exploration of DTCO goes beyond the scope of the present book. Some further optimization of the tile floorplan is possible, however, and we will introduce it in the next chapter, where it becomes relevant to designing high-performance switch-patterns. We also note that much like design constraints of modern FPGA architectures made the problem of channel segmentation design tractable, the gridded nature of FinFET layouts could allow for even more precise modeling than what we have developed in this chapter, without much loss in terms of evaluation speed. In the limit, since the layout solution space is so significantly restricted, compared to planar technologies, it may even be possible to completely automatically generate optimized full-custom layouts of each FPGA. For example, it has already been demonstrated that in a gridded context, it is possible to autogenerate custom standard cells for a given design on the fly, during physical implementation of an ASIC [Ryz11]. This, along with extending DTCO capabilities, could be an appealing avenue for future work on the topic presented in this chapter.

The source code used to produce the results of this study is available at https://github.com/EPFL-LAP/fpga21-scaled-tech.

References

[Abu14] Abusultan, M., & Khatri, S. P. (2014). A comparison of FinFET-Based FPGA LUT designs. In *Proceedings of the 24th Edition of the Great Lakes Symposium on VLSI*. Houston, TX (pp. 353–58).

[Ahm00] Ahmed, E., & Rose, J. (2000). The effect of LUT and cluster size on deep-submicron FPGA performance and density. In *Proceedings of the ACM/SIGDA International Symposium on Field Programmable Gate Arrays*. Monterey, CA (pp. 3–12).

[Ahm01] Ahmed, E. (2001). The Effect of Logic Block Granularity on Deep-Submicron FPGA Performance and Density. *Master Thesis*. University of Toronto.

[Ahm04] Ahmed, E., & Rose, J. (2004). The effect of LUT and cluster size on deep-submicron FPGA performance and density. *IEEE Transactions on Very Large Scale Integration (VLSI) Systems, 12*(3), 288–98.

[Alt01] Altera Corporation (2001). *FLEX 10K Embedded Programmable Logic Device Family*. ver. 4.1. 101 Innovation Drive, San Jose, CA 95134.

[Bet96] Betz, V., & Rose, J. (1996). Directional bias and non-uniformity in FPGA global routing architectures. In *Proceedings of International Conference on Computer Aided Design*. San Jose, CA (pp. 652–659).

[Bet98] Betz, V., & Rose, J. (1998). How much logic should go in an FPGA logic block? *IEEE Design & Test of Computers, 151*, 10–15.

[Bet99a] Betz, V., & Rose, J. (1999). FPGA routing architecture: segmentation and buffering to optimize speed and density. In *Proceedings of the 1999 ACM/SIGDA Seventh International Symposium on Field Programmable Gate Arrays*. Monterey, CA (pp. 59–68).

[Bet99] Betz, V., Rose, J., & Marquardt, A. (1999). *Architecture and CAD for deep-submicron FPGAs*. Kluwer Academic Publishers. ISBN: 0-7923-8460-1.

[Bha15] Bhanushali, K. N., & Davis, W. R. (2015). FreePDK15: An open-source predictive process design kit for 15 nm FinFET technology. In *Proceedings of the 2015 Symposium on International Symposium on Physical Design*. Monterey, CA (pp. 165–170).

[Boe17] Boer, M. (2017). *Altera Stratix II EP2S60 Die Shots*. https://www.flickr.com/photos/sic66/albums/72157667394503929/. Accessed on 03 Oct 2024.

[Cha01] Chang, Y.-W., Lin, J.-M., & Wong, M. D. F. (2001). Matching-based algorithm for FPGA channel segmentation design. *IEEE Transactions on Computer-Aided Design of Integrated Circuits and Systems, 20*(6), 784–791.

[Cha15] Chandrakar, S., Gaitonde, D., & Bauer, T. (2015). Enhancements in UltraScale CLB architecture. In *Proceedings of the 2015 ACM/SIGDA International Symposium on Field-Programmable Gate Arrays*. Monterey, CA (pp. 108–16).

[Cha22] Chang, C.-H., Chang, V. S., Pan, K. H., Lai, K. T., Lu, J. H., Ng, J. A., et al. (2022). Critical process features enabling aggressive contacted gate pitch scaling for 3 nm CMOS technology and beyond. In *2022 International Electron Devices Meeting (IEDM)*. San Francisco, CA (pp. 27.1.1–27.1.4).

[Che14] Chen, J. H.-C., Standaert, T., Alptekin, E., Spooner, T., & Paruchuri, V. (2014). Interconnect performance and scaling strategy at 7 nm node. In *IEEE International Interconnect Technology Conference*. San Jose, CA (pp. 93–96).

[Chi13] Chiasson, C. (2013). Optimization and Modeling of FPGA Circuitry in Advanced Process Technology. *MA Thesis*. University of Toronto.

[Chi13a] Chiasson, C., & Betz, V. (2013). COFFE: Fully-automated transistor sizing for FPGAs. In *Proceedings of the 2013 International Conference on Field-Programmable Technology*. Kyoto (pp. 34–41).

[Cho91] Chow, P., Seo, S. O., Au, D., Fallah, B., Li, C., & Rose, J. (1991). A 1.2μmCMOS FPGA using cascaded logic blocks and segmented routing. In *Proceedings of the International Workshop on Field Programmable Logic and Applications*. Oxford.

[Chr20] Chromczak, J., Wheeler, M., Chiasson, C., How, D., Langhammer, M., Vanderhoek, T., et al. (2020). Architectural enhancements in Intel®Agilex™FPGAs. In *Proceedings of the 2020 ACM/SIGDA International Symposium on Field-Programmable Gate Arrays*. Monterey, CA (pp. 140–149).

[Cio16] Ciofi, I., Contino, A., Roussel, P. J., Baert, R., Vega-Gonzalez, V., Croes, K., et al. (2016). Impact of wire geometry on interconnect RC and circuit delay. *IEEE Transactions on Electron Devices, 63*(6), 2488–2496.

[Cio17] Ciofi, I., Roussel, P. J., Saad, Y., Moroz, V., Hu, C.-Y., Baert, R., et al. (2017). Modeling of via resistance for advanced technology nodes. *IEEE Transactions on Electron Devices, 64*(5), 2306–2313.

[Cla16] Clark, L. T., Vashishtha, V., Shifren, L., Gujja, A., Sinha, S., Cline, B., et al. (2016). ASAP7: A 7-nm FinFET predictive process design kit. *Microelectronics Journal, 53*, 105–115.

[Com17] Computer History Museum (2017). *Oral History of Steve Trimberger*. https://archive.computerhistory.org/resources/access/text/2018/07/102740229-05-01-acc.pdf.

[DeH95] DeHon, A. (1995). *Entropy, counting, and programmable interconnect*. Transit Note 128. University of CA.

[Gai19] Gaide, B., Gaitonde, D., Ravishankar, C., & Bauer, T. (2019). Xilinx adaptive compute acceleration platform: Versal architecture. In *Proceedings of the 2019 ACM/SIGDA International Symposium on Field-Programmable Gate Arrays*. Seaside, CA (pp. 84–93).

[Gan16] Ganusov, I., & Devlin, B. (2016). Time-borrowing platform in the Xilinx UltraScale+ family of FPGAs and MPSoCs. In *2016 26th International Conference on Field Programmable Logic and Applications (FPL)*. Lausanne (pp. 1–9).

[Hua17] Huang, Z., Wei, X., Zgheib, G., Li, W., Lin, Y., Jiang, Z., et al. (2017). NAND-NOR: A compact, fast, and delay balanced FPGA logic element. In *Proceedings of the 2017 ACM/SIGDA International Symposium on Field-Programmable Gate Arrays*. Monterey, CA (pp. 135–140).

[Hut97] Hutton, M. (1997). Characterization and Parameterized Generation of Digital Circuits. *PhD Thesis*. University of Toronto.

[Koc21] Koch, D., Dao, N., Healy, B., Yu, J., & Attwood, A. (2021). FABulous: An embedded FPGA framework. In *The 2021 ACM/SIGDA International Symposium on Field-Programmable Gate Arrays*. Virtual Event (pp. 45–56).

[Lee06] Lee, E., Lemieux, G., & Mirabbasi, S. (2006). Interconnect driver design for long wires in field-programmable gate arrays. In *2006 IEEE International Conference on Field Programmable Technology*. Bangkok (pp. 89–96).

[Lem01] Lemieux, G., & Lewis, D. (2001). Using sparse crossbars within LUT clusters. In *Proceedings of the 2001 ACM/SIGDA Ninth International Symposium on Field Programmable Gate Arrays*. Monterey, CA (pp. 59–68).

[Lew03] Lewis, D., Betz, V., Jefferson, D., Lee, A., Lane, C., Leventis, P., et al. (2003). The stratix routing and logic architecture. In *Proceedings of the 2003 ACM/SIGDA Eleventh International Symposium on Field Programmable Gate Arrays*. Monterey, Calif (pp. 12–20).

[Lew05] Lewis, D., Ahmed, E., Baeckler, G., Betz, V., Bourgeault, M., Cashman, D., et al. (2005). The stratix II logic and routing architecture. In *Proceedings of the 2005 ACM/SIGDA 13th International Symposium on Field-Programmable Gate Arrays*. Monterey, CA (pp. 14–20).

[Lew09] Lewis, D., Ahmed, E., Cashman, D., Vanderhoek, T., Lane, C., Lee, A., et al. (2009). Architectural enhancements in stratix-III™and Stratix-IV™. In *Proceedings of the ACM/SIGDA International Symposium on Field Programmable Gate Arrays*. Monterey, CA (pp. 33–42).

[Lew12] Lewis, D., & Chromczak, J. (2012). Process technology implications for FPGAs. In *Proceedings of the 2012 International Electron Devices Meeting*. San Francisco, CA (pp. 25.2.1–4).

[Lew13] Lewis, D., Cashman, D., Chan, M., Chromczak, J., Lai, G., Lee, A., et al. (2013). Architectural enhancements in stratix V™. In *Proceedings of the ACM/SIGDA International Symposium on Field Programmable Gate Arrays*. Monterey, CA (pp. 147–156).

[Lew16] Lewis, D., Chiu, G., Chromczak, J., Galloway, D., Gamsa, B., Manohararajah, V., et al. (2016). The stratix™10 highly pipelined FPGA architecture. In *Proceedings of the 2016 ACM/SIGDA International Symposium on Field-Programmable Gate Arrays*. Monterey, CA (pp. 159–168).

[Lin10] Lin, M., Wawrzynek, J., & Gamal, A. E. (2010). Exploring FPGA routing architecture stochastically. *IEEE Transactions on Computer-Aided Design of Integrated Circuits and Systems, 29*(10), 1509–1522.

[Lin23] Lin, B. J. (2023). Immersion and EUV lithography: Two pillars to sustain single-digit nanometer nodes. In *Proceedings of the 2023 International Symposium on Physical Design*. Virtual Event.

[McM23] McMaster, J. (2023). *Xilinx Artix 7 XC7A50T Die Shots*. https://siliconpr0n.org/archive/doku.php?id$=$mcmaster:xilinx:xc7a50t. Accessed on 03 Oct 2024.

[Moh23] Mohaghegh, A., & Betz, V. (2023). Tear down the wall: Unified and efficient intra-and inter-cluster routing for FPGAs. In *Proceedings of the 33rd International Conference on Field-Programmable Logic and Applications (FPL)*. Gothenburg (pp. 130–136).

[Mor23] Moroz, V. (2023). Gate-all-around technology is coming.: What's next after GAA? In *Proceedings of the 2023 International Symposium on Physical Design*. Virtual Event.

[Mur13] Murray, K. E., Whitty, S., Liu, S., Luu, J., & Betz, V. (2013). Titan: Enabling large and complex benchmarks in academic CAD. In *Proceedings of the 23rd International Conference on Field programmable Logic and Applications*. Porto (pp. 1–8).

[Mur20] Murray, K. E., Petelin, O., Zhong, S., Wang, J. M., Eldafrawy, M., Legault, J.-P., et al. (2020). VTR 8: High-performance CAD and customizible FPGA architecture modelling. *ACM Transactions on Reconfigurable Technology and Systems (TRETS), 13*(2), 9:1–60.

[Nan11] ASU Nanoscale Integration and Modeling (NIMO) Group (2011). *Predictive Technology Model*. http://ptm.asu.edu/. Accessed: 26 Aug, 2020.

[Nik20] Nikolić, S., Zgheib, G., & Ienne, P. (2020). Straight to the point: Intra- and intercluster LUT connections to mitigate the delay of programmable routing. In *FPGA '20: The 2020 ACM/SIGDA International Symposium on Field-Programmable Gate Arrays, Seaside, CA, February 23–25, 2020*. Seaside, CA (pp. 150–160).

[Nik21] Nikolić, S., Catthoor, F., Tőkei, Z., & Ienne, P. (2021). Global is the new local: FPGA architecture at 5 nm and beyond. In *The 2021 ACM/SIGDA International Symposium on Field-Programmable Gate Arrays*. Virtual Event (pp. 34–44).

[Pet16] Petelin, O., & Betz, V. (2016). The speed of diversity: Exploring complex FPGA routing topologies for the global metal layer. In *Proceedings of the 26th International Conference on Field Programmable Logic and Applications*. Lausanne (pp. 1–10).

[Pet21] Petersen, M. B., Nikolić, S., & Stojilović, M. (2021). NetCracker: A peek into the routing architecture of Xilinx 7-series FPGAs. In *Proceedings of the 2021 ACM/SIGDA International Symposium on Field Programmable Gate Arrays*. Virtual Event (pp. 11–22).

[Pra19] Prasad, D., Nibhanupudi, S. S. T., Das, S., Zografos, O., Chehab, B., Sarkar, S., et al. (2019). Buried power rails and back-side power grids: Arm®CPU power delivery network design beyond 5 nm. In *Proceedings of the 2019 IEEE International Electron Devices Meeting*. San Francisco, CA (pp. 19.1.1–4).

[Qia21] Qian, J., Shen, Y., Shi, K., Zhou, H., & Wang, L. (2021). General routing architecture modelling and exploration for modern FPGAs. In *Proceedings of the 2021 International Conference on Field-Programmable Technology*. Auckland (pp. 1–9).

[Rub11] Rubin, R., & DeHon, A. (2011). Timing-driven pathfinder pathology and remediation: Quantifying and reducing delay noise in VPR-pathfinder. In *Proceedings of the ACM/SIGDA 19th International Symposium on Field Programmable Gate Arrays*. Monterey, CA (pp. 173–176).

[Ryz11] Ryzhenko, N., & Burns, S. (2011). Physical synthesis onto a layout fabric with regular diffusion and polysilicon geometries. In *2011 48th ACM/EDAC/IEEE Design Automation Conference (DAC)*. San Diego, CA (pp. 83–88).

[Sax03] Saxe, T., & Faith, B. (2003). *EFLX®eFPGA Resources*. https://www.eetimes.com/metal-layers-a-key-to-interconnect-delay/. Accessed on 24 April, 2023.

[Sch22] Schor, D. (2022). *IEDM 2022: Did We Just Witness The Death Of SRAM?* WikiChip, https://fuse.wikichip.org/news/7343/iedm-2022-did-we-just-witness-the-death-of-sram/. Accessed on 24 April, 2023.

[She18] Sherazi, S. M. Y., Chae, J. K., Debacker, P., Matti, L., Raghavan, P., Gerousis, V., et al. (2018). Track height reduction for standard-cell in below 5 nm node: How low can you go? In *Design-Process-Technology Co-optimization for Manufacturability* (Vol. 10588, pp. 1058809:1–13). SPIE.

[Shr23] Shrivastava, S., Nikolić, S., Ravishankar, C., Gaitonde, D., & Stojilović, M. (2023). IIBLAST: Speeding up commercial FPGA routing by decoupling and mitigating the intra-CLB bottleneck. In *Proceedings of the 2023 IEEE/ACM International Conference on Computer Aided Design (ICCAD)*. San Francisco, CA.

[Sti17] Stillmaker, A., & Baas, B. (2017). Scaling equations for the accurate prediction of CMOS device performance from 180 nm to 7 nm. *Integration, 58*, 74–81.

[Tok16] Tőkei, Z., Ciofi, I., Roussel, P., Debacker, P., Raghavan, P., van der Veen, M. H., et al. (2016). On-chip interconnect trends, challenges and solutions: How to keep RC and reliability under control. In *2016 IEEE Symposium on VLSI Technology*. Honolulu, Hawaii (pp. 1–2).

[TSM23] TSMC (2023). *Logic Technology*. https://www.tsmc.com/english/dedicatedFoundry/technology/logic. Accessed on 29 March, 2023.

[Tyh15] Tyhach, J., Hutton, M., Atsatt, S., Rahman, A., Vest, B., Lewis, D., et al. (2015). Arria™10 device architecture. In *Proceedings of the IEEE Custom Integrated Circuit Conference*. San Jose, CA (pp. 1–8).

[Wan14] Wang, C. C., Yuan, F.-L., Yu, T.-H., & Markovic, D. (2014). A multi-granularity FPGA with hierarchical interconnects for efficient and flexible mobile computing. In *2014 IEEE International Solid-State Circuits Conference Digest of Technical Papers (ISSCC)*. San Francisco, CA (pp. 460–61).

[Won00] Wong, S.-C., Lee, G.-Y., & Ma, D.-J. (2000). Modeling of interconnect capacitance, delay, and crosstalk in VLSI. *IEEE Transactions on Semiconductor Manufacturing, 13*(1), 108–111.

[Wu13] Wu, S.-Y., Lin, C. Y., Chiang, M. C., Liaw, J. J., Cheng, J. Y., Yang, S. H., et al. (2013). A 16 nm FinFET CMOS technology for mobile SoC and computing applications. In *Proceedings of the 2013 IEEE International Electron Devices Meeting*. Washington, DC (pp. 9.1.1–4).

[Wu16] Wu, S., Lin, C. Y., Chiang, M. C., Liaw, J. J., Cheng, J. Y., Yang, S. H., et al. (2016). A 7 nm CMOS platform technology featuring 4th generation FinFET transistors with a 0.027 um2 high density 6-T SRAM cell for mobile SoC applications. In *Proceedings of the 2016 IEEE International Electron Devices Meeting*. San Francisco, CA (pp. 2.6.1–4).

[Wu22] Wu, S.-Y., Chang, C. H., Chiang, M. C., Lin, C. Y., Liaw, J. J., Cheng, J. Y., et al. (2022). A 3 nm CMOS FinFlex™platform technology with enhanced power efficiency and performance for mobile SoC and high performance computing applications. In *2022 International Electron Devices Meeting (IEDM)*. San Francisco, CA (pp. 27.5.1–27.5.4).

[Wu20] Wu, T., Luo, H., Wang, X., Asenov, A., & Miao, X. (2020). A predictive 3-D source/drain resistance compact model and the impact on 7 nm and scaled FinFETs. *IEEE Transactions on Electron Devices, 67*(6), 2255–2262.

[Xil93] Xilinx Inc. (1993). *XC2000 Logic Cell Array Families*. San Jose, CA.

[Yan91] Yang, S. (1991). *Logic Synthesis and Optimization Benchmarks User Guide, Version 3.0*. Technical Report. Microelectronics Center of North Carolina.

[Yaz19] Yazdanshenas, S., & Betz, V. (2019). COFFE 2: Automatic modelling and optimization of complex and heterogeneous FPGA architectures. *ACM Transactions on Reconfigurable Technology and Systems (TRETS), 12*(11), 3:1–27.

[Ye06] Ye, A., & Rose, J. (2006). Using bus-based connections to improve field-programmable gate-array density for implementing datapath circuits. *IEEE Transactions on Very Large Scale Integration (VLSI) Systems, 14*(5), 462–473. ISSN: 1557-9999.

[Yea19] Yeap, G., Lin, S. S., Chen, Y. M., Shang, H. L., Wang, P. W., Lin, H. C., et al. (2019). 5 nm CMOS production technology platform featuring full-fledged EUV, and high mobility channel FinFETs with densest 0.021µm2 SRAM cells for mobile SoC and high performance computing applications. In *Proceedings of the 2019 IEEE International Electron Devices Meeting*. San Francisco, CA (pp. 36.7.1–4).

[You15] Young, S. P., Song, Y., & Chong, N. (2015). Two gate pitch FPGA memory cell. US 9177634 B1. Patent assignee: Xilinx Inc.

[Zgh16] Zgheib, G., Lortkipanidze, M., Owaida, M., Novo, D., & Ienne, P. (2016). FPRESSO: Enabling express transistor-level exploration of FPGA architectures. In *Proceedings of the 2016 ACM/SIGDA International Symposium on Field-Programmable Gate Arrays*. Monterey, CA (pp. 80–89).

[Zgh17] Zgheib, G., & Ienne, P. (2017). Evaluating FPGA clusters under wide ranges of design parameters. In *Proceedings of the 27th International Conference on Field Programmable Logic and Applications*. Ghent (pp. 1–8).

Chapter 5
Switch Presence Negotiation

Contents

© The Author(s), under exclusive license to Springer Nature Switzerland AG 2025
S. Nikolić, *Modern Programmable Interconnect Design*,
https://doi.org/10.1007/978-3-031-80629-2_5

In the last chapter, we have seen that in scaled technologies it is imperative to reduce the distance that signals traverse at the low, highly resistive metal layers. It is also necessary to reduce the capacitive load on channel wires, as well as the active area consumed by the routing multiplexers so that the lengths of different wires can be reduced. At the same time, not all connections will be used by the router to implement timing-critical signals of user circuits, meaning that while general improvement in the above three metrics is highly desirable, further benefits can be reaped if optimization specifically targets the connections commonly used by the critical signals. Needless to say, while designing the switch-block to meet these criteria, care must be taken that overall routability is not excessively damaged.

How does one simultaneously take into account (1) the influence of technology and physical implementation on the performance of different routing resources and (2) their utility for implementing different user circuits? Our premise is that automating the switch-pattern design process is the only way to reliably balance these two factors under the rapidly changing technological and application constraints. In this chapter, largely based on our prior work published in 2021 at the 31st International Conference on Field-Programmable Logic and Applications (FPL), under the title "Turning PathFinder Upside-Down: Exploring FPGA Switch-Blocks by Negotiating Switch Presence" [Nik21a], and especially its journal extension published in ACM Transactions on Reconfigurable Technology and Systems, under the title "Exploring FPGA Switch-Blocks without Explicit Pattern Listing" [Nik24], we present a novel algorithm for solving this design problem, which overcomes a fundamental scalability barrier of the prior approaches. Before describing the details of the algorithm, let us first take a brief look at how the problem was solved during the past three decades.

5.1 A Little Bit of History

In the early days of FPGA research, considerable attention was given to the problem of designing efficient switch-patterns. However, much like Dr. Trimberger recalls [Com17], the process was mostly manual, relying on careful observation and intuition about what may and may not work, derived from analyzing small examples that could be effectively reasoned about [Ros91, Wil97]. To validate the intuition, researchers typically relied on placing and routing a standard set of benchmark circuits using place and route tools, which produced performance metrics, such as the routed critical path delay [Bet99]. There were also some attempts to formulate fundamental properties that an optimal switch-pattern should satisfy and construct solutions accordingly. An example of that was the so-called *Universal* switch-pattern [Cha96]. Although this approach is very appealing, for it enables proofs about an architecture's general optimality, instead of relying on empirical demonstrations on a finite set of benchmark circuits, it quickly became obvious that definitions of optimality that one could actually reason about were far too local and simplistic to capture the needs of actual circuits and the capabilities of actual CAD tools; it turned out that different patterns, which were not optimal in the postulated sense, such as *Wilton* [Wil97], performed better in practice. By the end of the last century, three main switch-pattern designs emerged as the most popular: *Disjoint* (also called *Subset* or *Planar*) [Xil98], *Universal* [Cha96], and *Wilton* [Wil97]. Although industry has long since moved away from these designs [Pet21], they are still used in majority of academic FPGA architecture research produced today, with Wilton being the most popular choice [Tan19].

5.1.1 Importance of Parametric Patterns

These three classical switch-pattern designs had one very important characteristic: They were parametric in terms of the number of tracks in the routing channels; in fact, they could be thought of as concisely described procedures for constructing switch-blocks on an arbitrary number of tracks. Lemieux and Lewis even managed to formalize these succinct procedures as permutations of tracks entering the switch-block on one side and onto tracks exiting it from the three remaining sides [Lem02].

The importance of such parametric designs was vast. When routing channels were traced between logic blocks and not above them, each channel track directly contributed to the FPGA's area. Hence, determining the minimum channel width at which circuits were routable on a given FPGA was a critical metric [Ros89], and this would not have been possible without parametric switch-patterns like Wilton. In fact, parametric switch-patterns continue to be highly important even today. In Sect. 4.7.4, we were able to explore channel segmentations in an enumerative manner, precisely because the switch-pattern of Fig. 4.18 could automatically adapt to different segmentations.

5.1.2 Importance of Per-Segmentation Switch-Pattern Optimization

Once segmentation exploration is close to convergence and a handful of most promising segmentations is determined, it is necessary to optimize the switch-pattern for each of them individually, if the best results are to be obtained. For example, when discussing the improvements in Stratix II, Lewis et al. mention the following [Lew05]:

Beyond this, approximately a 20% reduction in routing capacity (normalized to logic density) compared to Stratix was implemented, despite the doubling of target logic density, due to increased ability to tune FMT[1] *routing patterns, and improvements in the production router, offering a further 6% improvement in overall area per unit logic.*

In this chapter, we present a new algorithm that enables automated design of switch-patterns optimized for any given channel segmentation. It is also interesting to note the importance of CAD tools when designing an interconnect architecture, which is exposed by the above remark of Lewis et al. Even if we could prove optimality of some pattern in any particular sense, this would mean little if no CAD tools existed which could efficiently implement circuits on an FPGA based on this pattern. Hence, the majority of prior work used CAD tools and predefined benchmark circuits to evaluate the quality of different candidate architectures [Tri97, Bet99]. We take this one step further and actually use the CAD tools themselves—in particular the PathFinder routing algorithm [McM95]—to design switch-patterns. This tight coupling between exploration and exploitation of switch-patterns ensures that the solution is appropriate for the given CAD tools implementing circuits similar to the ones used in exploration.

5.1.3 Danger of Neglecting Assumptions

As we have already mentioned, classical switch-patterns like Disjoint, Wilton, and Subset are still predominantly used to drive academic FPGA architectural exploration. However, these patterns were designed for and analyzed on architectures containing bidirectional wires spanning one tile. Without thorough reconsideration, conclusions about qualities of these patterns were carried over to more modern architectures, comprising unidirectional wires with a variety of different lengths that appeared toward the beginning of this century [Lew03, Tri15]. Almost 20 years after the inception of the Wilton switch-pattern [Wil97] and 12 years since the advantages of unidirectional wires were clearly demonstrated by Lemieux et al. [Lem04],

[1] FMT is an FPGA architecture modeling and evaluation tool developed from VPR [Bet98] and used by Altera/Intel for evaluating new FPGAs [Lew03].

Petelin and Betz concluded that many of the classical switch-patterns cannot even be implemented in these newer architectures while respecting the original construction procedures [Pet16]. Perhaps it should then not be too surprising that large disparities between the somewhat free interpretations of classical results that have been in use in academia for decades and actual commercial switch-patterns were observed even in mature FPGA families, such as the 28nm Xilinx/AMD 7-Series [Pet21].

Discrepancies are further increased in state-of-the-art FPGA architectures designed for the latest technologies with highly resistive lower metal layers, such as the Altera/Intel Agilex, where groups of wires are divided into largely disjoint *planes* [Chr20]: It is not possible to even locally [Pet16] construct a Wilton switch-pattern, as there are not enough wires of any individual type in a plane (namely, there are one or two [Chr20]) to implement the required permutation function [Lem02].

5.1.4 Importance of Considering Physical Implementation Aspects

Already this illustrates the importance of taking into account the underlying technology and physical implementation when designing the switch-pattern. However, since at the time when the classical switch-pattern designs were produced, the delays of connections implemented by the FPGA depended mostly on the number of hops through the switch-blocks [Gop06], little attention was given to the efficiency with which different switch-patterns could actually be laid out in silicon. One notable exception is the work of Schmit and Chandra who concluded that the Disjoint switch-pattern vastly outperforms Universal and Wilton, when layout is taken into account [Sch02], even though conventional wisdom of the time, based on experiments neglecting this important aspect, claimed its inferiority [Lem04a].

As we have mentioned in the last chapter, with technology scaling, it becomes imperative to reduce capacitive loading on wires and the distances traversed at lower metal layers. This can only be done by considering layout efficiency together with FPGA router requirements when designing the pattern. To this end, we leverage the physical modeling framework introduced in the previous chapter and augment it with optimization of placement of the individual multiplexers that constitute the switch-block, as described in Sect. 5.8.1.2.

5.2 Inaptness of the Black-Box Approach

In the previous section, we have identified that contrary to some other aspects of FPGA architecture design, such as LUT and cluster sizes, popular switch-pattern choices have not resulted from automated exploration. Let us now turn to analyzing why the main technique used for automating these other tasks—enumerative exploration—is not fit for the purpose of designing switch-patterns.

Most of the conclusions about what constitutes a good FPGA architecture reached in the past 30 years came from applying a variant of the approach which we also used in Chap. 4 to revisit the problem of optimal cluster sizing [Ahm00]:

1. Select an architectural parameter p, the influence of which is to be assessed and fix a range P for it.
2. For every value in P, create an architectural model with p taking that value, run the CAD flow on a number of preselected benchmark circuits, and record some performance metrics (e.g., critical path delay and area).
3. Choose the value (or a value range) of p which optimizes the performance metrics.

For instance, optimal ranges of LUT [Ahm00] and cluster [Bet98] sizes were discovered in this manner. Given that LUT area increases exponentially with input count and that the size of a crossbar with fixed sparsity increases roughly quadratically with cluster size, it is not surprising that this parameter-sweeping approach was highly successful: Reasonable ranges of these parameters are very small and can easily be exhaustively explored.

5.2.1 How Large is the Switch-Pattern Search Space?

To better illustrate why a brute-force exploration cannot be applied to designing switch-patterns, let us first try to make a quick assessment of how large the search space for switch-patterns could be. Let there be 10 wires exiting a switch-block from one side and 30 entering it on the three remaining sides. Assuming that each multiplexer driving an exiting wire should be able to select from six incoming wires, there are $\binom{30}{6}^{10} \sim 10^{57}$ ways to form the pattern. This includes pathological cases, where, e.g., some wires have no fanout, as well as isomorphic duplicates. Yet, among the ten wires per side, it is likely that most will be of different lengths or coming from neighboring planes [Chr20], which would require treating them as unique during enumeration. Such a large space clearly cannot be exhaustively explored.

5.2.2 Black Box in the Loop

A step toward more efficient exploration would be to iteratively improve a starting pattern based on postrouting performance metrics. This approach—illustrated in Fig. 5.1—has been successfully used by Lin et al. [Lin10] and subsequently by Shi et al. [Shi22]. Starting from some, perhaps arbitrary, switch-pattern, a search engine—typically based on simulated annealing—proposes a modification which is then evaluated by the router on a preselected set of benchmark circuits. Performance

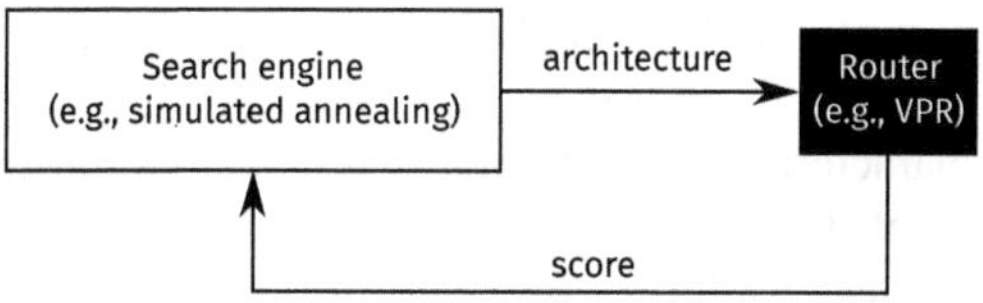

Fig. 5.1 An example of using the router as a "black box in the loop" to drive iterative switch-pattern improvement. At each iteration, a modification of the switch-pattern is proposed, and a complete architecture model is generated. The performance metrics obtained from the router are used to decide on accepting the proposed modification (based on an approach proposed by Lin et al. [Lin10])

metrics obtained from the router are then used to decide whether to accept the proposed modifications. While the exploration is no longer brute-force, it is important to note that the algorithm still explicitly constructs a pattern and then uses the router as a complete black box, merely to obtain the performance metrics. In this chapter we argue that this is on the one hand a fundamental limitation and on the other completely unnecessary. Given that routing a modern circuit even with a state-of-the-art router can easily take minutes, if not hours [Mur20], the number of modifications that can be evaluated in this manner is rather limited.

5.2.3 *Proxy Oracles*

To speed up the evaluation process that is the bottleneck of the black-box-in-the-loop approach, attempts have been made to substitute the router for a proxy oracle that tries to predict the score that the router would output, in a fraction of its time. An example of this was proposed by Petelin and Betz [Pet15]. Although appealing, proxy oracles can only speed exploration up by a constant factor; evaluating each switch-pattern might be significantly faster, but the number of switch-patterns that have to be explicitly listed to cover any sizeable fraction of the search space remains prohibitively large. Another downside is that while it is possible to assess how closely oracles mimic the router on a limited set of test architectures, it is difficult to ensure that they appropriately approximate the router for all architectures that may occur during the course of the exploration. Failure to do so could silently mislead the search.

An interesting approach to proxy design was introduced by Lemieux and Lewis [Lem02]. They first limited the set of switch-patterns to be explored to those which can be described using *permutation functions* [Lem02]. Then they conjectured that a certain characteristic of a switch-pattern that can be quickly measured (*diversity*) has an important influence on its routability. Instead of directly optimizing the performance metrics obtained from the router or some proxy trying to mimic it, they used this characteristic as the maximization objective. Given the fast computation of the objective and a search space significantly reduced by the initial constraints, it was possible to find solutions maximizing the chosen objective using randomized and even brute-force search. While such an approach can help to understand which characteristics lead to highly routable switch-patterns, proposing the characteristics is left to the human designer. Similarly, constraining the search space a priori can be very useful for allowing it to be searched in practice, but it is often difficult to ensure that the imposed constraints do not exclude promising solutions. For instance, permutation functions as defined by Lemieux and Lewis assume that each incoming wire has exactly one target at each of the three remaining sides of the switch-block. However, some newer industrial architectures do not meet this constraint [Pet21].

Instead of proposing another method to further reduce the time taken to evaluate each particular solution and thus enable exploration of yet more points of the search space, we propose one that altogether avoids the need to explore these individual points. This removes the fundamental scalability barrier of the previous approaches. Yet, even though the proposed method is free from fundamental theoretical limitations, for the reasons discussed at length in Sect. 5.14, it still cannot be directly applied to designing switch-patterns intended for use in large FPGAs. Section 5.14 will also offer some potential remedies.

5.3 A Brief Review of Negotiated-Congestion Routing

In this section, we give a brief review of *Negotiated-Congestion Routing*, as embodied in *PathFinder* [McM95]—a very important algorithm that forms the basis of most modern FPGA routers [Kap12], as well as the switch-pattern exploration method developed in this chapter. Only a simplified review of congestion negotiation is given here, focusing on aspects most relevant to the proposed algorithm. For a more in-depth discussion of the theoretical aspects of PathFinder, as well as the evolution of ideas behind it, interested readers may refer to the paper of McMurchie and Ebeling that first introduced it [McM95] and Chapter 17 of Hauck and DeHon [Hau07], in which the authors discuss these topics at greater length. For more practical aspects of engineering, an efficient implementation of PathFinder, interested readers may refer to the works of Betz et al. [Bet99] and Murray et al. [Mur20].

Once a user circuit is placed, physical locations of both endpoints of each of its edges are known and fixed. It is the duty of the *router* to connect these endpoints together by forming appropriate paths from wire instances connected by switch instances. Paths implementing edges with different tail nodes (different nets) must not intersect in a legal routing solution as that would constitute a short circuit. When two or more such paths do intersect on a wire u, u is said to be *congested* [Bet99]. The number of different nets using a wire u is typically called the *occupancy* of u, $O(u)$ [Bet99]. We can then express the magnitude of congestion on u as $C(u) = O(u) - 1$, since every wire can legally carry one signal.

Algorithm 5.1 Simplified PathFinder [McM95, Bet99]

Input: $G = (V, E)$—rr-graph, $E_c \subseteq V \times V$—all connections to be routed;
Output: A routing tree of each signal

1: **function** CONGESTION_COST(u, s) ▷ computes the congestion cost of node u when routing signal s
2: **if** $u \in RT(s)$ **then return** 0 ▷ if u is already used by one connection of s, others can freely use it
3: **return** $b(u) \times (1 + p_{fac} \times O(u)) \times (1 + h_{fac} \times C_h(u))$ ▷ otherwise, compute congestion
4: **for** $u \in V$ **do**
5: $O(u) = 0$; $C_h(u) = 0$ ▷ set occupancy and historical congestion of all nodes to 0
6: **if** $(\exists v \in V)\,((u, v) \in E_c)$ **then**
7: $RT(u) = \{u\}$ ▷ initialize the routing tree of each signal
8: $i = 0$; $p_{fac} = p_{fac}^{init}$
9: **do**
10: **if** $i \geq$ max_iter **then return** UNROUTABLE
11: ▷ no congestion-free routing was found in max_iter iterations
12: **for** $s \in \{u \in V : (\exists v \in V)\,((u, v) \in E_c)\}$ **do**
13: ▷ all signals are ripped up and rerouted in each iteration; modern incremental routers deviate from this [Mur20]
14: **for** $u \in RT(s)$ **do**
15: $O(u) = O(u) - 1$ ▷ reduce the occupancy of all nodes used by the signal s that is ripped up
16: $RT(s) = \{s\}$; $O(s) = O(s) + 1$ ▷ rip up the signal
17: **for** $t \in V : (s, t) \in E_c$ **do**
18: $P =$ SHORTEST_PATH($s, t, \forall u \in V : cong(u) =$ CONGESTION_COST(u, s))
19: ▷ (re)route the connection $s \rightarrow t$
20: **for** $u \in P$ **do**
21: **if** $\neg(u \in RT(s))$ **then**
22: $O(u) = O(u) + 1$ ▷ increase the occupancy of all nodes not already used by the signal s
23: $RT(s) = RT(s) \cup P$ ▷ add the connection route to the routing tree of s
24: **for** $u \in V$ **do**
25: $C_h(u) = C_h(u) + \max(0, O(u) - 1)$ ▷ update historical congestion
26: $i = i + 1$
27: $p_{fac} = p_{fac} \times p_{fac}^{mult}$ ▷ increase the penalty of using occupied nodes; $p_{fac}^{mult} > 1$
28: **while** $\exists u \in V : O(u) > 1$ ▷ finish if there is no congestion
29: **return** $\forall RT$ ▷ return all routing trees

Congestion negotiation was first introduced by McMurchie and Ebeling in their seminal paper which presented the *PathFinder* routing algorithm [McM95]. Likely the most popular open implementation of PathFinder is *VPR*, first developed by Betz and Rose [Bet97], which introduced several refinements to the original algorithm. A negotiated-congestion router operates on the so-called *routing-resource graph* (rr-graph). In an rr-graph, each wire and each pin (endpoint of one of the circuit's edges after placement) is represented by a node, while each switch is represented by an edge [McM95]. A simplified version of PathFinder is shown in Algorithm 5.1. The algorithm proceeds iteratively, by routing all connections of a circuit using the shortest path in the rr-graph between their respective endpoints (fixed during placement). This is designated by the loop starting at line 12, while the shortest path search itself is performed on line 18. All signals are routed independently, and hence their paths can intersect. As mentioned before, this constitutes a short circuit and must be avoided. The key to this lies in how PathFinder assigns costs to each rr-graph node. Namely, each node u has a *base cost*, $b(u)$, which determines how preferable it would be for any signal to use it, if congestion is entirely ignored. There are many ways to compute $b(u)$, some of which are discussed by Murray et al. [Mur20]. This base cost is then multiplied by a product [Bet99] of two additional costs: one directly related to *present* occupancy $O(u)$ [McM95] and another directly related to historical congestion $C_h(u)$ [McM95]. Computation of this *congestion cost* is performed on line 3. Occupancy is updated on lines 15 and 22: Whenever a signal's routing tree is ripped up, occupancy of all of the nodes of the tree is reduced by one; whenever an rr-graph node is added to a signal's routing tree, its occupancy is increased by one [Bet99]. Finally, historical congestion of each node is updated on line 25 [Bet99].

Because of the dependence of the cost of each node on its current occupancy, even though each signal is routed independently, it is motivated to deviate from its preferred path as determined by the base costs and avoid using the nodes already in use by other signals. Deviation from preferred paths happens only gradually, however: Initially, the present occupancy coefficient p_{fac} is made very small [Mur20] and increases exponentially with iterations (line 27); additionally, historical congestion is zero at first, but gradually increases, driving signals away from repeatedly congested areas [McM95]. Finally, to make the router timing driven, an additional term is added to the cost of each node:

$$cost(u)\Big|_{(s,t)} = crit(s,t) \times t(u) + (1 - crit(s,t)) \times cong(u). \tag{5.1}$$

Here $t(u)$ is the intrinsic delay of the node u, while $crit(s,t)$ is the timing criticality of the connection (s,t) of the circuit [McM95]. The first term attempts to route more critical connections through faster wires, whereas for others, the second term dominates causing them to release the congested wires to the more critical connections.

5.4 Main Idea

Before diving into technical details, including a formal definition of the switch-pattern design problem, let us first briefly explain the main idea behind the proposed approach.

5.4.1 Implicit Search Space Representation

An example portion of a simple FPGA is shown in Fig. 5.2a. The disjoint *switch-pattern* used in all switch-blocks is shown in the top-left corner of the figure. A switch-pattern describes which channel wires can be connected together: A line connecting the head of a wire W_I^i to the tail of a wire W_I^o means that W_I^i can drive

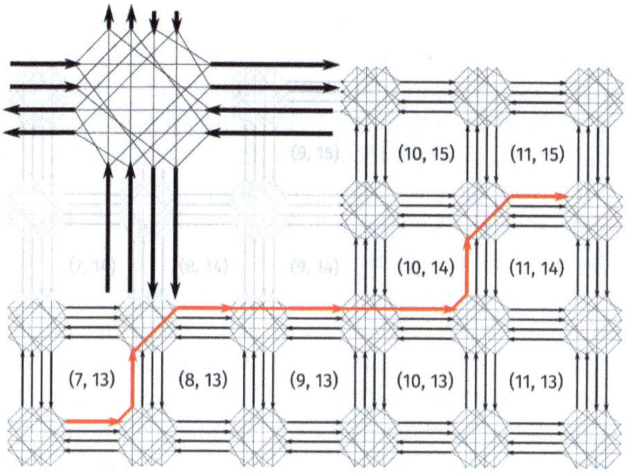

(a) Disjoint switch-pattern.

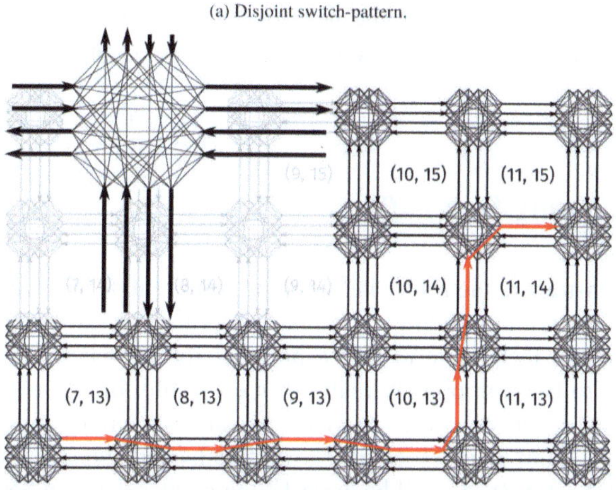

(b) All possible switches.

Fig. 5.2 Example of a simple architecture model with a disjoint switch-pattern (**a**) and all possible switches between the channel wires (**b**). One of the main downsides of the disjoint switch-pattern identified in prior work was the lack of a possibility to switch between different tracks in a channel [Wil97]. If track changes are needed for implementing a circuit, the router will be able to determine it in the architecture model that holds all possible switches (**b**)

W_l^o; otherwise, there is no such possibility. In practice, this means that one of the inputs to the multiplexer driving W_l^o is provided by W_l^i. However, we say more generically that there exists a *programmable switch* between W_l^i and W_l^o. Since the switch-pattern describes connectivity in *all* switch-blocks, we say that switch-blocks are *instances* of a switch-pattern or, in turn, that a switch-pattern specifies their *type*. Hence, we call a wire and a switch in a switch-pattern a *wire type* and a *switch type*, respectively, while we call a wire and a switch in one particular switch-block a *wire instance* and a *switch instance*, respectively. This will be formalized further in Sect. 5.5.

Depending on congestion, the shortest path connecting a source in tile (7, 13) to a sink in tile (11, 15) could be the one depicted in the figure. Figure 5.2b shows the same portion of almost the same simple FPGA architecture, with one major difference: Instead of containing exactly the switches corresponding to the disjoint switch-pattern, its switch-blocks contain all switches that could potentially be fabricated. When routing the same signal from tile (7, 13) to tile (11, 15), it is possible that the router discovers that changing tracks is beneficial to avoid congestion, as depicted by the shortest path in the figure. Note that if the router sees all switches that could be fabricated, there is no need for a designer to guess that track changing is useful and construct a switch-pattern that allows it, nor is there a need for some randomized exploration process to propose a modification which enables track changes; the router itself can select the appropriate switches and reap the benefits of track changes, where they exist. In other words, presenting the router with the entire search space embedded in the routing-resource graph lets it explore this space on its own, alleviating the need to explicitly construct switch-patterns during automated exploration and thus resolving the aforementioned scalability issues.

5.4.2 *Negotiating Switch Types*

Without any constraints, the router is free to use different switch types in different switch-blocks, while the switch-pattern that is fabricated must be common to all. Hence, it is crucial to be able to find a minimal set of switch types that allows all connections to be appropriately realized in every tile. In negotiated-congestion routers, evolving congestion costs allow the signals of a circuit to negotiate which ones will deviate from their respective shortest paths and *spread* to less congested wire *instances* (Sect. 5.3). As will be described in greater detail in Sect. 5.7, we use the same principle of evolving costs, only applied in reverse, to allow the signals of a circuit to negotiate which ones will deviate from their respective shortest paths and *concentrate* on a minimal set of switch *types* that will enter the final pattern.

5.5 Problem Definition

Let us now precisely define the problem tackled in the rest of the chapter. Since our goal is to design switch-patterns, we assume that the rest of the routing architecture—namely, the connection block, the intracluster interconnect, and the wires in the routing channels—is given and fixed. We have already provided an informal definition of a switch-pattern and a switch-block in Sect. 5.4.1. As we have already mentioned on several occasions, instead of the channel wires surrounding the logic clusters, they are actually traced above them [Lew13]. This is illustrated

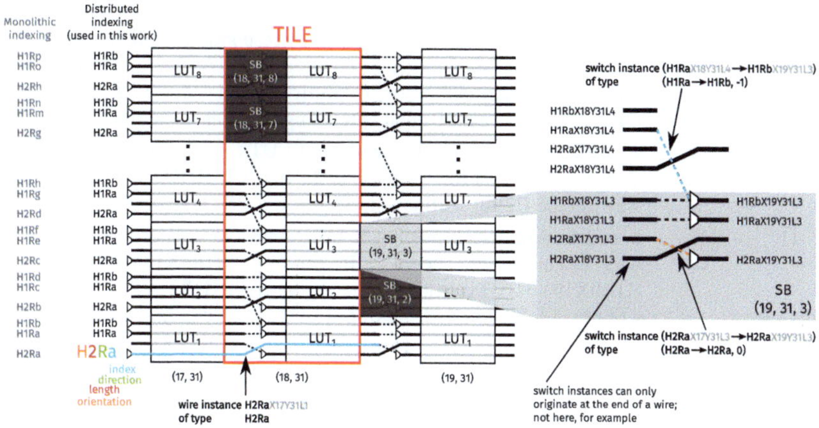

Fig. 5.3 Illustration of definitions. A switch-block (SB) is defined at the level of one LUT in one tile. All switch-blocks are identical, apart from those at the edges of a cluster, where inputs from neighboring tiles are omitted. In a real FPGA, these cross-cluster switches could also be implemented [Pet21], since consecutive tiles in a column are abutted with no empty space in between [McM23]. Yet, to simplify modeling in VPR, we do not explore this possibility

in greater detail in Fig. 5.3; for the sake of clarity, only horizontal wires going right are shown. In the most recent FPGA architectures, designed for scaled technologies for which this work is the most relevant, channels are composed in such a way that the same number of wires of the same length and direction start and end in the vicinity of each LUT [Chr20]. This is also illustrated in Fig. 5.3, where next to each LUT, wires H1Ra, H1Rb, and H2Ra start. These three wires, replicated at the height of every LUT together with the corresponding ones running leftward, give a combined effective channel width of $8 \times 2 \times (1 + 1 + 2) = 64$—the same as if the horizontal channel was composed of 16 H1R and 8 H2R wires specified in the figure under "Monolithic indexing." In such architecture, each wire *type* can be defined by its length, direction, and index within the LUT height (plane). Then, each wire *instance* can be defined by specifying its type along with coordinates of the tile and index of the LUT at which it originates. Similarly, a switch *type* can be defined by specifying the types of wires that it connects, along with the offset between their respective LUTs. Since in scaled technologies loading wires at nonterminal tiles damages performance too much and is thus no longer practiced [Chr20], there is no need to specify the offset between the origin tiles of the two wires connected by a switch type—it is already assumed in the length and direction of the driving wire.

This way of defining switch-patterns is very practical, since due to high resistance of lower layers of metal in scaled technologies, it is not feasible for switches to span a large number of LUTs. Let us now formalize these concepts that we draw from the *bundles* and *planes* of Agilex [Chr20] by introducing some notation which we will use throughout the chapter:

$OLDI$	a wire *type* with orientation $O \in \{H, V\}$, standing for *horizontal* and *vertical*, respectively; length $L \in \mathbb{N}$ corresponding to the number of tiles between its start and end; direction $D \in \{L, R, U, D\}$, standing for *left*, *right*, *up*, and *down*, respectively; and index $I \in [a..z]$. In Fig. 5.3, H2Ra is a horizontal wire going two tiles to the right.
$W_T XxYyLl$	a wire *instance* of type W_T, starting at LUT $l \in [0, N)$, in tile (x, y). Constant N stands for cluster size. In Fig. 5.3, H2RaX17Y31L1 is a wire of type H2Ra, starting at LUT 1 of tile (17, 31).
$(W_I^i \rightarrow W_I^o)$	a switch *instance*, providing a programmable connection between wire instances W_I^i and W_I^o. In Fig. 5.3, (H1RaX18Y31L4 $\rightarrow$ H1RbX19Y31L3) provides a connection from H1Ra starting at LUT 4 of tile (18, 31) to H1Rb starting at LUT 3 of tile (19, 31).
$(W_T^i \rightarrow W_T^o, d(l^i, l^o))$	a switch *type* connecting wires of type W_T^i and W_T^o, with the distance between their LUTs equal to $d(l^i, l^o)$. In Fig. 5.3, (H1Ra $\rightarrow$ H1Rb, -1) is the switch type of the previous switch instance example.
$SB(x, y, l)$	*switch-block*. The set of all switch *instances* driven by wire instances ending at LUT l of tile (x, y). The switch-block for $(x, y, l) = (19, 31, 3)$ is indicated in Fig. 5.3.
$SP(x, y, l)$	*local switch-pattern*. $SP(x, y, l) = \{(W_T^i \rightarrow W_T^o, l^o - l^i) : (W_T XxYyLl^i \rightarrow W_T XxYyLl^o) \in SB(x, y, l)\}$.
V	a set of available wire types.
$E = V \times V \times (-N, N)$	a set of all switch *types* that could exist in any local switch-pattern. Constant N stands for cluster size.

Definition 5.1 (Switch-Pattern) $E_a \subseteq E$, such that for each (x, y, l) in the FPGA, $SP(x, y, l) = E_a$.

Definition 5.2 (Usage, Denoted as $U(e)$) The number of switch-blocks in the FPGA in which the switch *type* e is used to route at least one connection of the given circuit.

Now we can define the problem itself:

Task 5.1 (Switch-Pattern Exploration) *Given a set of switch types E and a set of circuits of interest C, find the switch-pattern $E_a \subseteq E$, such that all circuits in C can be routed and their critical path delays minimized.*

5.6 Basic Algorithm

As mentioned in Sect. 5.4, our proposed method relies on implicitly representing the entire switch-pattern search space by embedding it in the rr-graph. It then simply observes the usage statistics of the different switch types across all switch-blocks and in all circuits used in exploration and constructs the pattern from the most used ones. This is more precisely defined by Algorithm 5.2. All switch types that can be fabricated are added to the rr-graph on line 1. Initially, the switch-pattern E_a is empty, and all switch types are allowed to enter it (line 3). The algorithm then

Algorithm 5.2 Simple greedy

Input: $\theta \in \mathbb{R}^+$—switch adoption threshold; **Output:** switch-pattern

1: Add all $e \in E$ to the rr-graph at cost $\varepsilon \in \mathbb{R}^+$
2: ▷ represent in the rr-graph all switch types that can be fabricated
3: $E_a = \{\}, E_p = E$ ▷ at the beginning, no switch type is in the pattern to be fabricated
4: **do**
5: Route the chosen benchmark circuits
6: $U_{max} = max(\{U(e) : e \in E_p\})$ ▷ find the maximum usage of any switch type not yet in the pattern
7: $E_a = E_a \cup \{e \in E_p : U(e) \geq U_{max}/\theta\}$
8: ▷ extend the pattern by all switch types with usage $\geq 1/\theta$ of the maximum
9: Set cost of all $e \in E_a$ to 0 ▷ switch types already in the pattern become free in subsequent iterations
10: $E_p = E \setminus E_a$ ▷ switch types already in the pattern cannot be added again
11: **while** $\exists e \in E_p : U(e) > 0$ ▷ if the router used only switch types already in the pattern, stop
12: **return** E_a

proceeds iteratively, first routing the benchmark circuits chosen for the exploration using PathFinder on line 5 (see Algorithm 5.1) and then adding to the pattern on line 7 all of the switch types with usage $1/\theta$ of the maximum over all that are not already in the pattern. Here, the *adoption threshold*, θ, is a parameter. Upon growing the switch-pattern, costs of all added switch types are reset to 0 on line 9, because once a switch type has been marked for fabrication, it can be used by the router for free in subsequent iterations of the algorithm—using it no longer implies any increase in pattern size. The algorithm stops when the router no longer uses any switch types that are not already in the pattern. We note here that once a switch type enters the pattern, it is never removed from it. It may be beneficial to revisit this in future work, but it guarantees that the algorithm always converges: In the worst case, all possible switch types are taken ($E_a = E$). Preventing this situation will be the main focus of the next section. We first further discuss the algorithm's general structure, however.

5.6.1 Benefits of Iteration

In principle, usage statistics obtained from a single run of the router could already provide valuable information about which switch types would be useful in the switch-pattern. In fact, prior research has successfully relied on usage to design novel interconnect architectures [Wan06]. Nevertheless, there are two important benefits of progressively growing the switch-pattern. First, after each run of the router, some switch types will have a significantly higher usage than others and can thus be clearly deemed useful. An example of this is illustrated by the orange curve in Fig. 5.6. Once these switch types are adopted at a lower cost (note the small initial cost used on line 1 to distinguish switch types not yet in the pattern), though, it may happen that more signals will use them instead of other switch types, thus leading to minimization of the entire pattern. Second, in between iterations, the switch-

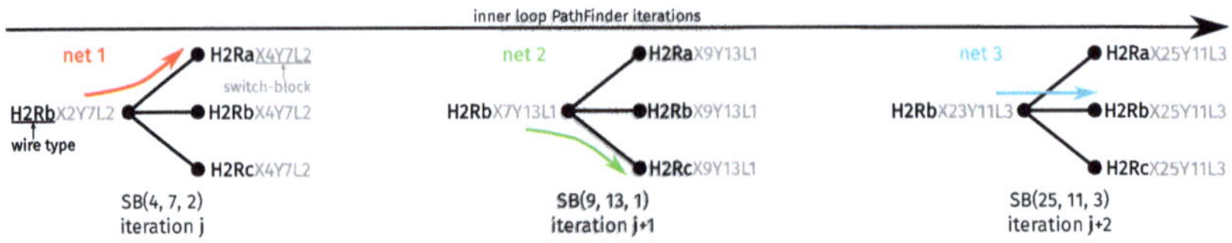

Fig. 5.4 An example of usage spreading over multiple switch types. Colored arrows mark the paths chosen by the router for three different nets passing through three different switch-blocks in three different regions of the FPGA. Each net uses an instance of a switch of a different type, even though this may not have been necessary. As a result, a switch-pattern common to all three switch-blocks would need to contain all three switch types, even if one would have sufficed

block can be optimized at the physical level, by changing the positions of different multiplexers, depending on which switch types were added to the pattern. Similarly, up-to-date implications on delay and area increase of choosing each switch type can be presented to the router in the next iteration. We will discuss this in more detail in Sect. 5.8.1.2.

5.6.2 *Shortcomings of Uncompressed Usage Statistics*

The idea of constructing the switch-pattern from the post-routing usage statistics relies on the intuition that the router itself will be able to best determine which switch types are useful for routing a given circuit. However, since it greedily routes each connection of the circuit using a shortest path in the rr-graph (line 18 of Algorithm 5.1), independent of others, by default it has no incentive to maximize the number of common switch types between the routes of different nets, which would lead to minimizing the switch-pattern size. Before suggesting a remedy to this problem, let us illustrate it on an example. Figure 5.4 depicts three different nets being routed through three different switch-blocks (note the different tile coordinates). As all three nets can arbitrarily choose the switch instances they take, for they all seem equally good, it is possible that usage is spread equally among the three switch types. Once at line 7, Algorithm 5.2 has to accept all of them. In other words, it is impossible to know if all three switch types are essential for routing the circuit, or the router used all of them equally often simply because it had no incentive to do otherwise.

5.7 Turning PathFinder Upside-Down

In this section, we introduce a remedy to the above problem: using the principles of *congestion negotiation* [McM95] to make the nets reach a consensus on which switch types are really important for routing a given circuit.

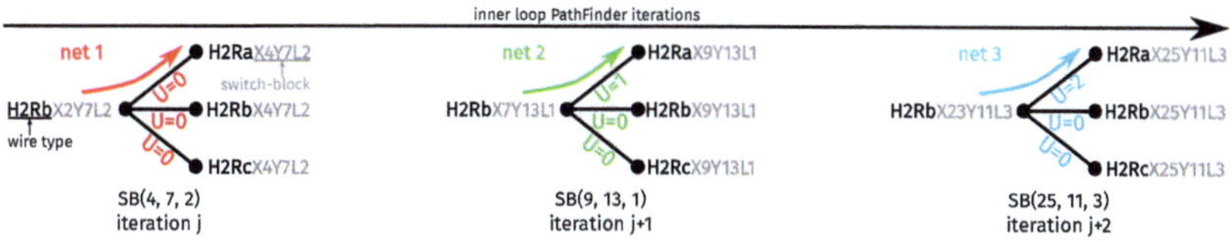

Fig. 5.5 Inversely relating switch instance cost to its type's usage across all switch-blocks motivates nets to concentrate on the same switch types. Current usage of each switch type at the time of routing of each net is depicted next to the corresponding edge

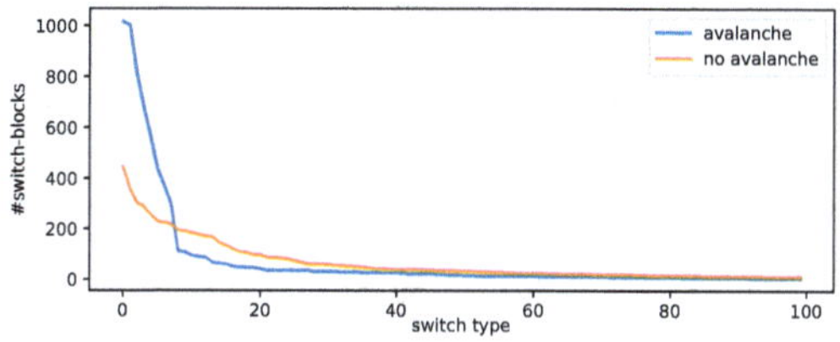

Fig. 5.6 Concentration effect achieved by the avalanche costs. Usage of the 100 most used switch types, out of the 564 available in one particular experiment, is shown for the case when avalanche costs are enabled and disabled, respectively. The area under the two curves is not identical, as switch type concentration also changes the total number of wires used for routing

5.7.1 Avalanche Costs

Figure 5.5 shows the same routing process as Fig. 5.4 with one important difference: Switch instance costs are no longer constant, but inversely related to their type's usage, indicated on the corresponding edge. For the first net that is routed nothing changes, it still sees the same cost at all three switch instances and freely chooses one of them. The second net, however, sees the switch instance of the type already used by the first net as cheaper, due to the inverse relationship between cost and usage. Hence, it is inclined to choose the same switch type. By the time the router starts processing the third net, the relative cost of (H2Rb $\rightarrow$ H2Ra, $+0$) becomes still smaller, so it is even more inclined to use it. In subsequent iterations, the router will rip-up and reroute nets, leading some of them to choose switch types that have in the meantime become cheaper than the ones they chose in the previous iterations when the cost differences among switch types may not have been as pronounced. This will create an avalanche effect, where the positive feedback keeps reducing the cost of switch types with large usage, increasing their usage even more. Evolving costs enable the nets to reach a consensus on which switch types are important for implementing the given circuit.

Figure 5.6 shows a concrete example of how the *avalanche costs* concentrate bulk of the usage in a limited subset of the available switch types, suppressing the long tail of others with moderate and low usage. It is interesting to note that if, for example, the cost of the switch type (H1La $\rightarrow$ H1Lb, $+0$) drops significantly below the cost of (H2La $\rightarrow$ H2La, $+0$), more noncritical nets may choose to use

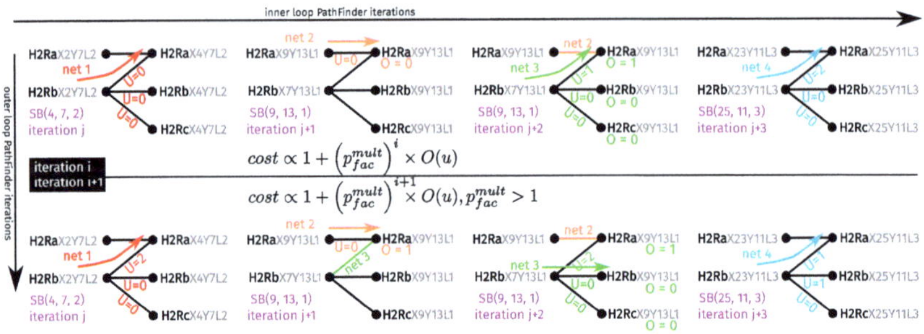

Fig. 5.7 Too much concentration on switch types can lead to congestion on wire instances. As the cost of using occupied wires (indicated by $O > 0$) rises over time (through exponential increase of p_{fac} [Bet99]), at some point, a net will choose a less used switch type. Eventually, the two effects balance out, producing a legal routing with a minimized number of switch types

four H1L wires instead of two H2La wires, thus increasing the total usage compared to the situation where the cost differences did not exist. This effect causes the area under the blue curve of Fig. 5.6 to be larger than that under the orange curve.

5.7.2 *Negotiating Both Congestion and Switch Presence*

In Sect. 5.3 we have seen that PathFinder gradually increases the cost of congested wire *instances*, pushing the nets toward a consensus on which ones will deviate from their desired paths, to spread congestion to other wires and eventually eliminate it [McM95]—making the same *instance* choices as other nets is *penalized.*

Inversely relating the cost of switch *types* to usage in avalanche costs makes the principle act in the opposite direction, causing a consensus on concentration, instead of spreading—making the same *type* choices as other nets is *rewarded.*

Let us see through the example of Fig. 5.7 how these two directions of the same principle naturally act together. At routing iteration *i*, *net 3* may choose to take the switch instance (H2RbX7Y13L1 $\rightarrow$ H2RaX9Y13L1), as it is cheaper because its type (H2Rb $\rightarrow$ H2Ra, +0) is already used by *net 1*. This causes congestion on wire H2RaX9Y13L1 already in use by *net 2*. However, in the subsequent iteration, p_{fac} will be increased by a factor of $p_{fac}^{mult} > 1$ (line 27 of Algorithm 5.1), in turn increasing the cost of congestion on any wire instance (line 3 of Algorithm 5.1). Hence, even though the switch instance (H2RbX7Y13L1 $\rightarrow$ H2RaX9Y13L1) will still be cheaper than (H2RbX7Y13L1 $\rightarrow$ H2RbX9Y13L1) in the next iteration, the router will choose the later to avoid congestion on H2RaX9Y13L1.

5.7.2.1 Can Congestion Elimination Be Guaranteed?

To guarantee that avalanche costs will never prevent congestion resolution, we must ensure that this tipping point when penalization of congestion on wire instances surpasses the reward of concentration on switch types always occurs. Given that congestion penalization is not bounded from above, due to the exponential increase of p_{fac} (line 27 of Algorithm 5.1), it suffices to ensure that the difference between the maximum and the minimum switch instance cost (which constitutes the maximum switch type concentration reward) is bounded. As we will see in Sect. 5.7.3, we make the maximum switch instance cost—corresponding to unused switch types—constant, while we prohibit the avalanche costs from dropping below zero. Hence, the above requirement is satisfied, and avalanche costs do not prevent congestion from being eliminated, though they may increase the number of iterations (line 9 of Algorithm 5.1) needed to achieve this. Further details on this problem will be provided in Sect. 5.14.

5.7.2.2 Is Congestion Elimination Always Necessary?

When PathFinder is being used for implementing a circuit on an existing FPGA (its usual intended use), it is necessary to eliminate all congestion—otherwise, the routing is illegal. However, here we are not using PathFinder to implement a circuit on an existing FPGA but to design a new switch-pattern, by observing which switch types are most useful for routing the circuits selected for use in exploration. This may become apparent long before congestion is fully eliminated. Especially in the early iterations of Algorithm 5.2, switch types that once surpass the adoption threshold are unlikely to drop below it again. As discussed further in Sect. 5.14, stopping PathFinder at this point could be used to greatly speed up the exploration process.

5.7.3 Functional Form of Avalanche Costs

As discussed in Sect. 5.7.1, avalanche costs should be high for switch types that are unused and drop in proportion with usage of the particular type. Additionally, not to prevent congestion resolution, they must be bounded from both below and above. To satisfy these criteria, we use a functional form similar to congestion costs of PathFinder (Sect. 5.3):

$$a(u) = max(0, s(u) - (a_p \times U(u) + a_h \times U_h(u))). \tag{5.2}$$

The $s(u)$ term in Eq. 5.2 is the starting cost assigned to the given switch type, which is also its maximum cost. Parameter a_p determines how quickly the avalanche cost drops as a function of the current usage of the switch type, $U(u)$, while a_h determines how quickly it drops as a function of its cumulative historical usage, $U_h(u)$. Usage tracking is completely analogous to occupancy tracking of Sect. 5.3 and $U(u)$ is updated each time a net is routed (lines 15 and 22 of Algorithm 5.1). Similarly, historical usage tracking is completely analogous to historical congestion tracking, and $U_h(u)$ is updated at the end of each routing iteration (line 25 of Algorithm 5.1). The main difference, however, is that unlike the occupancy trackers that are bound to individual nodes of the rr-graph (individual wire *instances*), the usage trackers $U(u)$ and $U_h(u)$ are shared between all nodes representing instances of switches of the same *type*. This allows for communicating switch type choices to nets using entirely different switch-blocks (Fig. 5.5) and eventually reaching a consensus on which switch types will enter the pattern that is common to them all.

We note here that there are many other functions which would satisfy the requirement of avalanche costs dropping in proportion with switch type usage and being bounded from both sides. Adding together the present and historical usage terms as in Eq. 5.2 has a benefit of providing a relatively easy way of tuning the coefficients. This will be discussed in Sect. 5.13.1. It also has a downside, compared to the analogous product used in the congestion costs of PathFinder: The historical term quickly dominates. This may lead to a reduced capacity for driving nets to common switch types. Aside from multiplying the two terms or making $a_p \gg a_q$, a possible remedy for this could be to update the historical usage using an exponential decay, giving higher importance to more recent history [Hau07]. Nevertheless, investigating the effectiveness of functional forms other than the one of Eq. 5.2 goes beyond the scope of the present work.

5.7.4 A Note on Implementation

Most implementations of PathFinder, including the one in VTR 8 [Mur20] that we use in this work, assign weights only to the nodes of the rr-graph. To retain the existing data structures, we simply split each edge that represents a potential switch instance by an additional node and assign the appropriate avalanche cost to this node. In particular, we compute the congestion cost of these virtual nodes as follows:

$$cong(u) = b(u) = a(u). \tag{5.3}$$

Splitting edges with additional nodes doubles the total edge count in the rr-graph and drastically increases its node count. As will be discussed in Sect. 5.14, this has a significant impact on the exploration time. However, implementational effort needed to adapt VTR's algorithms to accept both node and edge weights went beyond the scope of this work.

5.7.5 Respecting the Critical Paths

A good switch-pattern must enable the router to properly optimize the critical path of each circuit of interest. Hence, during the pattern search, critical connections must be able to route even through switch types with otherwise low usage. Critical

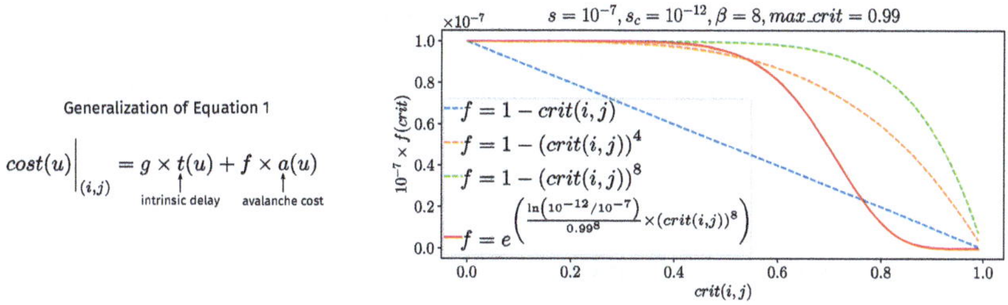

Fig. 5.8 Comparison of functions for criticality scaling of avalanche costs (generalizations of Eq. 5.1). The proposed function of Eq. 5.4 (solid red line) allows for precise tuning of the avalanche cost that the most critical nets perceive, so that the timing requirements are sufficient to motivate them to use switch types with otherwise low usage. It also creates a relatively flat region of low avalanche cost for a wider range of high criticalities, necessary for actually optimizing the critical path delay, given that the timing analysis during routing is done only infrequently. A relatively steep rise in cost ensues once criticality drops below the cut-off point, which is needed to discourage noncritical nets from increasing the switch-pattern size. The function of Eq. 5.1 and its exponentiated versions [Mur20] lack these features (dashed lines)

path delay of a typical circuit is on the order of 10^{-9}. In our experiments, we have determined that the starting avalanche cost is best set to the same order, or larger, even up to 10^{-7}. Under those circumstances, linearly scaling avalanche costs by $(1 - crit)$, like congestion costs in Eq. 5.1, would not give enough freedom to the critical paths to choose switch types with low usage; switch cost variations would simply overshadow the timing optimization. Another problem with linear scaling is that somewhat critical (e.g., criticality 0.5) nets are given unfair advantage in choosing switches compared to nets that are just slightly less critical. While exponentiating the criticality can help mitigate this second problem, it further worsens the first, as shown by the blue, orange, and green curves in Fig. 5.8.

To provide a solution to both problems, we designed a function represented by the red curve in Fig. 5.8. The curve shows a relatively wide flat range of very small values close to criticality of 1, which allows for the critical paths to actually be optimized. At the same time, there is a steep rise in the value of the function as criticality drops, which prevents the nets with comparatively low criticality from unnecessarily increasing the number of used switch types.

The combined timing and avalanche cost assigned to a node splitting an edge that represents a potential switch is

$$cost(u)\Big|_{(i,j)} = t(u) + e^{\left(\frac{ln(s_C/s)}{max\_crit^{\beta}} \times crit(i,j)^{\beta}\right)} \times a(u). \tag{5.4}$$

Here s_c is a parameter determining the perceived avalanche cost of a potential switch when routing the most critical possible net, with criticality max_crit (a standard parameter of VPR [Mur20]), and β is a criticality exponent used to tune the selectivity of the function. Approximate delay contribution of a switch to its driving wire is captured by the term $t(u)$, modeled as described in Sect. 5.8.1. We do not scale it by criticality of the net being routed, because all nets—regardless of criticality—should be aware of the implications of including a switch type in the pattern.

5.8 Completing the Algorithm

The complete algorithm is almost identical to Algorithm 5.2, apart from the fact that routing on line 5 is performed using a modified version of VTR 8 [Mur20], which incorporates the avalanche costs of Sect. 5.7. Another difference is that if there are switch types which got their avalanche cost reduced to zero in the current iteration, all of them are selected and the usage-threshold-based selection of line 7 is skipped. Nodes representing instances of the selected switch types are removed from the rr-graph, and their neighbors are connected directly. This is conceptually equivalent to resetting their costs to 0 (line 9) but has a practical benefit of reducing the size of the rr-graph.

5.8.1 Conveying Physical Information

Apart from the delays of the routing wires, necessary for proper timing optimization, the router must be aware of the implications on the architecture's performance of using switches of a type that is not yet in the pattern. Before each iteration of the algorithm, we run a physical modeling and optimization flow to provide these data for the modified switch-pattern. The impact that using each potential switch has on performance generally depends on which other potential switches are also used. However, if the adoption threshold θ (Sect. 5.6) is sufficiently small to prevent adoption of too many switches between reevaluations of the physical model of the switch-block, the simple approach of only informing the router about the impact of each switch in isolation, through the $t(u)$ term of Eq. 5.4, should suffice.

5.8.1.1 Modeling Flow

To extract delays of the channel wires, we rely on the physical modeling flow presented in Chap. 4. As we have already described in detail in Sect. 4.2.1, the flow assumes a floorplan similar to that of the Stratix FPGAs [Lew13], where LUTs are stacked on top of each other, while the routing multiplexers are arranged in columns padded to their left. Figure 5.9 depicts one such floorplan. The crossbar multiplexers are placed immediately next to the LUTs, followed by the connection-block multiplexers, and then finally the switch-block ones. Each multiplexer column

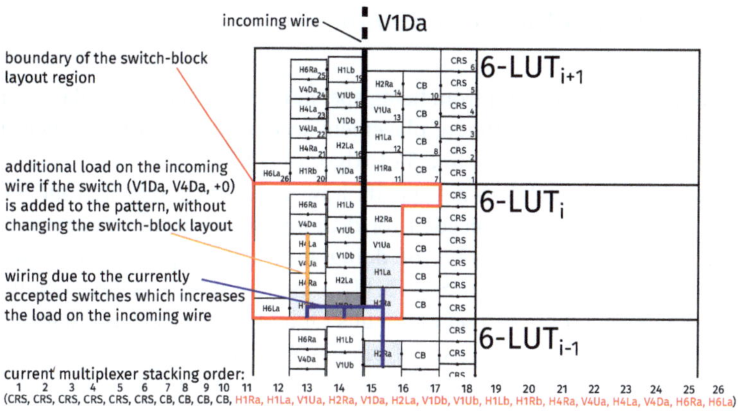

Fig. 5.9 Updated floorplan construction, not drawn to scale

is filled from the bottom up, until its height matches the height of the adjacent LUT. Only then a new column is started. Placement of multiplexers next to each LUT is identical.

5.8.1.2 Multiplexer Position Optimization

Precise positions of all multiplexers allow for accurate modeling of intra-switch-block wiring (depicted in blue in Fig. 5.9, for one source channel wire), which in turn allows for correctly taking into account the influence of this wiring on the delay of the channel wires. However, as the pattern evolves during the course of the avalanche search, positions of multiplexers in the tile floorplan may become suboptimal. In the previous chapter, multiplexers were stacked in a fixed order, derived from their input count [Nik21]. Now we adapt the order to the changing connectivity by performing a quick anneal of the stacking order. All moves represent swaps of two randomly selected multiplexers in the order, upon which a new floorplan is generated. For the cost function, we use a combination of the total intra-switch-block wirelength and a timing cost computed as a product of approximate routing wire delay and its exponentiated criticality extracted from the last routing run, summed over all routing wires. This cost function was adopted from VPR's timing-driven placer [Mar00]. Multiplexer position optimization and routing wire delay measurement only consider those switch types which have already been adopted to the final switch-pattern.

Routing wire delays reported to the router for the next iteration of the pattern search are obtained directly from SPICE simulations [Nik21]. However, to speed up optimization, annealing uses approximate delays obtained from a polynomial fitted to a set of SPICE simulations, which relates the total length of the intra-switch-block wiring that a routing wire drives to the increase in its delay. Output of the same model is used for timing costs of the potential switches during routing.

5.8.2 Preventing Overspecialization

To prevent the resulting pattern from being specialized to a particular placement, we replace the circuits using a different placement seed at the start of each iteration of the search algorithm (just before line 5 of Algorithm 5.2).

Line 5 of Algorithm 5.2 does not specify how multiple circuits are routed before measuring the usage statistics needed for deciding which switch types are added to the pattern. Two possible ways of doing this will be presented in detail in subsequent sections. Namely, in Sect. 5.11, a switch-pattern is first obtained by performing the exploration on a set of circuits C_1 and is then used as a starting point for a continued exploration on a set of circuits C_2. This can be highly beneficial for reducing the exploration runtime on large circuits, but it has a downside of giving preference to circuits from C_1 in deciding which switch types should enter the pattern. An alternative approach, presented in Sect. 5.13.2, is to route multiple circuits independently in parallel on line 5, combining the usage statistics observed on each of them before proceeding to switch type selection. While this avoids the problem of a priori favoring one set of circuits, it suffers from the usage spreading problem of Fig. 5.4: Introducing avalanche costs enabled nets passing through switch-blocks in different regions of the FPGA to negotiate which common switch types they will use; however, if the three nets in the example of Fig. 5.4 belong to three independently routed circuits, this negotiation is no longer possible, and they may again end up using different switch types even if a common one would have sufficed.

Note that these problems do not exist in the black-box-in-the-loop approach (Sect. 5.2), since in that case all circuits are routed on one and the same switch-pattern. To prevent the problems in the context of avalanche search as well, here we simply route multiple circuits simultaneously, allowing their nets to jointly negotiate the presence of different switch types. Implementational details of this are illustrated in Fig. 5.10. First, each circuit is packed and placed independently on the smallest FPGA that can fit it, as determined by VPR [Bet99], with its logical width and height adjusted to make the physical ones roughly equal. Then, a new large FPGA is created, so that its height equals the maximum height of all individual FPGAs and its width equals the sum of the widths of all individual FPGAs. The smaller individual FPGAs are placed on this larger one, much like if it were a passive interposer, but without any connections between dies (see Chap. 10 for more details on multi-die FPGAs). The packing and placement of each circuit are transferred to their respective isolated region on the new FPGA and assigned their own set of clock domains. The netlists are then merged and routed simultaneously, allowing them to share avalanche costs. Since different circuits are placed in isolated regions with no connectivity between them, nets of one circuit cannot cause congestion in others. Similarly, since each clock domain is individually optimized, timing characteristics of each circuit are preserved. On the other hand, shared avalanche costs enable joint switch-pattern negotiation.

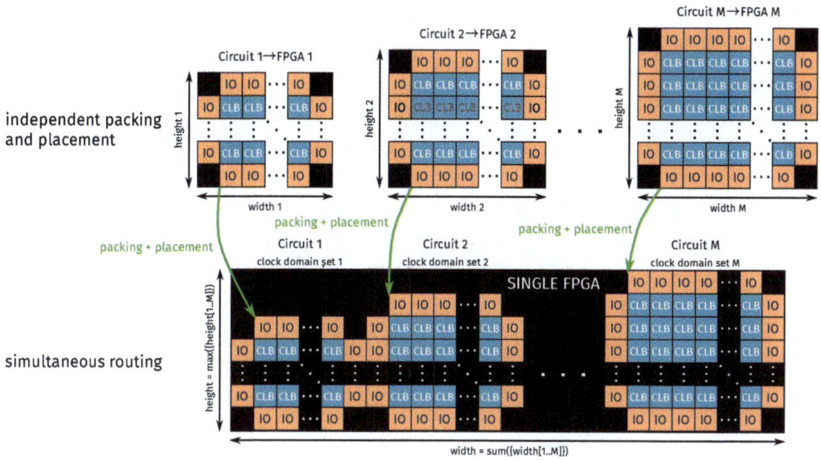

Fig. 5.10 To enable joint negotiation of switch-presence among multiple circuits, we route them simultaneously

5.9 Experimental Setup

All experiments are performed on an architecture with eight 6-LUTs in the cluster and a channel composition reminiscent of that of Agilex, but for the longest wires [Chr20]: $2 \times$ H1, H2, H4, H6, $2 \times$ V1, V4. These wires are repeated for each LUT of the cluster, leading to an equivalent width of a horizontal channel equal to $2 \times 8 \times (1 + 1 + 2 + 4 + 6) = 224$ and an equivalent width of the vertical channel equal to $2 \times 8 \times (1 + 1 + 4) = 96$. As mentioned in Sect. 5.5, wires are only allowed to drive other wires from their end tile. Without loss of generality, we consider only switch types with LUT offset $\in \{-1, 0, 1\}$ (Sect. 5.5) and prohibit switch types to a target wire going in the direction from which the driving one came [Pet21]. This results in 564 available switch types. The connection blocks and crossbars generated by the physical modeling flow are kept constant in all experiments, while delays are extracted from the 4 nm technology model of Chap. 4. Avalanche parameters are set as described in Sect. 5.13.1.

5.10 Effectiveness of Avalanche Costs

In this section, we assess the effectiveness of the proposed avalanche search method against the simple greedy algorithm of Sect. 5.6. Instead of introducing explicit ε costs without a physical meaning to the greedy algorithm, we use the timing costs of the switches equally visible to all nets, regardless of criticality (Eq. 5.4). Search was performed by simultaneously routing the *alu4*, *ex5p*, and *tseng* circuits. The

Table 5.1 Properties of the different patterns; *manual* and *annealed* will be discussed in Sect. 5.12

	Avalanche			Greedy			Trunc. greedy			Manual [Nik21]			Annealed		
# iterations	36			228			54						10,000 moves		
# sw. types	78			438			78			180			210		
Average	fi	fo	t[ps]	fi	fo	t[ps]	fi	fo	t[ps]	fi	fo	t[ps]	fi	fo	t[ps]
H1	5	3	13.9	31	25	23.1	6	3	13.2	10	10	16.0	13	13	19.6
H2	5	4	16.8	28	28	31.6	5	5	18.1	11	11	21.3	14	11	24.1
H4	4	7	27.4	21	27	43.2	4	6	25.7	11	11	30.8	16	12	32.1
H6	5	5	35.7	19	25	59.6	2	6	35.7	11	11	43.1	9	13	47.3
V1	7	6	21.8	38	31	35.5	8	7	22.1	12	12	24.6	14	15	29.2
V4	2	5	70.1	12	27	97.5	1	4	67.0	13	13	74.3	13	15	86.8
W(tile)	6792 nm			8904 nm			6816 nm			7464 nm			7488 nm		
CPD	1.38 ns			1.71 ns			1.38 ns			1.46 ns			1.55 ns		

switch adoption threshold θ was set to 1.1 for both algorithms. Final assessment of performance was done on all MCNC circuits, but for the pin-bound *dsip*, *des*, and *bigkey*. We note that the results reported here differ slightly from the ones reported in the original paper [Nik21a], because in the original paper, periodic rip-up did not affect all uncongested connections, leading to some missed opportunities for increased concentration. With periodic rip-up fully enabled (Sect. 5.14.3), a more compact pattern with better delay was obtained, at the expense of lower routability on circuits that are significantly more complex than the ones used in pattern search. This will be addressed in Sect. 5.11.

5.10.1 *Direct Comparison with Greedy*

Avalanche search converged after 36 iterations, accumulating 78 switch types, while greedy search converged only after 228 iterations, accepting 438 switch types (Table 5.1). This demonstrates that projected delay contributions of individual switch types alone are insufficient to deter the router from using them. The large number of switch types in the greedy pattern resulted in both a large increase of the tile width and the average fanin and fanout of channel wires. This in turn led to a large increase of average wire delays and the routed critical path delay (Table 5.1).

5.10.2 *Comparison with Truncated Greedy*

To better assess the differences in choices made by the two search methods, we truncated the greedy pattern after the 54th iteration, when it also contained 78 switch types. The exact distribution of fanouts and fanins enables a tighter packing of

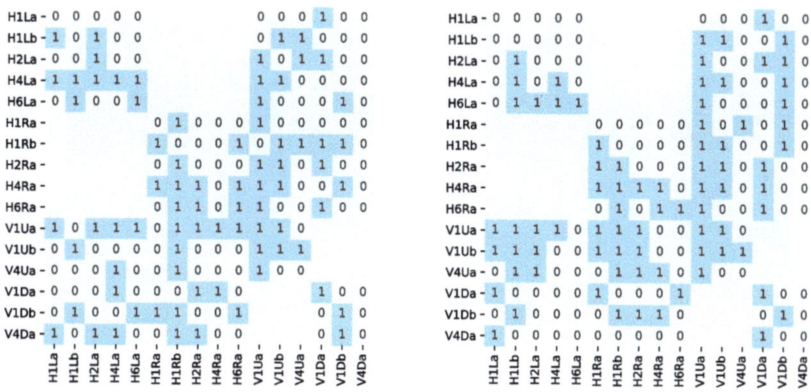

(a) Avalanche (b) Truncated greedy

Fig. 5.11 Adjacency of wire types: avalanche (**a**) and truncated greedy (**b**). Entries with no number are prohibited by construction. Rows correspond to drivers and columns to targets

multiplexers of the avalanche pattern, leading to a slightly lower tile width. Fanout and fanin sizes still predominantly determine the wire delays, however, which are very close between the two patterns and on average slightly lower for the truncated greedy (Table 5.1).

5.10.2.1 Adjacency

Adjacency between different wire types is illustrated in Fig. 5.11. Avalanche search resulted in more varied connectivity between wire types of different lengths. This can be seen by observing that, e.g., the fanouts of H1Ra and H1Rb are complementary in the avalanche pattern, whereas they have two switch types in common in the truncated greedy. Similarly, fanouts of V1Ua and V1Ub share three switch types in the avalanche pattern, whereas they share eight out of nine switch types in the greedy pattern. This suggests that the greedy search selects multiple switch types between the same lengths of wires, commonly connected by the router, where only a subset of them would suffice. As a result, with the same number of switch types, fewer different wirelengths can be connected.

5.10.2.2 Grid Distances

Consequences of selecting multiple switch types between the same wirelengths, instead of introducing more variety, can be seen in Fig. 5.12. Each entry of the matrices represents the minimum number of distinct channel wires needed to connect the center of the grid to the particular target, normalized by the minimum

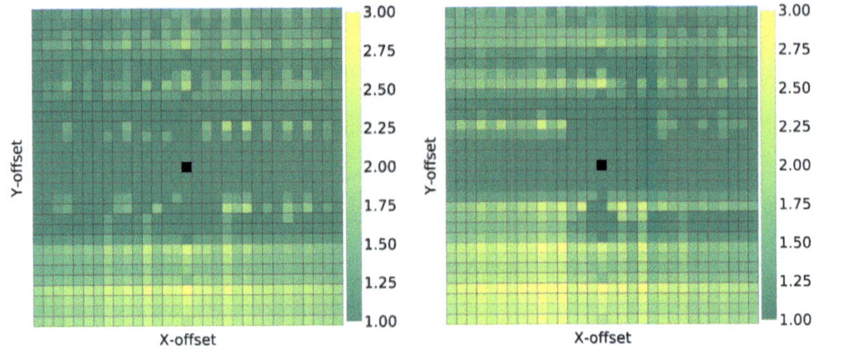

(a) (b)

Fig. 5.12 Hop-distances from the center of the FPGA to other tiles, normalized by the distances computed on a pattern containing all allowed switch types. Dark green is the best. (**a**) Avalanche. (**b**) Truncated greedy

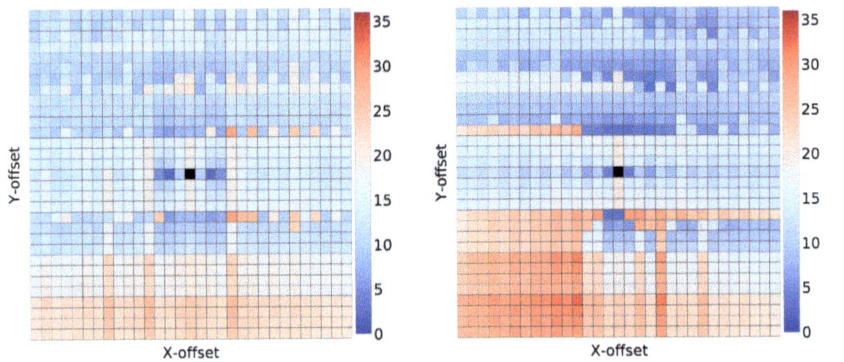

(a) (b)

Fig. 5.13 Percentage increase of the delay needed to reach other tiles from the center, compared to a hypothetical switch-pattern containing all allowed switch types with no impact on wire delay. Dark blue is the best. (**a**) Avalanche. (**b**) Truncated greedy

number of wires that would be needed if all switch types were available in the pattern. The avalanche pattern is closer to being optimal in this respect.

This is also reflected on the minimum delay distances, relative to an unrealistic fully connected pattern which disregards the impact of switch load on wire delay (Fig. 5.13). The relative inefficiency in connecting to the distant targets at the bottom of the grid was influenced by performing the search on small circuits requiring very small FPGAs. In a production setting, larger circuits should be used. We will discuss this further in Sect. 5.13.5.2.

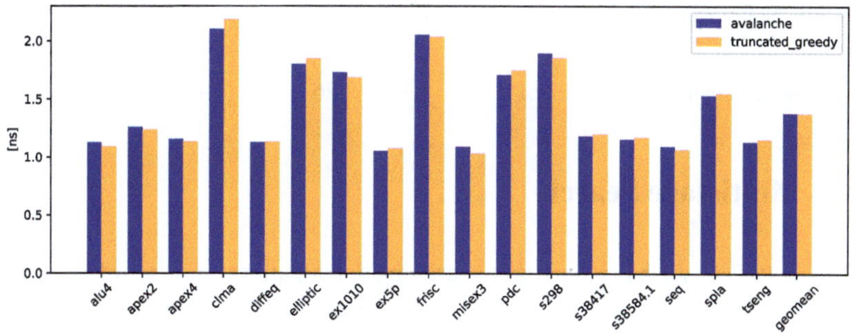

Fig. 5.14 Routed delays for the avalanche and the truncated greedy pattern

Table 5.2 Percentage of congested rr-graph nodes after 300 VPR routing iterations on Gnl circuits

Circuit	1	2	3	4	5	6	7	8	9	10
Avalanche	0.906%	0.274%	0%	0.521%	0.323%	0.187%	0.149%	0.262%	0.040%	0.002%
Trunc. greedy	1.061%	0.632%	0.484%	1.340%	0.597%	1.175%	0.982%	0.509%	0.275%	0.984%

5.10.2.3 Routed Delays

Despite the qualitative differences between the avalanche and the truncated greedy pattern, they are largely equivalent in terms of the routed critical path delays (Fig. 5.14). This could be due to the MCNC circuits imposing low stress on the routing architecture, making it easy to meet timing requirements. Another reason could lie in their large logic depth which, combined with oversimplified intracluster interconnect [Chr20], may make the delays inside the cluster dominant.

5.10.2.4 Routability

To see how the two patterns compare under increased stress, we generate ten synthetic circuits with about 10,000 LUTs using *Gnl* [Str99]. Rent's exponent was set to 0.7—close to the maximum used in the ISPD'16 routability-driven placement contest [Yan16]. We take the distribution of different LUT sizes in the circuits from [Hut04]. Then, we place the circuits on architectures based on the two switch-patterns and attempt to route them with a limit of 300 iterations. We neglect timing optimization since the circuits are synthetic.

Table 5.2 shows the percentage of congested rr-graph nodes at the end of the 300 routing iterations, for each circuit. The pattern obtained through avalanche search managed to legally route one of the ten circuits, while no circuit was routable on the greedy pattern. The difference in the percentage of remaining congested nodes also showcases the higher routability of the switch-pattern obtained through avalanche search. Nevertheless, not being able to route nine out of ten circuits is not acceptable for any meaningful pattern. We describe a remedy for this in the next section.

Table 5.3 Number of VPR iterations needed to route each of the ten Gnl circuits

Circuit	1	2	3	4	5	6	7	8	9	10
Gnl-extended avalanche	142	61	27	106	26	55	46	55	30	82

5.11 Multistage Search

In Sect. 5.10.2.4, we have seen that searching for a pattern on a set of benchmark circuits that has a lower connectivity demand than some circuits for which the final architecture is intended can result in those more complex circuits failing to route. However, switch types needed by the simpler circuits are most likely also needed by the more complex ones. Hence, running the search on the smaller circuits first and using the resulting pattern to initialize the search on more complex circuits is a reasonable way to reduce the search runtime. We demonstrate that in this section by presenting the results of continuing the search from the pattern of Fig. 5.11a, on the Gnl benchmarks described in Sect. 5.10.2.4.

5.11.1 *Convergence*

In each iteration, one of the ten Gnl circuits was routed to derive usage statistics, and the circuits were changed between iterations in a round-robin fashion. This additional run converged after three search iterations, adding six more switch types to the pattern. On this extended pattern, all ten Gnl circuits routed successfully in less than 300 router iterations (Table 5.3). Instead of running 39 search iterations on the more complex circuits which take about 7× more time to route, it was possible to run the majority of these iterations on the smaller circuits, drastically reducing the runtime. We will discuss runtime in more detail in Sect. 5.14. Here we would just like to note that even though the 10,000-LUT benchmarks are still rather small and running the entire search on them in a production setting would certainly be feasible, scaling the search to large modern circuits which can take several hours to route even in absence of avalanche costs could be more difficult without adopting this approach of gradually increasing the complexity of the used circuits.

5.11.2 *Pattern Changes*

The final pattern obtained after the additional search run on the Gnl benchmarks is shown in Fig. 5.15a, with switch types added on top of the pattern of Fig. 5.11a highlighted in red. It is interesting to note that besides providing access to V4Da that results in significant reduction in hop count needed to reach cells below the center (Fig. 5.15b), the new pattern also adds more options to switch between different

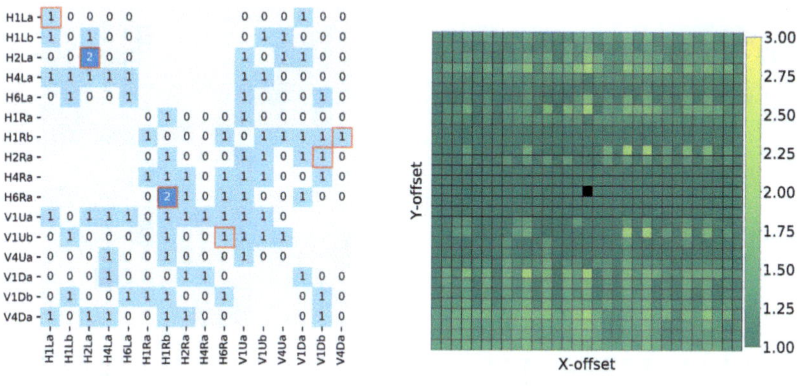

(a) (b)

Fig. 5.15 Pattern obtained after continuing the avalanche search on the Gnl benchmarks of Sect. 5.10.2.4. (**a**) Adjacency. Red switch types were added on top of those of the pattern of Fig. 5.11a. (**b**) Hop-distances. Addition of access to V4Da greatly improved the distances in the lower half-plane

LUT heights (entries with a value of 2; see Fig. 5.3). This may suggest that a few such *inter-plane* connections greatly help to reduce congestion, as mentioned by Chromczak et al. [Chr20]. The 8% increase of the pattern size led to a slight increase of the wire delays. This caused the geomean routed critical path delay of the MCNC benchmarks to rise by 0.6% to 1.39 ns.

5.12 Comparison with Simulated Annealing

In Sect. 5.13 we analyze in more detail various aspects of the avalanche search algorithm. However, before delving into details, we first compare avalanche search to a method inspired by previous work. Namely, Lin et al. successfully used simulated annealing for simultaneously optimizing channel composition and the switch-pattern [Lin10]. In this section, we investigate how a similar method compares with the proposed avalanche search.

5.12.1 Initial Pattern

We initialize the search with the default pattern produced by the physical modeling flow (see Sect. 4.7.6) [Nik21], which represents our best effort at manually capturing inter-wire-type connectivity of a modern tapless architecture [Pet21], with the constraint dictated by the high resistance of the lower metal layers that bulk of this connectivity is contained within wires starting and ending at the same LUT

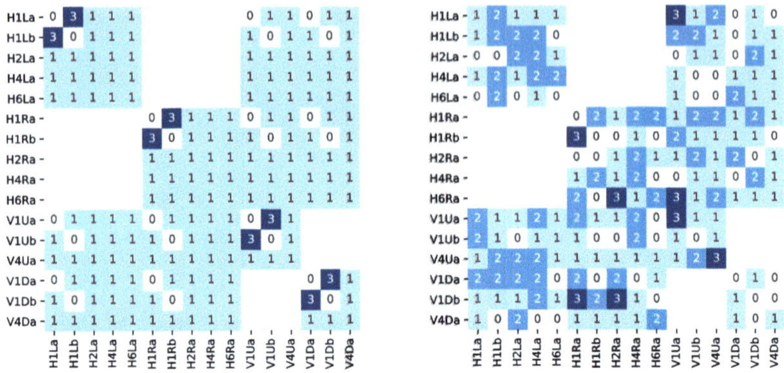

(a) Manual [Nik21] (b) Manual [Nik21] annealed

Fig. 5.16 Adjacency of wire types: initial manually designed pattern [Nik21] (**a**) and its annealed version (**b**)

height [Chr20]. The initial pattern contains 180 switch types organized as shown in Fig. 5.16a. The optimal hop-distances that it achieves are not sufficient to counter the wire delay increase due to a high load (Table 5.1). As a result, the geomean routed delay is 5.8% larger than for the avalanche pattern (Table 5.1).

5.12.2 Channel Segmentation Revisited

It is interesting to note that the 1.46 ns average routed delay is already 4.6% lower than the 1.53 ns that we observed in Fig. 4.19. The improvement comes primarily from the difference in channel segmentation, where the one used in this chapter has many fewer wires originating at each LUT: 16 compared to 36, on average for the three $N = 8$ 4 nm architectures used for measurements in Chap. 4. The reduced number of channel wires resulted in a reduced number of stored-select multiplexers that drive them, which in turn made both the horizontal channel wires and the wires connecting the cluster and the switch-block significantly shorter and faster.

This illustrates that while maximizing the utilization of the track space available above the active area of the cluster is likely useful, much like Lewis et al. mentioned [Lew13], care should be taken that this does not result in a large increase in the number of padded multiplexer columns, beyond the minimum which provides sufficient routability. We note that this does not prevent brute-force approach to exploring channel segmentations that was used in Chap. 4. Instead of simply padding the length-1 wires until all track space above the active area is filled, one could do a binary search for a minimal routable padding, as has been typically done when channel width was a major evaluation metric for FPGA architectures [Bet99]. This minimum produces the best delays of the routing resources, since the wires maintain their minimum physical length, under the constraint that all benchmark circuits are routable. When congestion is entirely neglected and no signals make detours (e.g., in the first iteration of VPR's router [Mur20]), we can measure minimum achievable critical path delays, using that physical implementation. The only real reason (i.e., when disregarding the CAD tool noise) why these delays could be exceeded in practice is that the minimum padding is not sufficient for resolving congestion without some timing-critical signals having to detour. By increasing the padding, these detours can be minimized, at the expense of increasing the delays of the routing resources. From the lower bound critical paths, we can compute the maximum elongation of horizontal wires which would make these lower bound delays inferior to the real ones measured at the end of routing the corresponding circuits on the minimum-padding architecture. These elongations bound the maximum number of multiplexer columns that can be added to trade-off routing resource speed for detour minimization. We can then sweep the padding in this additional range, to obtain optimal performance.

5.12.3 What About Floorplan Optimization?

Besides the difference in channel segmentations between the architectures using the manually designed parametric switch-pattern in Chap. 4 and the one reported here, what also differs is the position of switch-block multiplexers in the tile floorplan. While in Chap. 4, the multiplexers were stacked in size order, here their positions are optimized to reduce the overall intra-SB wirelength and delay. However, because the manually designed switch-pattern is highly symmetric, most reasonable placements of multiplexers result in similar wirelength and delay, making the optimization of Sect. 5.8.1.2 not particularly useful. Concretely, the difference in terms of total routing resource delay, between enabling and disabling the optimization when generating the floorplan for the manually designed architecture, was merely 0.15%. However, the impact of this optimization on the outcome of automated switch-pattern design is much more significant. Namely, when we disable floorplan optimization during avalanche search, the produced switch-pattern achieves 2.57% higher routed critical path delay, on average. This once more illustrates the importance of taking into consideration the physical implementation aspects while designing a programmable interconnect architecture.

5.12.4 Setup

Let us now return to the comparison of avalanche search to a simulated annealing-based method. We use two very simple moves generated with equal probability: including or removing one of the 564 considered switch types. The self-normalizing two-term cost function of Marquardt et al. [Mar00] is used, with tile area and the

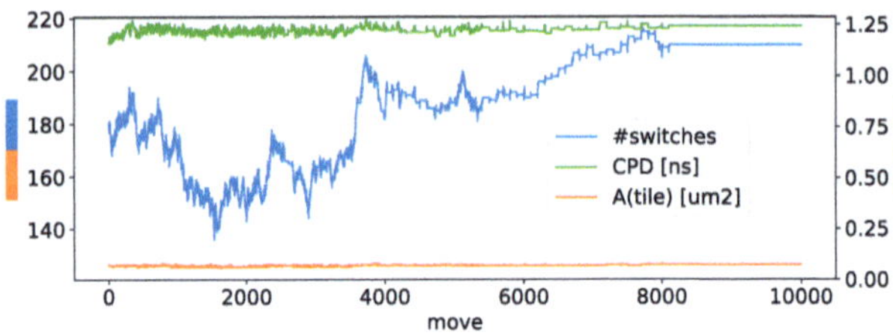

Fig. 5.17 Convergence of the simulated annealing optimization

geomean routed critical path delay of the circuits used in the search taken for the two terms, with equal contribution:

$$\Delta cost = 0.5\frac{\Delta A(tile)}{\text{prev. } A(tile)} + 0.5\frac{\Delta\text{geom. routed crit. path delay}}{\text{prev. geom. routed crit. path delay}}. \quad (5.5)$$

To save runtime, wire delays are measured only when the switch-pattern differs from that of the previously measured architecture in at least five switch types, while floorplan is optimized only on temperature change. The same three MCNC circuits driving the avalanche search of Sect. 5.10 are used again. The initial temperature is set to 0.02, and we perform 100 temperature changes, at the rate of 0.95, with 100 moves per temperature.

5.12.5 Results

Including or removing a single switch from the pattern most often has little influence on the critical path delay, or tile area, which only dramatically changes with a change in the number of columns needed to fit the multiplexers (Fig. 5.9). This makes convergence of the optimization difficult, as visible in Fig. 5.17. In the present experiment, 30 new switches were added, while both adjacency regularity (Fig. 5.16b) and hop-distance optimality were broken. Increased wire delays (Table 5.1) further increased the geomean routed delay by ~6%.

We conjecture that for Lin et al. annealing the switch-pattern proved valuable as during the optimization of the channel composition—likely causing larger and easier to capture changes in performance—the switch-pattern grew increasingly inappropriate for the new composition, and annealing it was just sufficient to rectify that. If applied to one fixed channel composition, success of the method seems less obvious. Of course, we do not claim that simulated annealing, or any other general optimization method, cannot be made to work for switch-pattern exploration, if extensive engineering of the cost function and move generation is performed. Nevertheless, much like the original PathFinder removed the need for elaborate ad hoc heuristics of early FPGA routers [Lem93], we believe that our avalanche search method—essentially relying on the same principles as PathFinder—removes the need for similarly elaborate heuristics to explore interconnect architectures.

5.13 Analysis of Some Further Aspects of Avalanche Search

In this section, we present results of several additional experiments which aim to increase the understanding of how avalanche search functions and what is the impact of different elements of the algorithm. Unless stated otherwise, an experimental setup of Sect. 5.9 was used in all experiments.

5.13.1 Parameters

The functional form of the avalanche costs (Eq. 5.2) involves three parameters: the starting avalanche cost, $s(u)$, and the two parameters dictating the rate of cost decrease with respect to usage, a_p and a_h. For the search method to be effective, these parameters must be assigned reasonable values. In this section, we present results of several experiments intended to help the understanding of the impact of parameter tuning on the effectiveness of the algorithm. We also give some remarks on how to choose good parameter values. Because different switch types are already distinguished by their timing cost, we chose to fix all $s(u)$ to a single parameter s.

5.13.1.1 Adaptive Tuning

The rate at which avalanche cost should drop with respect to usage depends fundamentally on the actual usage values attained during routing: A single fixed drop rate could be too high if many nets naturally tend to use the same switch types, whereas it could be too low if the number of nets that do so is very small. This depends on the size and structure of the circuits being routed, making it difficult to choose a single value for a_p and a_h. To resolve this issue, we first record the maximum usage during the first routing iteration, when the avalanche costs are temporarily reset to zero, much like VPR typically neglects congestion in the first iteration [Mur20]. This allows all nets to initially choose the timing-optimal resources. Let this maximum usage recorded in the first iteration be M_U^1. We compute a_p and a_h as

$$a_p = a_h = \frac{s}{M_U^1 \times (iter\_to\_zero + 1)}. \tag{5.6}$$

In other words, a_p and a_h are set to the value required for the avalanche cost to be reduced to zero in $iter\_to\_zero \in \mathbb{N}$ routing iterations, assuming a sustained usage of M_U^1. Thus we fix both a_p and a_h using a single metaparameter with a much more graspable meaning. Once computed in the first iteration of Algorithm 5.2, a_p and a_h do not change until the end of the search. Benefits of independently setting a_p and a_h are still to be investigated.

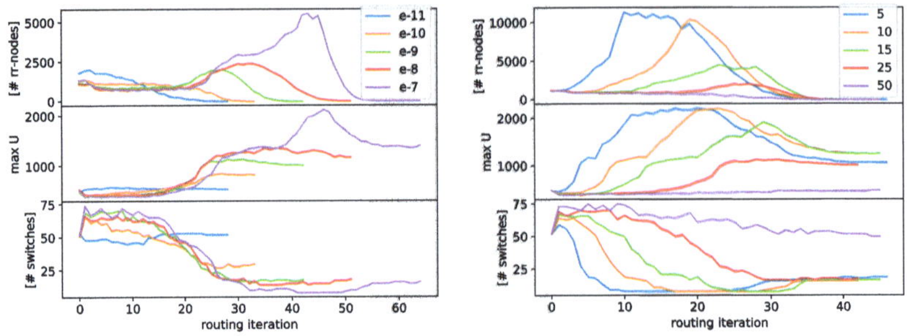

(a) Dependence on the starting avalanche cost.

(b) Dependence on the rate of avalanche cost decrease.

Fig. 5.18 Dependence of concentration on avalanche parameters. The top graph of Figure (**a**) shows the number of congested nodes in the rr-graph after each iteration of the router in the first iteration of the search algorithm, for *iter_to_zero* set to 25 and various starting avalanche costs. The middle graph shows the corresponding maximum usage, while the bottom one shows the number of switches with usage $\geq 0.05\times$ the current maximum. Graphs of Figure (**b**) are analogous to those of Figure (**a**), for the starting avalanche cost fixed at 10^{-9} and *iter_to_zero* $\in \{5, 10, 15, 25, 50\}$

5.13.1.2 Starting Cost

Figure 5.18a shows the effect of various starting costs on concentration and congestion resolving when simultaneously routing the *alu4*, *ex5p*, and *tseng* MCNC circuits [Yan91], with $iter\_to\_zero = 25$. In the first graph, we see that all explored values of s cause a rise in the number of congested nodes which disappears once congestion is penalized sufficiently for nets to move to switch types with lower usage and higher avalanche cost. Larger values of s lead to higher peaks of congestion occurring later in the routing process.

The middle graph clearly shows the correlation between rising concentration and congestion. Larger values of s initially make it less likely for nets to route through switch types with low usage, leading to larger peaks of maximum usage. However, excessive concentration is not sustainable because it prevents congestion resolution. The overshoot for $s = 10^{-7}$ depicts this clearly, and although its final maximum usage is also somewhat higher than for the other values of s, some routing iterations are inevitably wasted. Apart from the maximum usage, the number of switch types with significant usage (here set at $\geq 5\%$ of the current maximum) is also illustrative. As the bottom graph shows, all explored values of s—apart from 10^{-11} and 10^{-10} which are clearly too small to prevent nets from using switches of types not required by other nets—lead to very similar results in this respect, by the end of the routing process.

While larger values of s, such as 10^{-7}, may lead to additional reduction of the obtained switch-pattern size, in the experiments in this chapter, we use $s = 10^{-9}$ since it provides a reasonable trade-off between concentration and runtime.

5.13.1.3 Rate of Decrease

Figure 5.18b shows the results of sweeping *iter_to_zero* under the setup of Sect. 5.13.1.2, but with s fixed at 10^{-9}. Smaller values quickly reduce the cost of switch types which are intrinsically in high demand (usage close to M_U^1), causing an early concentration and congestion increase. Upon congestion resolution, however, different explored values converge to very similar results. The exception is 50, which results in too slow drop in avalanche costs that does not allow the higher usage switch types to attract nets to route through them. It appears that a good trade-off between concentration and runtime is given by values corresponding to about half the total number of routing iterations taken to achieve a congestion-free routing. In all experiments presented in this chapter, we use $iter\_to\_zero = 25$.

More comprehensive analyses could lead to parameter values that produce better quality solutions or reduce runtime. Nevertheless, at the moment it does not seem that avalanche search is particularly sensitive to the values of the parameters.

5.13.2 *Circuit-Level Parallelization*

In Sect. 5.8.2, we proposed to route multiple circuits at once, so that usage information and avalanche costs can be shared among them. The rationale was that different nets of multiple circuits can together negotiate a more compact pattern than when routed individually. We test that hypothesis in this section.

5.13.2.1 Average Normalized Usage

We run two experiments. In the first one, each circuit is routed independently, using individually autotuned avalanche parameters. Once all circuits are routed, the final avalanche costs of all switch types are averaged out among all circuits, while the usages are first normalized by the total usage in the particular circuit and then averaged. In this way, large circuits are no longer given an unfair advantage over the small ones in determining switch type selection. The average costs and average normalized usages are then used to determine which switch types are adopted in the pattern, as described in Sect. 5.8. The results of this experiment are shown in Figs. 5.19 and 5.20.

The search relying on independently routing the *alu4*, *ex5p*, and *tseng* circuits converged in 57 iterations (as opposed to 36 when routing all circuits together) and accumulated 87 switch types (as opposed to 78 when routing all circuits together). The smaller size of the pattern obtained through routing all circuits at once demonstrates the benefit of sharing the usage and cost information between circuits during routing. However, as Fig. 5.20 shows, the routed critical path delays on the larger pattern are only negligibly (0.25%) larger. This means that when running avalanche search on larger modern circuits where even routing one circuit may be

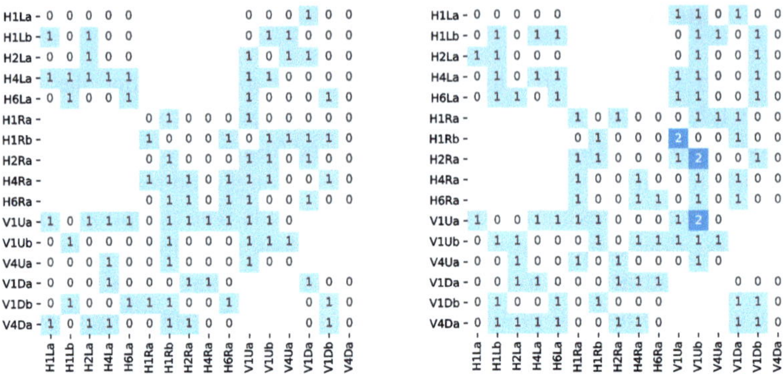

(a) Simultaneous routing (same as Figure 5.11a). (b) Independent parallel routing.

Fig. 5.19 Adjacency of wire types: simultaneous routing (**a**) and independent parallel routing (**b**). When routing all circuits at once and using the total usage to evolve the avalanche costs and select the switch types to enter the pattern, large circuits may have an unfair advantage. Routing many circuits at once may also not be feasible due to exceeding runtime. Figure (**b**) shows a pattern obtained from an avalanche search where all circuits are routed independently in parallel, with independently autotuned avalanche parameters. After each iteration, usages are normalized for each circuit, and their average is taken as a basis for switch type selection

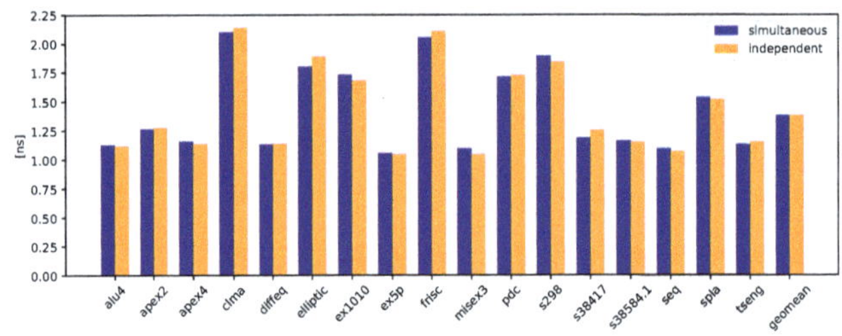

Fig. 5.20 Influence of simultaneously routing multiple circuits during pattern search

a challenge (see Sect. 5.14), let alone multiple of them at the same time, combining normalized usages can be a reasonable alternative. Of course, as an intermediate step, providing some mechanism to (periodically) pass cost information between different parallel threads routing different circuits independently could be useful.

It is interesting to note that even though the potential dominance of the larger circuits is now less likely, that did not improve the geomean routed delay of the three circuits used for the search; it is in fact negligibly higher (by 0.28%).

5.13.2.2 Total Usage

The second experiment attempts to more closely approximate the behavior of simultaneously routing all circuits while still routing them independently. Namely, the avalanche parameters are taken from the search which routes all circuits together, and after every iteration of the search, the current and historical usages of each switch type are summed among all circuits. This total usage and the costs computed from it are then used for selecting the switch types. Hence, the only difference between this approach and routing all circuits simultaneously is that usage information is not shared between the circuits during the search iterations.

This search converged in 51 iterations, accumulating 92 switch types. For the interest of space, we do not show the resulting pattern or plot the routed delays, which are on geomean 2.63% worse than when routing all circuits at once.

5.13.3 Sensitivity to Circuit Choice

One of the main advantages of benchmark-driven FPGA architecture design is that the obtained architecture can be tailored to some extent to the circuits of interest, represented by the selected benchmark set. However, this is also one of the main disadvantages of the approach, since if the benchmark set does not appropriately represent the intended use of the architecture, the architecture will either completely fail to implement some circuits of interest or fail to do so with appropriate performance. We have seen an instance of that already in Sect. 5.10.2.4. In this section, we attempt to provide a deeper understanding of dependence of switch type usage statistics—which forms the basis for the avalanche search algorithm introduced in this chapter—on the choice of the circuits used to derive these statistics.

In Sect. 5.13.2, we have determined that average normalized usage is an effective metric for choosing the switch types to adopt in the pattern. We use it here to assess the impact of circuit choice on usage statistics. The rationale is that the less dependent usage is on exact choice of the circuits on which it is observed, the closer the observation made on any individual circuit will be to the average of several circuits.

Let S_c be the set of all switch types, ordered by the decreasing usage achieved on circuit c. Let S_m be the set of all switch types, ordered by the decreasing average normalized usage on a set of circuits C. Furthermore, let S_c^n and S_m^n be the subsets of the aforementioned sets containing the first n of their elements. As a measure of similarity between usage statistics obtained on individual circuits from C and their average, we use the size of the intersection $\left|S_c^n \cap S_m^n\right|$, for various subset sizes n. We plot this metric for seven different individually routed MCNC circuits and n ranging between 1 and 100 in Fig. 5.21a. Usage statistics come from a single avalanche search iteration. For most values of n, the curves for most circuits are close to $y = x$. This means that there is significant similarity in the sets of most used switch types chosen by the router on different circuits. Hence the sensitivity of the search outcome to the circuits used to run it is not particularly high. However, we can see that as n increases, most curves start moving away from $y = x$. This is because larger subset sizes capture switch types with significantly lower usage,

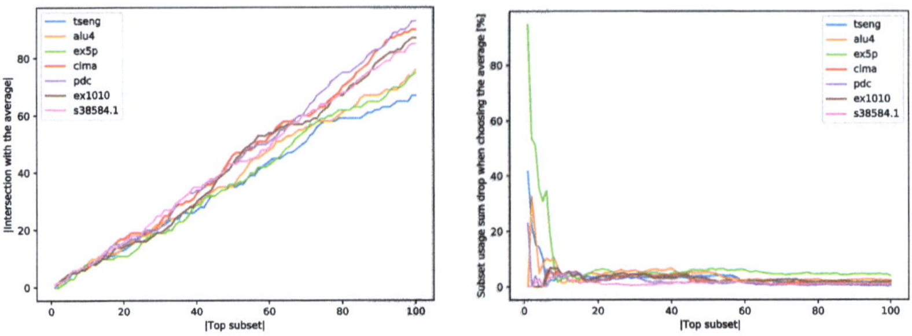

(a) Intersection of the max-usage subset and the average.

(b) Usage drop when choosing the average.

Fig. 5.21 Maximum-usage subset overlaps with the average of all circuits. Figure (**a**) shows the size of the intersection of the maximum-usage subset of each individual circuit and the same-size subset with maximum normalized average usage over all circuits, for subset sizes ranging between 1 and 100. Overlap between the two subsets is reasonably high (line close to $y = x$) for all circuits, with some local variation. This variation increases with subset size, since larger subsets capture more switch types with low usage, mostly used by the few critical paths. Figure (**b**) shows the relative total usage drop if the average subset is chosen instead of the subset determined for each circuit individually. Since the top few switches differ among the circuits, large drops are observed for very small subset sizes. However, for subsets larger than $\sim$10 switch types, cumulative usage drop is consistently below 10% for all circuits

mostly catering to the needs of the critical paths, where similarity between the circuits is lower.

Let $U(S) = \sum_{e \in S} U(e)$ be the total usage of all switch types in a set S. We have seen that a maximum-usage subset of each circuit in Fig. 5.21a significantly overlaps with the maximum-average-usage subset. Now we would like to quantify how large a drop in total usage each circuit $c \in C$ would experience if S_m^n is chosen for it in place of S_c^n. Figure 5.21b plots $\frac{U(S_c^n)-U(S_m^n)}{U(S_c^n)} \times 100\%$, for the same circuits and values of n used in Fig. 5.21a. The total usage of both subsets is computed solely with usage information of the respective circuit. For very small n, the drop is significant, because the few most used switch types differ between circuits. However, as n increases beyond 10, the total usage drop is consistently below 10%. This experiment further suggests that dependence of the usage statistics on the circuits on which they are observed is not particularly high and that the major differences could arise only in the long tail of switch types with low usage.

5.13.4 Influence of Approximate Switch Type Delay on Adoption

As was described in Sect. 5.8.1, different switch types are differentiated in the rr-graph by a delay which strives to approximate the impact of adding the particular switch type to the pattern on loading the wire that is driving it. While this is a way

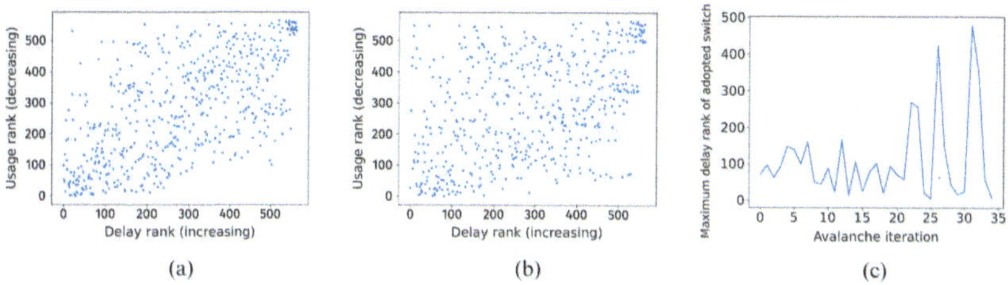

Fig. 5.22 Low correlation between switch type adoption and its approximate delay. (**a**) Low correlation between switch rank in sorting by increasing approximate delay and decreasing usage after avalanche iteration #1. (**b**) Still lower correlation between switch rank in sorting by increasing approximate delay and decreasing usage after avalanche iteration #16. (**c**) Maximum rank of an adopted switch type in sorting by increasing approximate delay, for each avalanche iteration

of informing the router on how expensive it would be to add a particular switch type to the pattern, potentially motivating it to use another, less detrimental switch type, we must confirm that the choices made by the router are not predominantly based on this delay penalty. If that were the case, avalanche costs would become irrelevant as the pattern would largely be determined by the a priori assigned delays.

In Fig. 5.22a, we plot the correlation between switch type usage and its intrinsic delay reported to VPR in the first iteration of avalanche search. The x-coordinate of each point representing a switch type is determined by its delay rank, with switch types with lower delay being closer to the origin. The y-coordinate of each point is determined by the switch type's usage rank, with higher usage being closer to the origin. Ties are broken by names.

If switch type adoption was solely dependent on switch type delay, all points would be along the $y = x$ line. We can clearly see that this is not the case and that the router is capable of escaping any potential local minima set by the preassigned delay penalties. However, there is some correlation between delay and usage, since points are scattered more closely to the $y = x$ line than to $y = 564 - x$; this was the intention of conveying the physical information through the approximate delays.

The correlation weakens as the iterations of the avalanche search progress, which can be seen by the larger spread away from $y = x$ in Fig. 5.22b, depicting the same situation at iteration 16. Perhaps more illustrative of this phenomenon is Fig. 5.22c, depicting the maximum delay rank (i.e., the slowest) of the switch types adopted in each iteration of the avalanche search. We can see that some of the slowest switch types are adopted toward the end of the search. This could be because some critical path needed the particular switch types to form a connection between wires that optimally implement it, or the particular switch type was needed to resolve some outstanding congestion hotspot.

5.13.5 Routability-Driven Search

In Sect. 5.7.5, we commented on the importance of allowing the router to route critical paths with switch types that otherwise have low usage and would not be adopted in the final pattern. We updated the node cost function in PathFinder to enable better selectivity of the critical paths and to allow them to see the high-cost switch types with low usage as sufficiently cheap. In this section, we inspect whether this was successful and whether the switch types that entered the pattern of Fig. 5.11a were not merely required for making it possible to route the circuits used in the search.

To check this, we reran the search neglecting the timing information (insofar as it is not embedded in the base costs of the wires which are dependent on the delay [Mur20]) by setting the *max_crit* parameter of VPR to 0. This search converged in 23 iterations, which is significantly less than the 36 that were needed when timing was considered. More importantly, only 27 switch types entered the pattern, as opposed to 78 when timing was considered. This significantly smaller pattern is shown in Fig. 5.23a. As can be expected, most of the connectivity incident to the long wires has disappeared, because the benchmarks used in the search are small enough that their congestion can largely be resolved while using only the short wires. Under these circumstances, long wires serve only to enhance the speed of the implemented circuits and can thus be disregarded if timing is not of concern. For instance, H6Ra had a fanout of six in the pattern obtained with timing considered (Fig. 5.11a), whereas in this smaller pattern it has no fanout at all.

It is interesting to note that between the eight length-1 wires, there are 17 switch types, making their average fanout (F_S [Bet99]) very close to 2, which Lemieux and Lewis have determined to be the minimum required when two different switch-patterns are used in the FPGA grid in a checkered fashion [Lem04]. Additional switch types may have been chosen to compensate for the lack of two different patterns or simply because the pattern was not fully minimized.

5.13.5.1 Impact on Performance

Routing results obtained on the smaller pattern are shown side by side with the results of the pattern of Fig. 5.11a at the bottom of Fig. 5.23b. The smaller pattern does not provide enough connectivity to successfully route all MCNC circuits. For the subset that can be routed, the geomean critical path delay is 13.86% higher than on the pattern of Fig. 5.11a, despite the fact that the lower capacitive load and smaller tile area resulted in wires being faster. This clearly demonstrates the utility of considering the critical paths when performing the search.

There could be two main reasons why the delays are so significantly higher on the smaller pattern: (1) Pairs of wires needed to optimally implement critical connections of the circuits cannot be connected due to the lack of the appropriate switch and (2) due to lower routability, nets need to detour more for the congestion to be resolved. To determine which issue contributes more to delay deterioration, we also report the lower bound critical path delays obtained after the first iteration of VPR at the top of Fig. 5.23b. Significant deterioration is present there as well, but it is less pronounced: The geomean is 4.84% higher on the smaller pattern. Hence, it is the poor routability that is the main culprit.

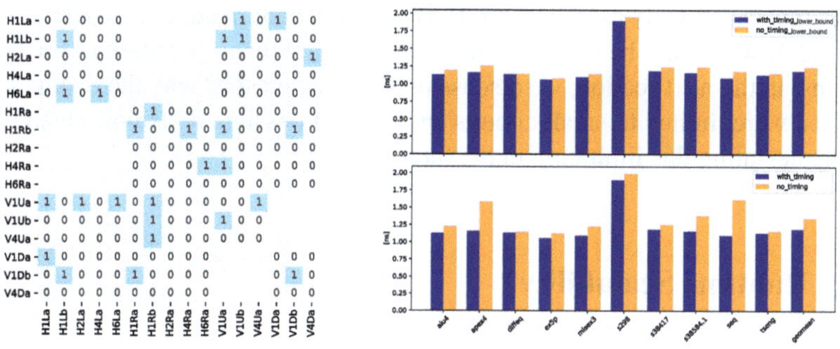

(a) Adjacency.

(b) Routed delays.

Fig. 5.23 Results of an avalanche search ignoring timing information. (**a**) Adjacency. (**b**) Routed delays

The question of how much freedom should be given to the critical paths to enlarge a minimal routable pattern remains to be answered in future work. In this section we demonstrated that at least some level of freedom is highly useful, but a more optimal point may lie somewhere in between. One possible solution would be to first find a minimal pattern that supports all of the routability requirements, using a routability-driven search, and then extend it in a subsequent timing-driven search, until a predefined budget of additional switch types or exploration iterations is surpassed.

5.13.5.2 What Do the Results Tell Us?

Compaction of the switch-pattern presented in this section, which resulted in some of the longer wires being unable to connect to other channel wires, once more points to the fact that the circuits used in the exploration are not large and complex enough to saturate the channel capacity of a modern plane-based FPGA. This makes it hard to derive general rules of the sort "an FPGA implemented in a 4nm technology should have between 24 and 32 switches per 16 wires in a plane" from the currently available results. Nevertheless, the compaction also demonstrates the effectiveness of the proposed exploration method in minimizing the pattern size where an opportunity for that exists. This was indeed the main intention of this work—to develop a method for automatically designing switch-patterns that are appropriate for the conditions created by the underlying technology and the requirements of the target circuits, even in situations when general design rules with which a human designer is familiar no longer hold. As we have seen in Sect. 5.12, simulated annealing does not possess this feature—at least not in the adopted implementation. Similarly, the stark difference between the performance of the patterns obtained in routability- and timing-driven search demonstrates the capacity of the method to select switch types important for delay optimization.

Oracles that can quickly assess routability of a given pattern but not its performance (Sect. 5.2.3) would not be useful in this context, other than for pruning away some solutions. In the next section, we discuss in detail the reasons why the results so far presented were limited to small circuits and suggest possible remedies which could help in alleviating this limitation in the future.

5.14 Runtime Scalability

Implicitly representing the entire search space in the rr-graph during avalanche search removes the need to route thousands of explicitly constructed solutions. This means that increased runtime of each routing run can be tolerated. Nevertheless, it is important to assess how large this increase is and where it comes from, so that it can be mitigated when necessary. The total routing time spent in the single 37-iteration exploration run of Sect. 5.10 was about 5 hours. Of that, about 4.5 hours were spent by the actual PathFinder extended with avalanche costs, while the remaining half an hour was taken by lookahead computation and allocating the data structures in VPR. When combined with SPICE simulations, rr-graph generation, packing, placement, and placement manipulation, the entire search took about 10 hours. While this may not seem like an exceedingly long time, it is important to give it a context: routing the same three circuits used in exploration (which have a combined size of about 2700 LUTs), but on an architecture containing only the final pattern, with no switch-splitting nodes and no usage tracking, took a mere 5.5 seconds. This $\sim$80$\times$ average runtime increase per iteration is very significant. For instance, each iteration of avalanche search gives a possibility to evaluate 80 different patterns in the black-box-in-the-loop approach, although, as we have seen, even 10,000 moves were not sufficient for simulated annealing-based exploration of Sect. 5.12 to converge to results comparable to those obtained after 37 iterations of the avalanche search (giving a budget of $\sim$3000 move evaluations).

What is more important, however, is that with such a large increase in routing runtime (likely to deepen further with growing circuit size), it is infeasible to use in exploration circuits that normally take even minutes and let alone hours or days to route. Hence, in this section we give a detailed analysis of the origins of this runtime increase and suggest several remedies for which we believe that they could be successful in alleviating the problem.

5.14.1 Routing Graph Size

Intuitively, increasing connectivity in the rr-graph could be expected to reduce the routing time, as more flexibility makes it easier to eliminate congestion. This is indeed true, but only up to a certain point: If the rr-graph already offers sufficient flexibility to eliminate congestion, any further increase in its size only leads to deterioration in runtime, as it takes longer to find a shortest path for each connection. This was observed by Moctar et al. who determined that routing circuits on an architecture with a fully populated cluster input crossbar represented within the rr-graph takes about 2.5× more time than routing the same circuits on an architecture where this crossbar is 40% populated (Figure 4 in Moctar et al. [Moh12]). Complexity of Dijkstra's shortest path algorithm is $\Theta(|E| + |V| \log |V|)$ [Cor09]. Assuming that the size of these rr-graphs was dominated by the number of edges describing the intracluster interconnect and assuming that the population of both architectures was sufficient to easily eliminate congestion, we could expect that runtime is roughly linear in $|E|$. Indeed, the ratio between the edge counts of the fully and the 40%-populated crossbars ($1/0.4 = 2.5$) aligns with the runtime ratio observed by Moctar et al.

Since avalanche search relies on embedding the entire search space in the rr-graph (Sect. 5.4), it is inevitable that the rr-graph used in exploration is much larger than the final one. For example, the final rr-graph obtained from the exploration of Sect. 5.11 contained in each switch-block 16 nodes (representing the 16 channel wires originating from it) and 84 edges. The graph representing the entire design space, on the other hand, contained $16 + 564$ nodes and $2 \times 564 = 1128$ edges—a 36× increase in $|V|$ and a 13× increase in $|E|$. Using the same rough assessment based on the complexity of Dijkstra's shortest path algorithm as above, a runtime increase between 13× and 44× can be expected. To measure this impact we performed an experiment where we recorded the time taken to complete the first iteration of the inner loop of PathFinder (line 12 of Algorithm 5.1), when congestion costs are neglected, in three different cases: (1) on the final pattern, (2) on an rr-graph containing all possible switches, but without nodes to split them, and (3) on an rr-graph with all switches split by additional nodes. The runtimes were 1.8 s, 9.7 s, and 21.9 s, leading to a runtime increase of 12.2× and 5.4× when additional switches are and are not split by nodes, respectively. This suggests that the increased edge count is the dominant problem and that providing support for weighting both nodes and edges, so as to avoid the need to split switches, could lead to a ~2× runtime improvement.

5.14.1.1 Possible Remedy: Suppressing Low-Usage Switches

One way to reduce the rr-graph size is to suppress low-usage switch types after the usage is first measured in the first iteration of PathFinder, before the avalanche costs are initialized. For example, if it is known that the size of all multiplexers should be some fixed number m, as is often the case in commercial architectures [Pet21], then selecting the intrinsically most used $k \times m$ switch types for each multiplexer, where k is some small constant, can lead to a drastic reduction in rr-graph size. The remaining switch types can periodically be brought back, to ensure that some of them did not become more preferable, due to prior adoption decisions.

5.14.1.2 Possible Remedy: Randomized Instance Sparsification

Another orthogonal approach could be to remove switches of a certain type from some switch-blocks but retain them in others. Provided that a sufficient number of instances are available for each type, it is plausible that the collected statistics would show little change compared to the situation when each switch type is represented in every switch-block. Removing switches from one region would likely alter the paths in another, but the hope is that if sparsification is done in such a way as to ensure that each multiplexer receives a sufficient number of inputs, average behavior would be similar to representing all switches everywhere.

5.14.1.3 Possible Remedy: Partition, Sample, and Mix

The method of simultaneously routing multiple circuits depicted in Fig. 5.10 may seem completely impractical following the analysis of this section. However, it also offers a possible remedy for the problem of rr-graph size increase. Rather than combining multiple complete circuits together on a common FPGA, placements of large circuits could be partitioned into manageable pieces, a certain subset of these pieces could be selected through random sampling, possibly across multiple circuits, and then combined on a common FPGA. In this way, switch types necessary for short-haul connections could be quickly determined, before proceeding recursively, as in Sect. 5.11, to extend the pattern with others required by the long-haul connections. Alternatively, long-haul connections passing over a particular piece could be determined through the use of a global router and approximated in the piece by appropriate input-to-output connections.

Besides the rr-graph size, there are several other reasons why routing with avalanche costs is significantly slower than usual. We list them in the following sections.

*5.14.2 A**

FPGA routers typically use A^* to speed up the shortest path finding [Mur20]. The idea is that when a node u is being pushed to the heap, it could be easy to obtain a lower bound on the cost of the path needed to reach the target t from u. Then, instead of pushing u with the known cost f needed to reach u from the source s, it is pushed with the cost $f + g$, where g is the estimate for reaching the target. If g is really a lower bound, the algorithm will never fail to miss the shortest path. At the same time, though, the addition of g will prevent some nodes from ever being popped from the heap, thus often drastically reducing the portion of the rr-graph that needs to be explored. The closer the chosen lower bound to the actual cost of the remaining path is, the more effective this pruning will be.

5.14.2.1 Congestion Lookahead

In case of congestion costs, the minimum value of congestion is 0. Hence, a lower bound estimate (lookahead) can only be computed when congestion itself is completely neglected and only base costs of routing resources are taken into account [Mur20]. This means that pruning will be more effective toward the beginning of the routing, because there the actual congestion costs are closer to 0. Fortunately, the number of congested nodes generally drops as the routing iterations progress (Fig. 5.25), so only a relatively small portion of the rr-graph and the circuit's nets will be affected by the poor performance of the lookahead.

5.14.2.2 Avalanche Lookahead

Similarly to congestion costs, the lowest value that avalanche costs can attain is 0. Hence, when computing a lookahead that stores the values of g in a fixed look-up table, as is typically done by VTR 8 [Mur20], we have to set all avalanche costs to 0; otherwise it would not be a true lower bound (*admissible*). However, contrary to congestion costs that are (close to) 0 in the beginning of the routing process, avalanche costs drop to zero toward the end of it. For most iterations, such a lookahead is ineffective in presence of avalanche nodes.

Let us illustrate this by the example of Fig. 5.24, showing a portion of the rr-graph with the source and the sink nodes designated as s and t, respectively. There are two possible paths between them, each composed of two wires. The path on the left, composed of the nodes $u1$ and $u3$, has a total cost of $20e - 12$, when there is no congestion present, while the path on the right, composed of $u2$ and $u4$, has a total cost of $10e - 12$. In the absence of avalanche nodes and congestion (left figure), the lookahead is exact and only the nodes of the shorter path get popped. Once the avalanche nodes are inserted (right figure), their cost overshadows the admissible lookahead, and nodes from both paths need to be popped. The increase in the number of nodes along the shortest path necessarily increases the number of pops needed to find it. However, this alone would cause an increase from 4 to 7 pops ($1.75\times$), while the actual number of pops required to find the shortest path in the right figure is 11 ($2.75\times$ more). The difference comes from lookahead ineffectiveness. This is only a toy example to illustrate the mechanism of pruning. In practice, when nodes have higher out-degrees, the difference between having an effective pruning function g and not is much higher. For example, Swartz et al. have demonstrated that A^* can produce a speedup of about $50\times$ on MCNC circuits [Swa98]. In larger circuits with longer average paths, the impact could be even higher. For example, as we have already mentioned in the previous section, it took PathFinder 21.9 s to complete the first iteration of routing of one of the Gnl circuits, under the setting in which all avalanche nodes were in place, but their cost was reset to zero, which made the admissible lookahead effective. Raising the avalanche costs up to $1e - 9$ increased this time to 3899 s—an almost $180\times$ increase. After the first iteration of avalanche search, when 12 switch types have

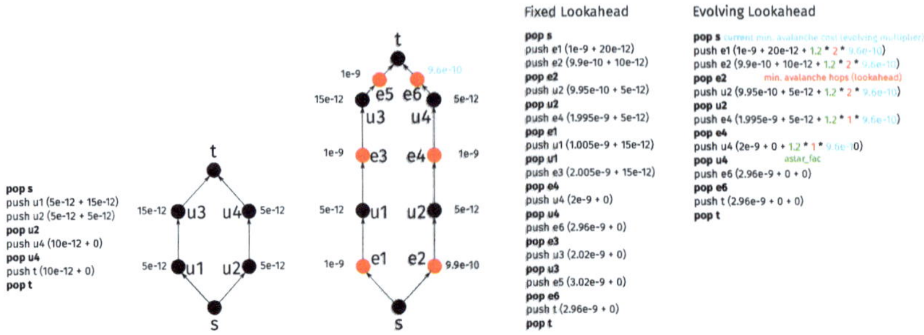

Fig. 5.24 Example of lookahead ineffectiveness. The left figure shows a portion of the rr-graph without avalanche nodes and a sequence of heap operations needed to find the shortest path from s to t, while the right one shows the same portion of the graph with avalanche nodes (red), along with the corresponding heap operation sequence. Costs are annotated next to nodes. Ratios between the avalanche costs and the base costs of the wires are realistic. Lookahead values are on the right side of the "+" signs. "Fixed Lookahead" is the usual *map* lookahead used by VTR [Mur20], while "Evolving Lookahead" is the modification proposed in Sect. 5.14.2.3

been adopted to the pattern, the same runtime reduced to 319 s, bringing the lookahead ineffectiveness gap down to ~15×. Once the cost of the adopted switch types drops to zero, lookahead becomes effective for paths routed through them. Hence, it is imperative to improve lookahead effectiveness in the initial iterations with few adopted switch types.

5.14.2.3 Possible Remedy: Evolving Lookaheads

One idea that could help in resolving this issue is to store in an additional look-up table the minimum number of avalanche nodes needed to connect a node of given type to the target tile. Then, when pushing nodes onto the heap, their cost could be increased by the appropriate entry from this table, multiplied by the cost of the currently cheapest avalanche node. This is illustrated on the right of Fig. 5.24, for the *astar_fac* parameter multiplying g set to 1.2 (VTR 8 default [Mur20]). The proposed lookahead now finds the shortest path with seven pops, which is the minimum in presence of switch-splitting nodes. Note that this approach is not easily applicable to congestion lookahead, because with very high probability, some node will always remain unused, thus reducing the lowest congestion cost to zero. Avalanche nodes, on the other hand, have the highest cost when their type is unused.

Another potential advantage of avalanche costs is that they are tied to types and not instances, unlike congestion costs. This could perhaps allow runtime improvements by increasing preprocessing effort, as information would need to be stored only about relationship between the few types, rather than the numerous instances. At any rate, for the proposed method to be truly scalable, A^* must be made effective.

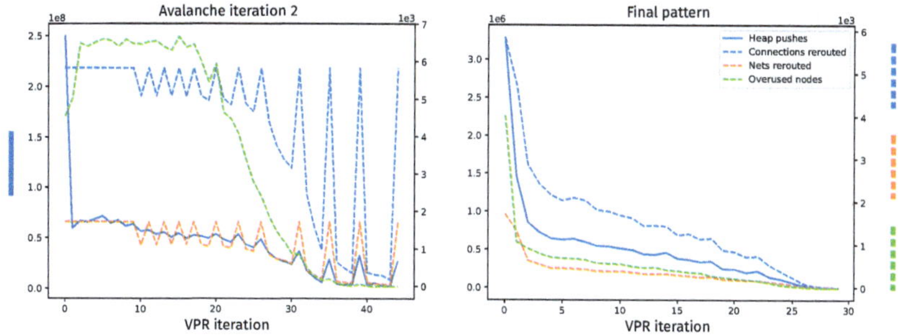

Fig. 5.25 Heap operations and rip-ups during the second iteration of the avalanche search (most congested). Bars on the sides of the graphs designate which y-axis corresponds to which curve(s) (right for the dashed and left for the solid)

5.14.3 Periodically Forcing Rip-Up

Figure 5.25 plots rip-up and congestion statistics of routing in two extreme iterations of the avalanche search: (1) the second iteration where the highest congestion was observed (left figure) and (2) the last iteration where the switch-pattern has been finalized and no potential switches and in turn no avalanche costs are present. Dashed orange and blue lines represent the number of nets and connections, respectively, which were ripped-up and rerouted in the corresponding router iteration. We can see that in the right figure where avalanche costs are not present, these curves almost monotonically decrease. This comes from the incremental-rerouting capabilities of VTR 8 [Mur20], where only connections that use congested nodes or fail to meet timing are ripped up, contrary to the original VPR PathFinder implementation which ripped up all connections in every routing iteration [Bet99, Luu14]. The corresponding curves in the figure on the left contain a number of peaks at an increasing distance from one another, as the routing iterations progress. These peaks represent iterations where rip-up of all connections has been forced. If this had not been done, it would have been possible for some switch types to be used in the initial iterations where the avalanche cost differences were still small, without the paths using them later moving to a cheaper switch type. This would then potentially cause some nonessential switch types to be included in the final pattern. As the routing progresses, avalanche cost changes become smaller, so it becomes less useful to rip up the legal connections, and hence this forcing is done less frequently.

5.14.3.1 Possible Remedy: Path-Cost Bounding

Peaks in the number of heap pushes coincide with the forced rip-up peaks, which is expected. However, this increase in the amount of work could potentially be reduced. Namely, if a legal connection is being routed, its pre-rip-up avalanche cost is also an upper bound on the avalanche cost of the new shortest path that can implement it, because avalanche costs are monotonically nonincreasing. This upper bound could be used for more efficient pruning when searching for the new shortest path.

5.14.4 Congested Nodes

The green dashed curve in Fig. 5.25 depicts the number of congested nodes in each PathFinder iteration. In the figure on the right, we see that this curve is again almost monotonically decreasing, with the peak being in the first iteration which neglects congestion altogether. In the figure on the left, however, the situation is drastically different. Since in the first PathFinder iteration avalanche costs are also neglected, congestion is comparable to that of the first iteration in the right figure. As soon as the avalanche costs start to be considered, they create concentration on switch types which drives the congestion up. This congestion starts to get resolved only after the avalanche costs drop sufficiently for a large enough number of switch types, and at the same time, congestion costs increase enough to outweigh the avalanche costs. This occurs roughly around the middle of the routing run, which coincides with the intended avalanche cost reduction to zero for the most used nodes (see Sect. 5.13.1.1). As discussed in Sect. 5.7.2.2, fully resolving congestion may not always even be necessary, and early stopping could benefit the runtime.

5.15 Conclusions and Future Work

In this chapter, we introduced a new method for automated exploration of FPGA switch-patterns, which removes the fundamental limitation of prior techniques: necessity to explicitly list and test numerous architectures in a place and route flow. The proposed method achieves this by leveraging the router itself to perform the exploration, instead of perceiving it merely as a black box used for evaluation of explicitly listed solutions. We hope that this will open up new interesting possibilities in the FPGA architecture domain. One of the most exciting aspects of the method is that it represents the design space implicitly in the routing-resource graph, meaning that it could be useful for exploring many other aspects of programmable interconnect. In the limit, recalling the discussion of Sect. 3.6 which postulated that the perception of programmable interconnect as a collection of programmable switches connecting prefabricated channel wires is somewhat dated and could be better replaced by a periodic graph of stored-select multiplexers, we could even think of using avalanche search to design the entire interconnect architecture at once.

Of course, for this to be possible in practice, it is first necessary to solve the runtime issues identified in Sect. 5.14, perhaps with the help of possible remedies

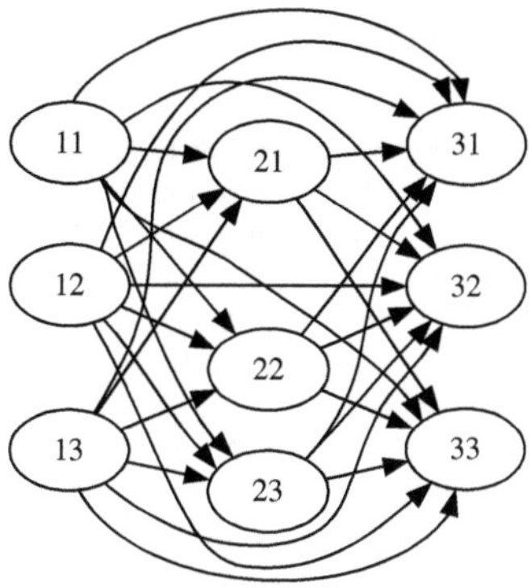

Fig. 5.26 Example of low-cost-path issue in exploring multilevel multiplexing structures

suggested in that section as well. Additionally, support for architectures containing hardened blocks should be introduced by extending the physical modeling framework of Chap. 4. Hardened blocks may pose different routing requirements than the LUT-based logic clusters, which could change the effectiveness, though not the generality of avalanche search. Nevertheless, applying the method to a more complex design task introduces other interesting challenges. For the sake of illustration, let us assume that we are exploring sparsifications of the graph shown in Fig. 5.26. How do we ensure that, for instance, edge (11, 33) is used by the router only if it is really needed for congestion resolution or delay optimization, and not simply because it represents a path of lower cost than any going through the second layer as well? One way could be to assign edge weights such that (11, 33) is more costly than the indirect alternatives. However, in a more general case, this would be tedious and prone to inadvertently skewing the outcome. A much better solution would be to dynamically order the potential drivers of each multiplexer by their usage and scale their cost by the position in this order. This would prevent any multiplexer from growing disproportionately large and thus eliminate the aforementioned problem.

Given that in most commercial FPGAs, sizes of multiplexers are a priori known [Pet21], rank-scaling of switch costs could even be nonlinear, such that addition of drivers to a multiplexer beyond its target size is penalized more. Automated design of switch-patterns that respect various regularity constraints that may be required to make (full-custom) layout feasible, or CAD tools more scalable, is a very interesting topic in its own right. We dedicate the next chapter to extending the avalanche search method that we have just presented so that it can produce solutions which respect arbitrary forms of regularity.

The source code used to produce the results of the study presented in this chapter is available at https://github.com/EPFL-LAP/fpl21-avalanche.

References

[Ahm00] Ahmed, E., & Rose, J. (2000). The effect of LUT and cluster size on deep-submicron FPGA performance and density. In *Proceedings of the ACM/SIGDA International Symposium on Field Programmable Gate Arrays* (pp. 3–12). Monterey.

[Bet97] Betz, V., & Rose, J. (1997). VPR: A new packing, placement and routing tool for FPGA research. In *Proceedings of the 7th International Workshop on Field-Programmable Logic and Applications* (pp. 213–22). London.

[Bet98] Betz, V., & Rose, J. (1998). How much logic should go in an FPGA logic block? *IEEE Design & Test of Computers, 15*(1), 10–15.

[Bet99] Betz, V., Rose, J., & Marquardt, A. (2012). *Architecture and CAD for deep-submicron FPGAs*. Kluwer Academic Publishers. ISBN: 0-7923-8460-1.

[Cha96] Chang, Y.-W., Wong, D. F., & Wong, C. K. (1996). Universal switch modules for FPGA design. *ACM Transactions on Design Automation of Electronic Systems, 1*(1), 80–101.

[Chr20] Chromczak, J., Wheeler, M., Chiasson, C., How, D., Langhammer, M., Vanderhoek, T., et al. (2020). Architectural enhancements in Intel®Agilex™FPGAs. In *Proceedings of the 2020 ACM/SIGDA International Symposium on Field-Programmable Gate Arrays* (pp. 140–49). Monterey.

[Com17] Computer History Museum. (2017). *Oral history of Steve Trimberger*. https://archive.computerhistory.org/resources/access/text/2018/07/102740229-05-01-acc.pdf

[Cor09] Cormen, T. H., Leiserson, C. E., Rivest, R. L., & Stein, C. (2022). *Introduction to algorithms* (3rd ed.). The MIT Press. ISBN: 0262033844.

[Gop06] Gopalakrishnan, P., Li, X., Pileggi, L. (2006). Architecture-aware FPGA placement using metric embedding. In *2006 43rd ACM/IEEE Design Automation Conference* (pp. 460–65). San Francisco.

[Hau07] Hauck, S., DeHon, A. (Eds.) (2007). *Reconfigurable computing: The theory and practice of FPGA-based computation*. Morgan Kaufmann. ISBN: 9780123705228.

[Hut04] Hutton, M. D., Schleicher, J., Lewis, D. M., Pedersen, B., Yuan, R., Kaptanoglu, S., et al. (2004). Improving FPGA performance and area using an adaptive logic module. In *Proceedings of the 14th International Conference on Field Programmable Logic and Application* (pp. 135–44). Leuven.

[Kap12] Kaptanoglu, S. (2012). Pathfinder: A negotiation-based performance-driven router for FPGAs. In *FPGA and Reconfigurable Computing Hall-of-Fame Endorsement*.

[Lem93] Lemieux, G. , & Brown, S. (1993). A detailed router for allocating wire segments in FPGAs. In *ACM/SIGDA Physical Design Workshop* (pp. 215–26). Lake Arrowhead.

[Lem04] Lemieux, G., Lee, E., Tom, M., & Yu, A. (2004). Directional and single-driver wires in FPGA interconnect. In *Proceedings of the 2004 IEEE International Conference on Field-Programmable Technology* (pp. 41–48). Brisbane.

[Lem04a] Lemieux, G., & Lewis, D. (2004). *Design of interconnection networks for programmable logic*. Kluwer Academic Publishers. ISBN: 1402077009.

[Lem02] Lemieux, G. G., & Lewis, D.M. (2002). Analytical framework for switch block design. In *Proceedings of the 12th International Conference on Field-Programmable Logic and Applications* (pp. 122–31). Leuven.

[Lew05] Lewis, D., Ahmed, E., Baeckler, G., Betz, V., Bourgeault, M., Cashman, D., et al. (2005). The Stratix II logic and routing architecture. In *Proceedings of the 2005 ACM/SIGDA 13th International Symposium on Field-Programmable Gate Arrays* (pp. 14–20). Monterey.

[Lew03] Lewis, D., Betz, V., Jefferson, D., Lee, A., Lane, C., Leventis, et al. (2003). The Stratix routing and logic architecture. In *Proceedings of the 2003 ACM/SIGDA Eleventh International Symposium on Field Programmable Gate Arrays* (pp. 12–20). Monterey.

[Lew13] Lewis, D., Cashman, D., Chan, M., Chromczak, J., Lai, G., Lee, A., et al. (2013). Architectural enhancements in Stratix V™. In *Proceedings of the ACM/SIGDA International Symposium on Field Programmable Gate Arrays* (pp. 147–56). Monterey.

[Lin10] Lin, M., Wawrzynek, J., & El Gamal, A. (2010). Exploring FPGA routing architecture stochastically. *IEEE Transactions on Computer-Aided Design of Integrated Circuits and Systems, 29*(10), 1509–1522.

[Luu14] Luu, J., Goeders, J., Wainberg, M., Somerville, A., Yu, T., Nasartschuk, K., et al. (2014). VTR 7.0: Next generation architecture and cad system for FPGAs. *ACM Transactions on Reconfigurable Technology and Systems, 7*(2), 1–30. ISSN: 1936-7406.

[Mar00] Marquardt, A., Betz, V., & Rose, J. (2000). Timing-driven placement for FPGAs. In *Proceedings of the 2000 ACM/SIGDA Eighth International Symposium on Field Programmable Gate Arrays* (pp. 203–13). Monterey.

[McM23] McMaster, J. (2023). *Xilinx Artix 7 XC7A50T die shots*. Retrieved October 3, 2024, from https://siliconpr0n.org/archive/doku.php?id$=$mcmaster:xilinx:xc7a50t

[McM95] McMurchie, L., & Ebeling, C. (1995). PathFinder: A negotiation-based performance-driven router for FPGAs. In *Proceedings of the 1995 ACM Third International Symposium on Field-Programmable Gate Arrays* (pp. 111–17). Monterey.

[Moh12] Moctar, Y. O. M., Lemieux, G. G., & Brisk, P. (2012). Routing algorithms for FPGAs with sparse intra-cluster routing cross-bars. In *Proceedings of the 22nd International Conference on Field Programmable Logic and Applications (FPL)* (pp. 91–98). Oslo.

[Mur20] Murray, K. E., Petelin, O., Zhong, S., Wang, J. M., Eldafrawy, M., Legault, J.-P., et al. (2020). VTR 8: High-performance CAD and customizible FPGA architecture modelling. *ACM Transactions on Reconfigurable Technology and Systems (TRETS), 13*(2) , 9:1–60.

[Nik21] Nikolić, S., Catthoor, F., Tőkei, Z., & Ienne, P. (2021). Global is the new local: FPGA architecture at 5 nm and beyond. In *The 2021 ACM/SIGDA International Symposium on Field-Programmable Gate Arrays* (pp. 34–44). Virtual Event.

[Nik21a] Nikolić, S., & Ienne, P. (2021). Turning pathfinder upside-down: Exploring FPGA switch-blocks by negotiating switch presence. In *31st International Conference on Field-Programmable Logic and Applications* (pp. 225–33). Virtual Event.

[Nik24] Nikolić, S., & Ienne, P. (2024). Exploring FPGA switch-blocks without explicitly listing connectivity patterns. *ACM Transactions on Reconfigurable Technology and Systems, 17*(1), 1–39. ISSN: 1936-7406.

[Pet15] Petelin, O., & Betz, V. (2015). Wotan: A tool for rapid evaluation of FPGA architecture routability without benchmarks. In *Proceedings of the 25th International Conference on Field Programmable Logic and Applications* (pp. 1–4). London.

[Pet16] Petelin, O., & Betz, V. (2016). The speed of diversity: Exploring complex FPGA routing topologies for the global metal layer. In *Proceedings of the 26th International Conference on Field Programmable Logic and Applications* (pp. 1–10). Lausanne.

[Pet21] Petersen, M. B., Nikolić, S., & Stojilović, M. (2021). NetCracker: A peek into the routing architecture of Xilinx 7-series FPGAs. In *Proceedings of the 2021 ACM/SIGDA International Symposium on Field Programmable Gate Arrays* (pp. 11–22). Virtual Event.

[Ros91] Rose, J., & Brown, S. (1991). Flexibility of interconnection structures for field-programmable gate arrays. *IEEE Journal of Solid-State Circuits, 26*(3) , 277–282

[Ros89] Rose, J., Francis, R. J., Chow, P., & Lewis, D. (1989). The effect of logic block complexity on area of programmable gate arrays. In *1989 Proceedings of the IEEE Custom Integrated Circuits Conference* (pp. 5.3/1–5.3/5). San Diego.

[Sch02] Schmit, H., & Chandra, V. (2002). FPGA switch block layout and evaluation. In *Proceedings of the 2002 ACM/SIGDA Tenth International Symposium on Field-Programmable Gate Arrays* (pp. 11–18). Monterey.

[Shi22] Shi, K., Zhou, X., Zhou, H., & Wang, L. (2022). An optimized GIB routing architecture with bent wires for FPGA. *ACM Transactions on Reconfigurable Technology and Systems, 16*(1), 1–28. ISSN: 1936-7406.

[Str99] Stroobandt, D., Depreitere, J., & Van Campenhout, J. (1999). Generating new benchmark designs using a multi-terminal net model. *Integration, 27*(2), 113–29.

[Swa98] Swartz, J. S., Betz, V., & Rose, J. (1998). A fast routability-driven router for FPGAs. In *Proceedings of the 1998 ACM/SIGDA Sixth International Symposium on Field Programmable Gate Arrays* (pp. 140–49). Monterey.

[Tan19] Tang, X., Giacomin, E., Alacchi, A., & Gaillardon, P. E. (2019). A study on switch block patterns for Tileable FPGA routing architectures. In *2019 International Conference on Field-Programmable Technology (ICFPT)* (pp. 247–50). Tianjin.

[Tri15] Trimberger, S. M. (2015). Three ages of FPGAs: A retrospective on the first thirty years of FPGA technology. *Proceedings of the IEEE, 103*(3), 318–331. ISSN: 1558-2256.

[Tri97] Trimberger, S., Duong, K., Conn, B. (1997). Architecture issues and solutions for a high-capacity FPGA. In *Proceedings of the 1997 ACM Fifth International Symposium on Field-Programmable Gate Arrays* (pp. 3–9). Monterey.

[Wan06] Wang, G., Sivaswamy, S., Ababei, C., Bazargan, K., Kastner, R., & Bozorgzadeh, E. (2006). Statistical analysis and design of HARP FPGAs. *IEEE Transactions on Computer-Aided Design of Integrated Circuits and Systems, 25*(10), 2088–2102.

[Wil97] Wilton, S. J. E. (1997). Architectures and Algorithms for Field-Programmable Gate Arrays with Embedded Memory. PhD Thesis. University of Toronto.

[Xil98] Xilinx Inc. (1998). *The programmable logicdata book.*

[Yan91] Yang, S. (1991). *Logic synthesis and optimization benchmarks user guide, version 3.0.* Technical Report. Microelectronics Center of North Carolina.

[Yan16] Yang, S., Gayasen, A., Mulpuri, C., Reddy, S., & Aggarwal, R. (2016). Routability-driven FPGA placement contest. In *Proceedings of the 2016 on International Symposium on Physical Design* (pp. 139–43). Santa Rosa.

Chapter 6
Searching for Regular Switch-Patterns

Contents

The switch-pattern exploration method that we presented in the last chapter suffers from the same downside as most of the previous attempts to automate solution of this design problem [Lem00, Lin10]: It produces highly irregular switch-patterns. In this chapter, largely based on our paper previously published at the 2023 ACM/SIGDA

© The Author(s), under exclusive license to Springer Nature Switzerland AG 2025

S. Nikolić, *Modern Programmable Interconnect Design*,
https://doi.org/10.1007/978-3-031-80629-2_6

International Symposium on Field-Programmable Gate Arrays, under the title "Regularity Matters: Designing Practical FPGA Switch-Blocks" [Nik23], we extend the method by combining it with *Integer Linear Programming* (ILP) so that it produces only those solutions that conform to arbitrary definitions of regularity. The avalanche search method of the previous chapter is a perfect candidate for such extension, since the total usage of switch types included in the switch-pattern constructed by solving the appropriate ILP is a natural maximization objective, capturing the requirements of the router when routing typical circuits. Nevertheless, other ways of assigning importance to different switch types could yield good results as well; for example, the switch-type-importance vector could be derived from some machine learning model or through black-box optimization. Hence, the contents of this chapter can be understood as a general method for constructing regular switch-patterns from a collection of switch types with some measure of importance assigned to them. Moreover, by using edit distance as a minimization objective, the proposed ILP-based techniques could be used to regularize any input architecture. All this makes the present chapter largely orthogonal to the last one, although at times this may not appear to be so. Before describing the method in detail, let us first briefly turn to why switch-pattern regularity is even of interest.

6.1 Who Cares About "Regularity"?

Figure 6.1 shows wire adjacency of two switch-blocks: one from a 7-Series FPGA [Pet21] (Fig. 6.1a) and the other resulting from avalanche search (Fig. 6.1b).

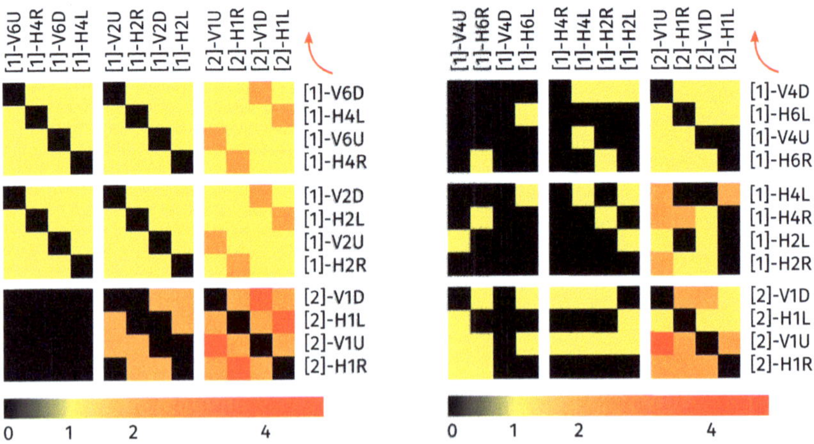

(a) Commercial switch-block.

(b) Result of automated search.

Fig. 6.1 Regularity of a commercial switch-block contrasted with an irregular solution obtained through automated search. Figure (**a**) shows simplified wire adjacency in a Xilinx/AMD 7-Series FPGA [Pet21]. An analogous plot of a switch-block obtained from avalanche search of the previous chapter is shown in figure (**b**). Although wire sets differ between the two switch-blocks and they can thus not be directly compared, a stark difference in "regularity" is readily observed

Although the matrices are not directly comparable as their wire sets differ, one thing immediately draws attention: The commercial switch-block is perfectly "regular," while the one resulting from automated exploration is very "irregular" in comparison. Could it be that imposing regularity is simply too constraining for the switch-block to reach peak performance? Or could there perhaps be other solutions with comparable performance that are also regular? Are there any downsides of irregularity, apart from the evident increase in the already high layout effort needed to produce an FPGA? If significantly different circuits are implemented on the optimized architecture, will it still outperform the regular commercial one, or will it become unroutable? All these are very interesting questions that, to the best of our knowledge, have not been systematically answered until now. In fact, most published search methods would produce highly irregular solutions (see Sect. 6.2) without much regard to whether they can be fabricated while retaining the observed performance gains. In this chapter, we attempt to correct this. First we propose a method for ensuring that the constructed switch-pattern conforms to any arbitrary definition of regularity encodable in terms of ILP constraints. Then, we compare the performance of patterns produced when regularity is imposed and when it is not in order to obtain answers to the above questions.

6.2 Related Work

As we have mentioned in the previous chapter, besides avalanche search [Nik21a], several other methods for automating switch-block exploration have been proposed [Lin10, Qia21], with simulated annealing forming the basis for majority of them. Due to randomized search and/or noise generated by the CAD tools used in the exploration loop, the architectures they produce tend to be highly irregular.

Nevertheless, techniques that enable quick architectural optimization can be very useful, even if the solutions they produce may not be easily fabricated in actual products. A notable example is found in the work of Lemieux et al., which demonstrated that sparse crossbars can bring large area reductions without sacrificing routability [Lem00]. While the key idea of sparsifying the crossbar was carried over to Stratix, for different layout reasons, the actual implemented pattern was considerably more regular [Lew03]. This also enabled simple LUT output swapping by the router without having to reconsider the local routing of the input signals: an interesting example of the case when constraints imposed on the interconnect architecture lead to both an efficient layout and a possibility to improve CAD efficiency.

When an automated search method is not intended to test a groundbreaking hypothesis, such as the one in the work of Lemieux et al., but to further optimize architectures from a known design space, inability to easily fabricate the solutions may question the actual utility of the method. Unfortunately, all of the papers cited at the beginning of this section propose methods that fall in this second category. In this chapter, we build on top avalanche search of Chap. 5—the most recent of the aforementioned algorithms—to produce practical switch-blocks.

6.3 Summary of Avalanche Search

In this section, we briefly review the avalanche search algorithm once more, presenting it in a slightly different manner than in the last chapter, so that the modifications introduced here can be distinguished more clearly. The goal of the avalanche search algorithm is to select a subset E_a of manufacturable switch types E, which connect wire types from a fixed set V, such that E_a is as small as possible but still allows the router to route all circuits of interest, while meeting timing constraints. The switch types entering E_a constitute a *switch-pattern* which is repeated in every FPGA tile, next to every LUT [Lew13]. Each switch type from E is designated by an ordered triplet (u, v, d_L), where $u \in V$ is the driving wire type, $v \in V$ the driven wire type, and d_L the distance between the end of the driving wire and the start of the driven wire in terms of the number of LUTs (see Fig. 5.3).

The algorithm is outlined in Fig. 6.2. Since in its extended version we will solve an ILP to construct the switch-pattern, we shall henceforth use SOL in place of E_a; this will make the essential differences between the two algorithms clearer. Avalanche search iteratively builds a switch-pattern by (1) routing circuits and observing in how many switch-blocks each switch type was used (*usage*, $U((u, v, d_L))$; box ②) and (2) extending the pattern by a certain number of most used switch types (box ⑤), until there are no more switch types used by the router which are not already in the pattern (boxes ③ and ④). To be able to measure usage of switch types that are not yet part of the pattern, each switch type that could be fabricated is represented in the routing-resource graph of box ②. These switch types, however, are represented with a high initial cost which drops in proportion to usage, whereas the switch types that are part of the pattern are assigned zero cost and their usage is no longer measured. As a result, the set of all switch types used by the router (*tentative solution*, $tSOL$, in box ③) slowly converges to the growing set of switch types accepted into the pattern (*partial solution*, $pSOL$).

6.4 Regularization Algorithm

The algorithm of Fig. 6.2 provides no mechanism for controlling the structure of the solution. We now extend it to impose arbitrary constraints while retaining tight coupling with the router.

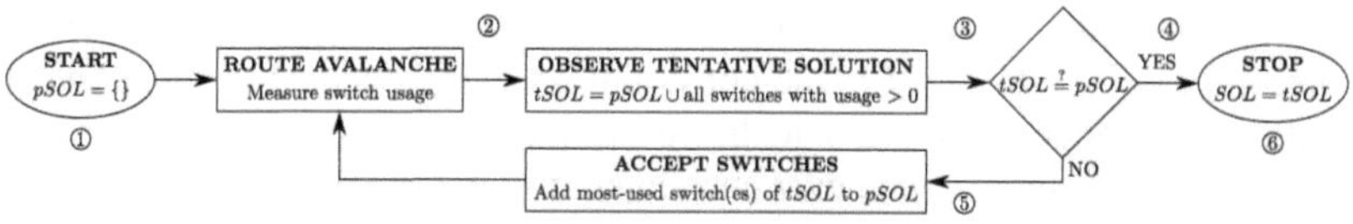

Fig. 6.2 Avalanche search algorithm of the previous chapter

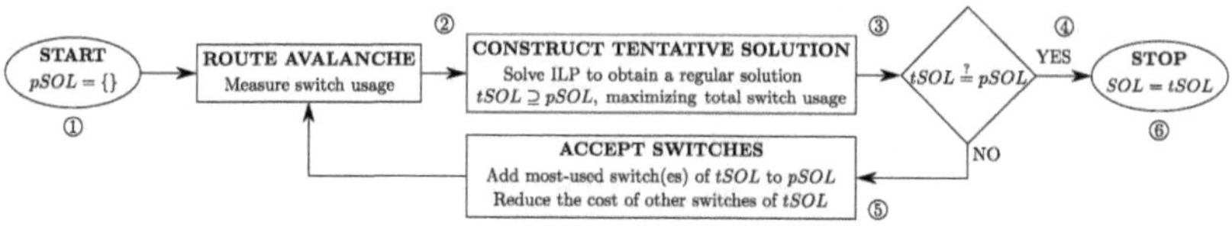

Fig. 6.3 Enhancements to the avalanche search algorithm, which enable enforcing arbitrary regularity constraints on the obtained solution

6.4.1 General Flow

The extended algorithm is shown in Fig. 6.3. It retains the structure of Fig. 6.2, with two important modifications. First, the tentative solution, $tSOL$, is not formed by merely observing which switch types were used by the router. An ILP problem is constructed instead, such that all of its feasible solutions satisfy all specified regularity constraints. Maximizing affinity of the router toward the switch types that enter $tSOL$, represented by their usage, increases the likelihood that the final solution is routable and appropriate for critical path optimization of the routed circuits.

Second, the selection of the most used switch types from $tSOL$ which enter the final pattern in box ⑤ is identical to that of Fig. 6.2, but the initial avalanche cost of the remaining switch types in $tSOL$ is set to a lower value in the next iteration of the algorithm. The rationale is as follows: In the first iteration, the router uses switches unaware of the constraints imposed on the switch-pattern. This may result in the set of used switch types being very far from meeting the constraints. The ILP solver legalizes the solution, respecting the decisions of the router as much as possible by maximizing the total usage. In the second iteration, those switch types that are known to be part of at least one legal solution ($tSOL$) are offered to the router at a reduced price, making it less likely for the router to violate the constraints again. Over time, the router and the ILP solver converge toward a common solution.

In order to guarantee that the algorithm ends, once a switch type enters the partial solution, $pSOL$, it is never removed from it; this is the same as in Fig. 6.2. Moreover, since $pSOL$ is always extended only by the switch types from $tSOL$, produced by the ILP solver and hence legal by construction, the algorithm is always guaranteed to end with a solution that satisfies all of the imposed constraints. It could happen, however, that the final solution does not encompass all of the switch types that the router used in the last iteration, because adding them to $tSOL$ would make it violate the constraints. These switch types may either slightly improve the delay of some (near-)critical paths or be necessary for complete removal of congestion. The second case would result in the constraint set being deemed *unsatisfiable* and it would have to be relaxed. We note, however, that in all experiments performed in preparation of this chapter, we never encountered such a situation.

6.4.2 Base ILP Problem

We first describe a skeleton ILP problem which completes the algorithm of Fig. 6.3. Different types of regularity constraints will be added to it in subsequent sections.

Given a switch type e connecting a wire type u to a wire type v at a LUT distance d_L, we designate its presence in the switch-pattern by the following binary variable:

$$x_{u,v,d_L} \in \{0, 1\}, \qquad \forall (u, v, d_L) \in E. \tag{6.1}$$

The corresponding switch type is part of the switch-pattern iff the variable is 1. To specify that the accepted switch types must be part of $tSOL$, we simply set the appropriate presence variables to 1:

$$x_{u,v,d_L} = 1, \qquad \forall (u, v, d_L) \in pSOL. \tag{6.2}$$

As mentioned in Sect. 6.4.1, the basic objective function strives to maximize the total usage of the switch types entering the solution:

$$\max \sum_{(u,v,d_L)\in E} U((u, v, d_L))x_{u,v,d_L}. \tag{6.3}$$

The usage of each switch type is observed from the router and is thus a constant in the ILP problem. To prevent selection of all switch types in the absence of further constraints, each ILP imposes an upper bound M on switch-pattern size:

$$\sum_{(u,v,d_L)\in E} x_{u,v,d_L} \leq M. \tag{6.4}$$

6.5 Experimental Setup

All of the subsequent sections that progressively introduce different kinds of regularity constraints share the experimental setup of Sect. 5.9. The only additional parameter introduced in Sect. 6.4 is the cost reduction factor for switch types participating in the ILP solution ($tSOL$) that have not yet been accepted into the partial solution ($pSOL$). We fix this to 0.9, which was experimentally determined to yield slightly better results than other tested values. A more thorough exploration of how this parameter impacts the quality of the obtained patterns is due.

In order to suppress the effects of experimental noise that could lead to false conclusions about which trade-offs a certain set of constraints brings, we need to compare sets of feasible architectures, rather than single points. To this end, we leverage the well-known fact that permuting the order in which the nets of a circuit are routed can have a significant impact on the outcome of the routing process [Rub11, Zha22] and hence also on the usage of different switch types. Permutation is achieved by performing 100 random swaps in the default-sorted netlist [Mur20]. For each constraint set, we construct five different architectures, by permuting the netlist using five different random number generator seeds.

The switch-pattern size is bounded by 96 in all experiments, which allows solutions to be found for all constraint types introduced in the subsequent sections, yet, does not make the patterns excessively larger than the ones produced by pure avalanche search in the previous chapter. We again note that our definition of switch-patterns includes only the switches between the channel wires and not the switches connecting CLB outputs to them; these are added on top of the budget of 96 switch types, as described in Sect. 4.7.6.

6.6 Limiting Multiplexer Size Variation

The first form of regularity apparent in industrial architectures is the very limited number of multiplexer sizes in their switch-patterns—often only one [And06, Pet21, Wol23a]. Limiting the number of different multiplexer sizes is a very logical step when full-custom layout is employed in implementing the FPGA. This could be especially useful for quick customization of programmable interconnect for different applications, discussed in Chap. 2. Similarly to implementing a design on an MPGA, a custom FPGA could be obtained by customizing only several layers of metal while retaining the common FEOL and the lowest few BEOL layers that correspond to a pre-laid-out collection of multiplexers. Reduction of the design and fabrication cost that could result from this is likely to be even higher with the rise of chiplet adoption [San23]. Additional reasons for multiplexer size uniformity observed in current architectures likely lie in the efficiency with which multiplexers of certain sizes can be laid out [You98, Pet21], as well as in increased routability and reduced router runtime. In this section, we investigate the benefits and downsides of limiting the number of multiplexer sizes. When the size of the multiplexer driving each wire v is known in advance to be some constant $\varphi(v)$, extending the ILP problem of Sect. 6.4.2 to respect this size distribution is trivial:

$$\sum_{u,d_L:(u,v,d_L)\in E} x_{u,v,d_L} = \varphi(v), \qquad \forall v \in V. \tag{6.5}$$

However, we would like to measure the impact of limiting the number of multiplexer sizes in general, without a priori assigning a size to any particular multiplexer, since this unrelated decision could influence the conclusions.

6.6.1 Encoding

Given a maximum allowed size for any multiplexer, M_φ, for each size $\varphi \in [0, M_\varphi]$, and each wire v, we introduce another binary variable $x_{v,\varphi}$, which is 1 iff the size of v's fanin is φ:

$$\sum_{u,d_L:(u,v,d_L)\in E} x_{u,v,d_L} = \sum_{\varphi\in[0,M_\varphi]} \varphi x_{v,\varphi}, \qquad \forall v \in V, \tag{6.6}$$

$$\sum_{\varphi\in[0,M_\varphi]} x_{v,\varphi} = 1, \qquad \forall v \in V. \tag{6.7}$$

For each allowed multiplexer size φ, we introduce another binary variable, x_φ, indicating that there is at least one wire in the switch-pattern driven by a multiplexer of that size:

$$x_\varphi = \bigvee_{v\in V} x_{v,\varphi}, \qquad \forall \varphi \in [0, M_\varphi]. \tag{6.8}$$

For each φ, the above disjunction is linearized in a standard way [Wil13]:

$$\sum_{v\in V} x_{v,\varphi} \geq x_\varphi, \tag{6.9}$$

$$x_{v,\varphi} \leq x_\varphi, \qquad \forall v \in V. \tag{6.10}$$

Finally, we need to limit the number of different multiplexer sizes present in the solution to the desired constant N_φ:

$$\sum_{\varphi\in[0,M_\varphi]} x_\varphi \leq N_\varphi. \tag{6.11}$$

If distribution of a specific set of sizes Φ (e.g., $\Phi = \{6, 20, 25\}$ [Pet21]) is sought, $[0, M_\varphi]$ is simply replaced by Φ in the above equations.

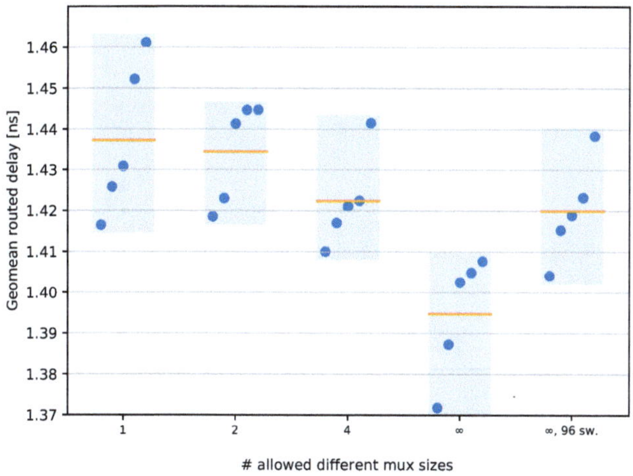

Fig. 6.4 Geomean critical path delays after routing the MCNC benchmarks. Each point represents one architecture. Average delay of architectures within the same constraint set is marked by an orange horizontal line. Blue rectangles are used merely for visual separation of architectures obtained for different constraint sets. Unconstrained architectures (∞) clearly have a performance advantage

6.6.2 Results

We investigate the difference in performance between architectures with up to one, two, four, or any number of arbitrary different multiplexer sizes up to 20:1 ($N_\varphi \in \{1, 2, 4, \infty\}$, $M_\varphi = 20$).

Critical Path Delay Figure 6.4 shows the critical path delays when routing MCNC circuits on the obtained architectures. Each point is a geometric mean of 17 circuits, and each circuit is in turn represented by the median over five different placements. Although there are significant differences between architectures within each group, we can observe that allowing two rather than a single multiplexer size provides slightly more stable results, while increasing the number to four leads to a clearer advantage. Architectures obtained when the number of multiplexer sizes is not bounded clearly outperform the others, with a close to 3% average critical path delay reduction over the case when only one size is allowed. This demonstrates that there is a performance advantage to be reaped when it is possible to tailor the multiplexer sizes specifically to the needs of the circuits of interest.

Architectures labeled as "∞, 96 sw." also do not limit the number of multiplexer sizes, but instead of considering $M = 96$ an upper bound on switch-pattern size, they all have exactly 96 switch types. In comparison, the ∞-labeled architectures have 72–86 switch types. Higher capacitive loading on wires largely negates the advantage of freely choosing multiplexer sizes.

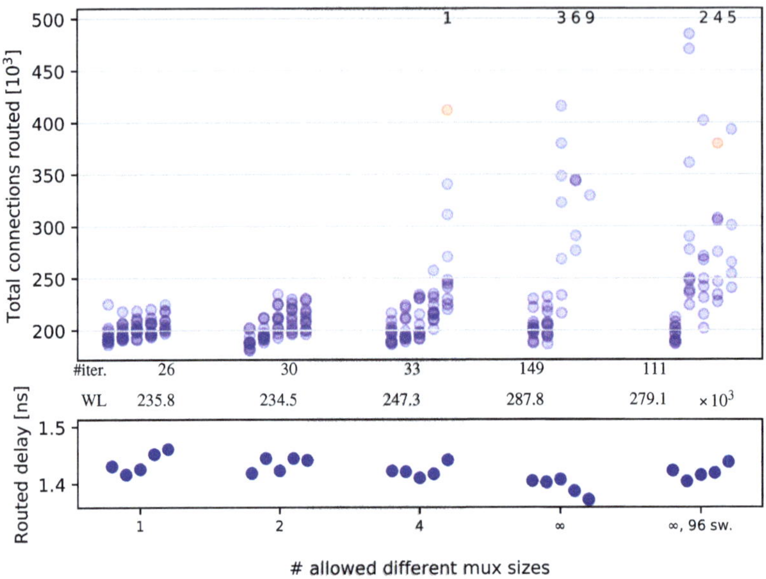

Fig. 6.5 Routability on Gnl circuits. Each point represents the total number of connections that were routed (including rip-ups) when routing a single Gnl circuit on particular architecture. Circuits that failed to route are shown in red. Since for many of the unroutable circuits more connections were rerouted than can fit in the plot, the number of failed circuits for each architecture is written at the top. All numbers are single digit (there were ten circuits in total). The plot below shows the average critical path delay on the MCNC circuits for the particular architecture as a reference (see the plot of Fig. 6.4). While unconstrained architectures were better optimized to achieve good performance on circuits from the set used to construct them, they struggle to route significantly more complex circuits. Multiplexer size regularity has a clear advantage in routability of more complex circuits

Routability Only three circuits were used in the search (see Sect. 5.9), yet all 17 circuits on which performance comparisons are based were routable. Nevertheless, the remaining 14 circuits come from the same benchmark set, so it is of interest to see how routable the patterns are when routing circuits from a more complex set. To this end, we attempt routing ten 10,000 LUT circuits with a Rent's exponent of 0.7, generated by Gnl [Str99, Nik21a]. Results of these experiments are shown in Fig. 6.5. Since modern routers, including VTR-8, are incremental [Mur20], we plot the total number of routed connections taken to complete the routing of each circuit on each architecture, as we believe that this is a more telling metric of the difficulty which the router faces than the traditionally more common iteration count. Geometric means of iteration count and wirelength of all circuits, averaged over all five architectures of the given constraint set, are shown in rows labeled as "#iter." and "WL," respectively. There is little difference in the number of connections that have to be routed between the architectures which have one or two multiplexer sizes. However, as soon as N_φ increases to four, there is a very clear drop in routability,

Table 6.1 Generalization

N_φ	Delay [ns]	# iter.	# routed con. [10^3]	WL [10^3]
Exploration circuits				
1	1.143	29	15.82	6.35
2	1.147	29	15.97	6.08
4	1.133	29	14.68	5.86
∞	1.113	29	16.72	6.01
∞, 96 sw.	1.131	29	16.22	6.00
Other MCNC circuits				
1	1.510	30	63.02	22.53
2	1.505	30	63.87	21.42
4	1.493	31	61.16	20.83
∞	1.464	31	66.50	21.48
∞, 96 sw.	1.491	31	68.10	21.50
Gnl circuits				
1		26	202.14	235.84
2		30	204.37	234.46
4		42	219.86	247.25
∞		149	581.91	287.80
∞, 96 sw.		111	343.75	279.13

with one architecture even failing to route all circuits. This trend further increases when the number of multiplexer sizes is unbounded, regardless of the switch-pattern size constraint: In both cases, three of the five unbounded architectures fail to complete some of the circuits within 300 iterations.

Table 6.1 shows the average routability metrics of architectures obtained for different constraint sets on circuits split into three groups: (1) those used in exploration, (2) the remaining circuits from the MCNC set, and (3) the more complex circuits from the Gnl set. Failed circuits enter the geometric mean for given architecture with 300 iterations, which was the limit used in the experiments. We can observe that the highly delay-optimized architectures with no constraint on multiplexer size count are somewhat less routable (albeit with a competitive wirelength) even on the circuits used for exploration. This trend is maintained on the remaining MCNC circuits, while the advantage of the regular architectures becomes really significant on the larger Gnl circuits. It seems quite clear that having uniform multiplexer sizes makes it much more likely that the architecture will support circuits with characteristics not captured by those used during the search. This is an important aspect for FPGA vendors, as it is often better to give up some of the potential performance benefits that the less regular architecture could bring, in order to make it significantly more general.

6.7 Limiting Fanout Size Variation

In the previous section, the number of different multiplexer sizes that occur in the switch-pattern was bounded, but there were no constraints on the fanout size variation. However, large variations in fanout size may make it more difficult to optimize the critical path delay due to differences in capacitive loads on wires, unless the router is able to effectively exploit them [Chr20]. We now assess the effect of constraining fanout and multiplexer size count together. Constraints that are needed for this are dual to the ones of Sect. 6.6.1, with edges being listed with respect to the tail node.

6.7.1 Results

All experiments are set up exactly as in Sect. 6.6. We bound the number of multiplexer and fanout sizes by the same constant.

Critical Path Delay Influence of regularizing fanout sizes on routed critical path delay is shown in Fig. 6.6. More balanced capacitive loads and more options that the router has to avoid congestion bottlenecks lead to improved results when the number of multiplexer sizes is very limited ($N_\varphi \in \{1, 2\}$). However, this improvement is still not sufficient to make the regularized architectures outperform the unconstrained ones, where the router can freely select the switch types that form the pattern.

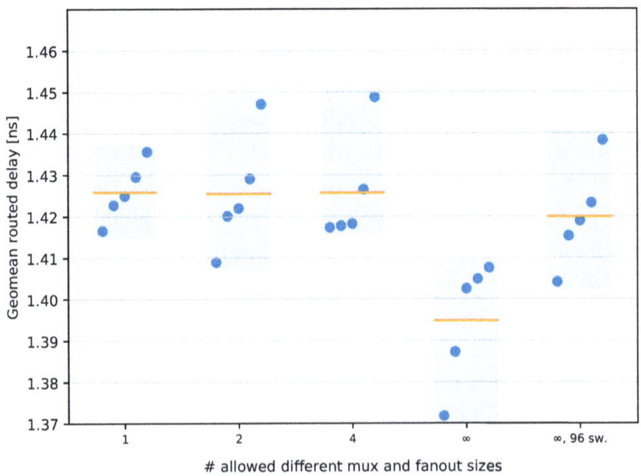

Fig. 6.6 Influence of bounding the number of different multiplexer and fanout sizes on the critical path delay of the implemented MCNC circuits. Regularizing fanout sizes balances capacitive loads and reduces the negative impact of multiplexer size regularization on delay

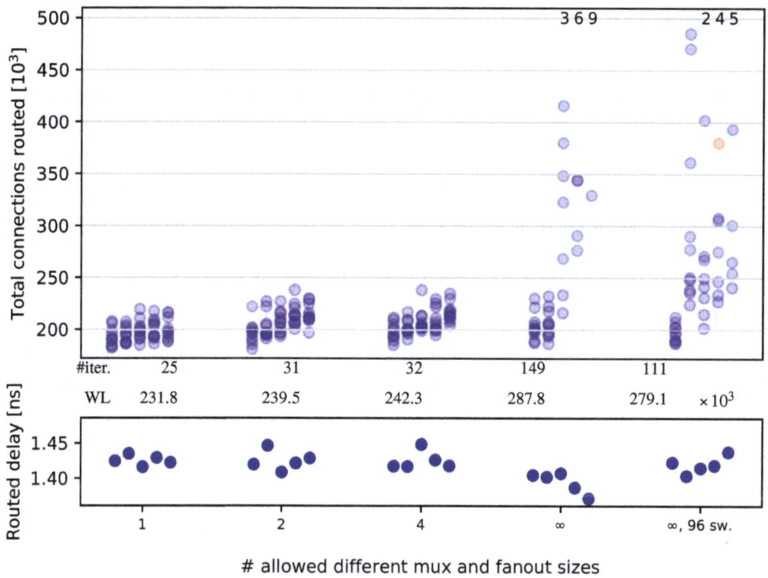

Fig. 6.7 Influence of bounding the number of different multiplexer and fanout sizes on routability. Regularizing fanout sizes further improves it

Routability Influence that regularizing fanout sizes has on routability is shown in Fig. 6.7. When compared with Fig. 6.5, we can see that regularization of fanout sizes contributes to additional improvement of routability. This is the most apparent when the number of sizes is bounded by four, where now all architectures manage to route all circuits, but it also brings improvement to the architectures with completely uniform fanins. Balancing the fanout sizes likely helps because it removes the situation when certain wires with a comparatively large fanout get too congested as others do not provide enough possibilities to complete the paths.

6.8 Multiplexer Input Sharing

Certain commercial FPGAs employ switch-patterns where pairs of multiplexers share a large fraction of inputs. For example, Young showed that forming a pair of two 6:1 multiplexers that share 5/6 inputs allows for significant area gains due to diffusion sharing [You98, Pet21]. Similarly, Chromczak et al. state that "adjacent muxes aggressively share input pins" to reduce vias in Intel Agilex FPGAs [Chr20]. In this section, we measure the impact of this approach.

6.8.1 Encoding

First we track the number of inputs shared by each pair of wires, v_1 and v_2,

$$S_{v_1,v_2} = \sum_{u,d_L:(\forall v \in \{v_1,v_2\})((u,v,d_L)\in E)} x_{u,v_1,d_L} \wedge x_{u,v_2,d_L}, \quad \forall v_1, v_2 \in V, \tag{6.12}$$

where S_{v_1,v_2} is an additional variable that counts the number of inputs shared between the wires v_1 and v_2. To linearize the conjunction, we need to introduce another binary variable x_{u,v_1,v_2,d_L} for each pair of wires v_1 and v_2 and each input switch that they could share (each addend of the above sum). The conjunction is then linearized in a standard way by the following constraints [Wil13]:

$$x_{u,v_1,v_2,d_L} \geq x_{u,v_1,d_L} + x_{u,v_2,d_L} - 1, \tag{6.13}$$

$$x_{u,v_1,v_2,d_L} \leq x_{u,v,d_L}, \quad \forall v \in \{v_1, v_2\}. \tag{6.14}$$

Next, for each pair of wires v_1 and v_2, we introduce a binary variable F_{v_1,v_2} which, when 1, will force v_1 and v_2 to share at least ξ inputs, where ξ is the specified constant:

$$\xi F_{v_1,v_2} \leq S_{v_1,v_2}, \quad \forall v_1, v_2 \in V. \tag{6.15}$$

Finally, we need to partition the wires into pairs, by specifying that each wire is forced into exactly one pair:

$$\sum_{v_2 \in V} F_{v_1,v_2} = 1, \quad \forall v_1 \in V. \tag{6.16}$$

6.8.2 Results

In these experiments, we force all multiplexers to take six inputs from other wires and sweep the minimum sharing bound ξ between 0 and 5. Since there are 16 wires in the switch-pattern, the total number of switches is exactly 96, and the multiplexer and fanout sizes correspond to the uniform case ($N_\varphi = 1$) of Sect. 6.7. The multiplexers are not exactly 6:1 like those of Young [You98], since each takes two additional inputs from the LUTs, as was described in Sect. 4.7.6.

Critical Path Delay Influence of input sharing between pairs of multiplexers on critical path delay of routed circuits is shown in Fig. 6.8. No significant critical path delay increase is apparent. In fact, there even appear to be some gains from applying these constraints, likely because wiring load is reduced. This can be observed in

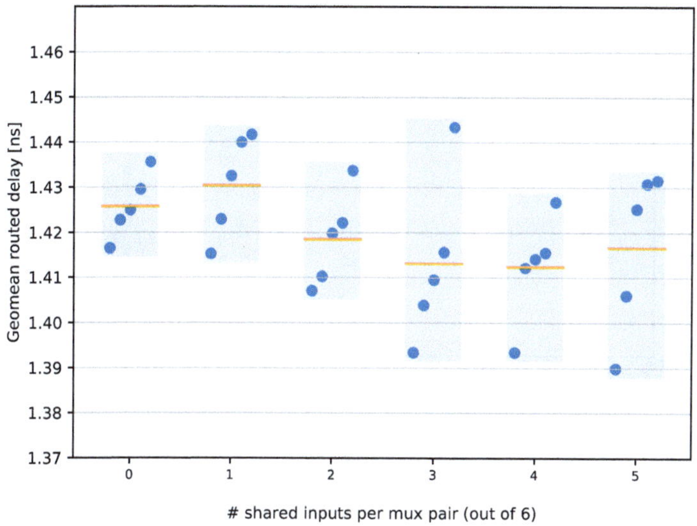

Fig. 6.8 Influence of sharing inputs between pairs of multiplexers on critical path delay. Input sharing shows no significant negative impact on routed delay; it sometimes even slightly improves it

Table 6.2 Average total wire delay with respect to the number of inputs shared

# inputs shared	0	1	2	3	4	5
t [ps]	223.2	221.6	220.6	220.4	220.4	218.6
Δ [%]	0.00	−0.72	−1.16	−1.25	−1.25	−2.06

Table 6.2 which contains average total wire delays for architectures with different amounts of input sharing. It is important to note that the models that we use (see Chap. 4) are not capable of accounting for area reduction obtained through diffusion sharing or for the reduced via use, which were the primary reasons for input sharing stated by Young [You98] and Chromczak et al. [Chr20]. Since input sharing appears to make no negative impact on the routed critical path delays, even with the conservative models that only take into account reduction in length of wires feeding multiplexer inputs, we hypothesize that FPGA vendors are reaping significant gains from this technique when its impact on layout is fully taken into account.

Routability One could expect that sharing majority of inputs between pairs of multiplexers may cause a drop in routability. However, the results of Fig. 6.9 show that this is not the case. Combining this observation with the expected performance gains, area reduction, and easier layout that were previously discussed, we conclude that input sharing is likely a very effective optimization technique.

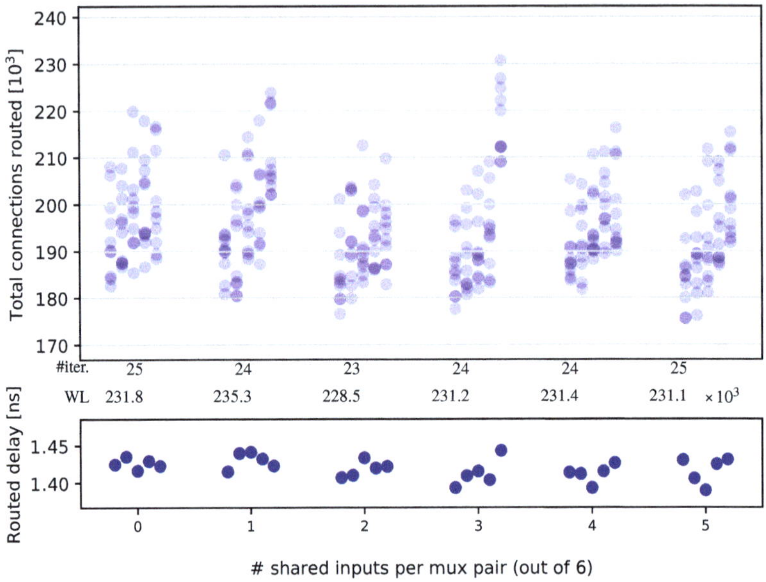

Fig. 6.9 Influence of sharing inputs between pairs of multiplexers on routability. Input sharing shows no negative impact on routability

6.9 Minimizing Wirelength

The only objective in the ILP problems that we have introduced so far was the total switch type usage. However, it may happen that some of the switch types are added to the switch-pattern solely to make it regular and that they otherwise have very low and even zero usage. In fact, in the experiments presented here, due to small size of the circuits on which the search is conducted, it was not uncommon to observe >20 of the 96 switch types entering the pattern exclusively for regularization reasons. In such cases, it seems reasonable to have a complementary objective to guide regularization. A good candidate for this is minimizing the total length of wires that provide inputs to the multiplexers within the switch-block. In this section, we describe how that objective can be modeled and what impact it has on performance and routability of the obtained solutions.

6.9.1 Encoding: Modeling Wirelength

Without loss of generality, we assume that all multiplexers are of the same size and that their physical width and height are α and β, respectively. For the solving approach discussed shortly, which formulates and solves multiple ILP problems with fixed multiplexer placement, relaxing this constraint is trivial. The same-size multiplexers form a uniform grid of n columns and m rows. We represent the placement of the different multiplexers in a similar way as Mihal and Teig [Mih13]. For each wire $v \in V$ and each location on the uniform grid (x, y), a binary variable $x_{v,(x,y)}$ indicates that the multiplexer driving v is placed at location (x, y). The following constraints make sure that each multiplexer is assigned a unique location:

$$\sum_{x,y} x_{v,(x,y)} = 1, \qquad \forall v \in V. \tag{6.17}$$

Overlaps between multiplexers are removed as follows:

$$\sum_{v \in V} x_{v,(x,y)} \leq 1, \qquad \forall (x, y). \tag{6.18}$$

We describe the total wirelength as the sum of Manhattan lengths of individual switch types:

$$\sum_{\substack{(u,v,d_L) \in E \\ x',x'' \in [0,n) \\ y',y'' \in [0,m)}} \lambda(x', y', x'', y'', d_L)(x_{u,v,d_L} \wedge x_{u,(x',y')} \wedge x_{v,(x'',y'')}), \tag{6.19}$$

where $\lambda(x', y', x'', y'', d_L)$ is a precomputed constant Manhattan distance between locations (x', x'') and (y', y''). The conjunction is linearized as in Sect. 6.8.1. To reduce solution time, instead of allowing the ILP solver to assign the locations of the multiplexers, we fix them a priori, then solve the simpler ILP problems, and reiterate the process several times (five in the experiments presented here), optimizing the placement between iterations using simulated annealing. During placement optimization, switch type selection is considered fixed, as determined by the ILP solution. We use the annealing schedule of Betz et al. [Bet99]. Tighter formulation of multiplexer placement—essentially a *Quadratic Assignment Problem* with Manhattan distance—is possible [Law63, Gue15] and could potentially allow the ILP solver to simultaneously select switch types and place multiplexers. However, this exceeds the scope of the present work. We note that multiplexer position optimization is performed in every iteration of the algorithm of Fig. 6.3, prior to SPICE simulations (Sect. 5.8.1.2). Additional optimization performed here is only intended to allow the ILP solver to form better decisions with respect to wirelength minimization and does not constitute an advantage in its own right.

6.9.2 Encoding: Combined Objective

To combine wirelength minimization with usage maximization, we adapt the *auto-normalizing* function of Marquardt et al. [Mar00]:

$$\min\left(\Lambda\frac{\text{total WL}}{\text{previous total WL}}-(1-\Lambda)\frac{\text{total U}}{\text{previous total U}}\right). \tag{6.20}$$

Objectives *total WL* and *total U* are determined by Eqs. (6.19) and (6.3), respectively, while *previous total WL* and *previous total U* designate the total wirelength and usage of the ILP solution from the previous iteration. Before solving the first ILP problem, a greedy solution is formed by selecting the M (upper bound on switch-pattern size, see Sect. 6.4.2) most used switch types, from which the initial *previous total WL* and *previous total U* are determined. Constant Λ sets the *wirelength trade-off*.

6.9.3 Results

In this section we present the results of experiments where all wires were constrained to have the same fanout (six, disregarding the connection block) and the same fanin (six, disregarding the LUT drivers). No other constraints were imposed, and the wirelength trade-off constant, Λ, was swept from 0 (wirelength completely disregarded) to 1 (usage completely disregarded).

Critical Path Delay Figure 6.10 shows the impact of minimizing wirelength on critical path delay of the implemented circuits. We see by comparing the edge cases ($\Lambda = 0$ and $\Lambda = 1$) that when the multiplexers are reasonably large and fanouts and fanins are uniform, which helps eliminate any pathological solutions, optimizing wirelength is about as important as respecting the affinity of the router toward certain switch types. The best results arise when both objectives are combined, however. The exact value of the trade-off constant Λ does not seem to matter too much, but the plot suggests that setting it to 0.5 yields the best results.

Routability Figure 6.11 indicates that taking wirelength into account also slightly improves routability. One hypothesis is that better distribution of capacitive loads alters the base costs of routing resources assigned by VPR [Mur20], thus reducing congestion. Since the differences are not as obvious as with delays, we did not attempt to put this hypothesis to test yet.

6.10 Enforcing Turns and Symmetries

So far, all the constraints that we have presented were motivated by layout considerations. In this section we investigate several other constraint types that should not directly impact the ease with which the switch-blocks can be laid out, but they do reflect features commonly observed in both industrial and academic architectures and may impact CAD tool performance.

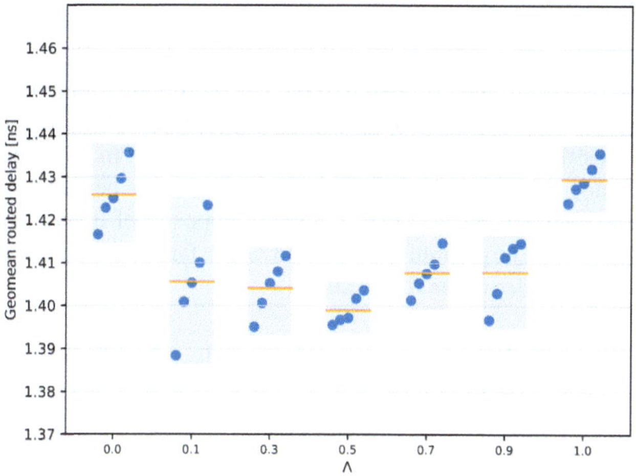

Fig. 6.10 Influence of wirelength optimization on critical path delay. $\Lambda = 0$ (=1) only maximizes usage (minimizes wirelength). Joint optimization of both objectives yields best results

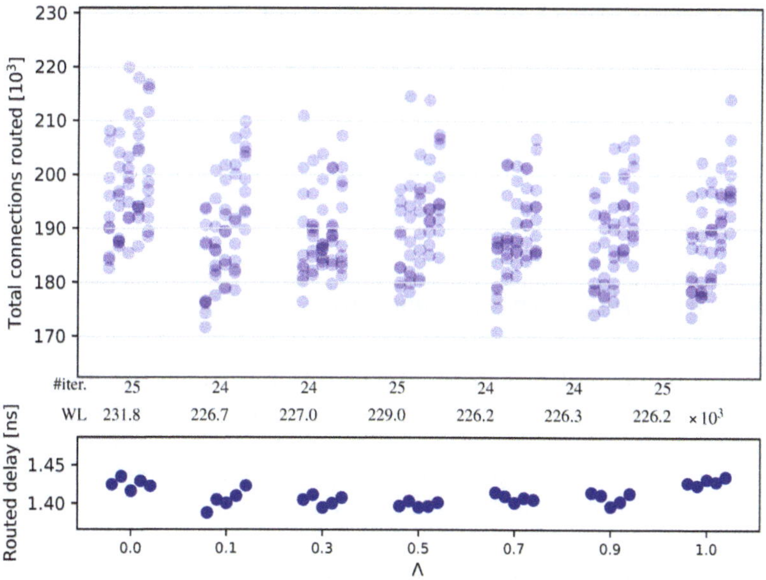

Fig. 6.11 Influence of wirelength optimization on routability. Joint optimization of wirelength and usage also leads to a slight increase in routability

6.10.1 Encoding: Turns

The first set of constraints is inspired by the assumption that has been very popular among the academic architectures since it was first formulated by Rose and Brown [Ros91]: namely, that each wire should fan out to one wire on each of the tree remaining sides of the switch-block; the two that are perpendicular to where the wire came from and the one that is directly opposite of it. We generalize this constraint to allow for larger multiplexers and larger fanouts, by turning the *exactly one* requirement to *at least one* requirement:

$$\sum_{v,d_L:(u,v,d_L)\in E \wedge DIR(v)=f_L(DIR(u))} x_{u,v,d_L} \geq 1, \qquad \forall u \in V, \tag{6.21}$$

$$\sum_{v,d_L:(u,v,d_L)\in E \wedge DIR(v)=f_R(DIR(u))} x_{u,v,d_L} \geq 1, \qquad \forall u \in V, \tag{6.22}$$

$$\sum_{v,d_L:(u,v,d_L)\in E \wedge DIR(v)=DIR(u)} x_{u,v,d_L} \geq 1, \qquad \forall u \in V, \tag{6.23}$$

$$f_L = \begin{pmatrix} R & L & U & D \\ U & D & L & R \end{pmatrix}, \qquad f_R = \begin{pmatrix} R & L & U & D \\ D & U & R & L \end{pmatrix}. \tag{6.24}$$

Equations (6.21), (6.22), and (6.23) specify that each wire must turn left, turn right, and continue in the same direction, respectively. Function $DIR(v)$ returns the direction of wire v, while f_L and f_R are mappings between directions that constitute left and right turns.

6.10.2 Encoding: Fanout Symmetries

The second set of constraints forces wire fanouts to be symmetric. There are two kinds of symmetry that we explore. The first one that we call *internal*, illustrated in Fig. 6.12a, specifies that whenever a wire u drives a wire v perpendicular to it, it must also drive a wire v' going in the opposite direction from v. We call the second kind *external*, and it is illustrated in Fig. 6.12b. It specifies that two wires that differ only in direction, such that they are opposing, must have identical sets of perpendicular wires to which they fan out and that the sets of parallel wires they fan out to must differ only in direction (i.e., the directions of the respective wires must be opposing). Both internal and external symmetries are trivially encoded by equating the switch-type-presence variables of symmetric pairs of switch types (e.g., $x_{H2La,V1Ua} = x_{H2La,V1Da}$ in Fig. 6.12a or $x_{H2La,V1Ua} = x_{H2Ra,V1Ua}$ in Fig. 6.12b). In the interest of space, we do not present them in their general form.

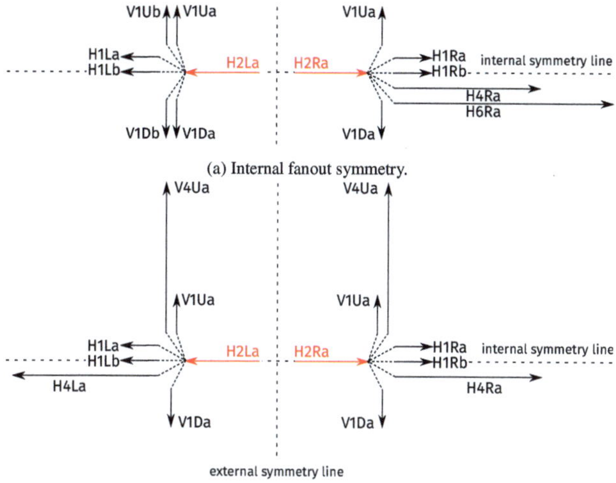

(a) Internal fanout symmetry.

(b) External fanout symmetry.

Fig. 6.12 Fanout symmetry illustration. Fanouts of the *H2Ra* and *H2La* wires (red) are shown for two different patterns. (**a**) Each of the wires has a symmetric fanout: There is a mirror symmetry between the vertical wires in the fanout, but the fanouts of the two wires are not mutually symmetric. (**b**) The two wires have mutually symmetric fanouts: The set of vertical wires they fan out to is identical, while there is mirror symmetry between the horizontal wires in the fanout. However, neither of them has an internally symmetric fanout

6.10.3 Results

All experiments in this section were conducted using an objective that combines usage maximization and wirelength minimization with a trade-off of 0.5, since this was previously shown to yield the best results. Fanins and fanouts of all wires were fixed at 6 (disregarding LUT-related ones), and no additional constraints were applied other than the ones indicated in the respective plot.

Critical Path Delay As shown by Fig. 6.13, enforcing any of the constraints from this section brings no benefit to the critical path delay and even slightly deteriorates it.

Routability Figure 6.14 shows that most constraints do not bring a tangible improvement of routability either. Hence, we conclude that there is no benefit in explicitly enforcing them, unless maybe when CAD tools specifically assume that they are satisfied.

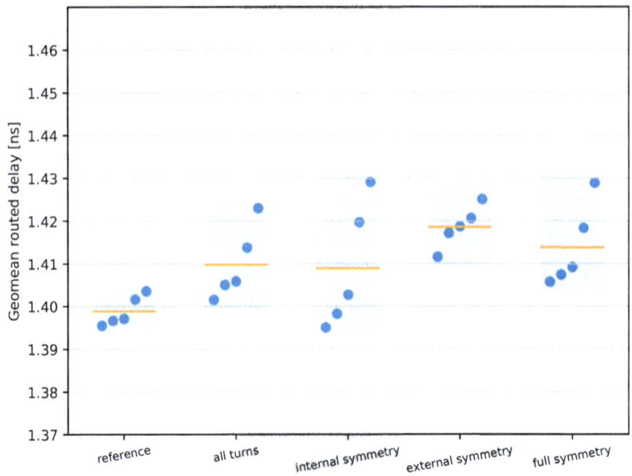

Fig. 6.13 Influence of enforcing different topological features on delay. Enforcing any of these features leads to a deterioration in performance

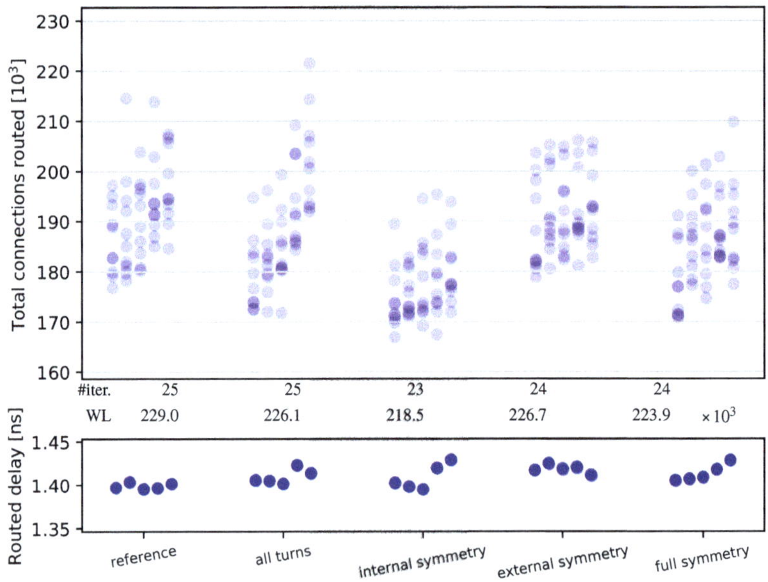

Fig. 6.14 Influence of enforcing different topological features on routability. Enforcing most of the considered features brings little change

6.11 Enforcing Hop-Distance Optimality

The final set of constraints that we consider is also related solely to the performance of CAD tools and was inspired by our previous observation in Sect. 5.10.2.2 that performance of a switch-pattern might be related to the lower bound on the number of hops needed to connect any two locations on the FPGA grid. It could be tempting to search for only those solutions for which this lower bound is the same as if all switch types were present in the pattern.

6.11.1 Encoding: Proof Grid

We start by computing the length of the shortest path from the center of the FPGA grid to all locations that are at most 12 tiles away from it horizontally and at most 8 tiles away from it vertically, when all switch types are present in the switch-pattern. The dimensions are dictated by twice the length of the longest horizontal (M_H) and vertical (M_V) wires, respectively (Fig. 6.1b). Without attempting a formal proof, we assume that proving that the same shortest path lengths are attainable by a pattern SOL on this grid suffices to prove that they are attainable by it on an infinite grid, which would render it *hop-distance optimal*. We experimentally confirmed hop-distance optimality of all solutions on a 100×100 grid.

6.11.2 Encoding: Shortest Paths

For the switch-pattern to be optimal, for each offset (x_δ, y_δ), the length of the shortest path connecting the tiles at this offset must equal the value precomputed above, $L(x_\delta, y_\delta)$. Hence, we need to encode an existence of a path of length $L(x_\delta, y_\delta)$ such that the sum of offsets of all horizontal (vertical) wires on it equals x_δ (y_δ). We do this similarly to how Hamiltonian Path can be encoded in SAT [Pap94]: For each position on the path, $p \in [1, L(x_\delta, y_\delta)]$, and each wire v, a binary variable $x_{(x_\delta,y_\delta),p,v}$ is 1 iff v is the pth node on the path. Each position on the path must be occupied by exactly one wire:

$$\sum_{v \in V} x_{(x_\delta,y_\delta),p,v} = 1, \qquad \forall p \in [1, L(x_\delta, y_\delta)]. \tag{6.25}$$

For two wires to be on consecutive positions in the path, there needs to be a switch between them:

$$x_{(x_\delta,y_\delta),p,u} \wedge x_{(x_\delta,y_\delta),p+1,v} \leq \sum_{d_L} x_{u,v,d_L},$$
$$\forall p \in [1, L(x_\delta, y_\delta)), \forall u, v \in V. \tag{6.26}$$

It only remains to sum up the horizontal wire offsets along the path and force them to x_δ (vertical offsets are analogous):

$$\sum_{p\in[1,L(x_\delta,y_\delta)],v\in V} \chi(v)x_{(x_\delta,y_\delta),p,v} = x_\delta, \tag{6.27}$$

where $\chi(v)$ is the horizontal offset of the wire v. To speed up the solution process, we add some additional constraints. Most importantly, if the number of shortest paths for a particular offset in the presence of all switch types is less than 100, we enumerate all of them and assign each another binary variable. The sum of all these path variables for the given offset is set equal to 1, which forces the solver to select exactly one of the paths. Then, the corresponding $x_{(x_\delta,y_\delta),p,v}$ variables are set by appropriate implication constraints.

6.11.3 Results

The experimental setup is the same as in Sect. 6.10, but because hop-distance optimality constraints make the problem significantly more complex and increase the solution time from the order of seconds to the order of minutes, we decided to set the wirelength trade-off to 0, thus avoiding the need to solve multiple ILP problems per iteration, due to multiplexer placement optimization.

Critical Path Delay Figure 6.15 indicates that enforcing hop-distance optimality slightly improves the performance of the obtained architectures, although the difference is rather insignificant.

Routability Similar conclusions about the impact that hop-distance optimality has on routability can be drawn from the results shown in Fig. 6.16. Combining the two results, we conclude that imposing these constraints is not detrimental; the guaranteed optimality could perhaps be useful for simplifying the CAD tools.

6.12 ILP Complexity

Most problem instances that we encountered are solved within second. A notable exception is the problems that enforce hop-distance optimality that sometimes take minutes and even tens of minutes to solve. Comparatively difficult instances also

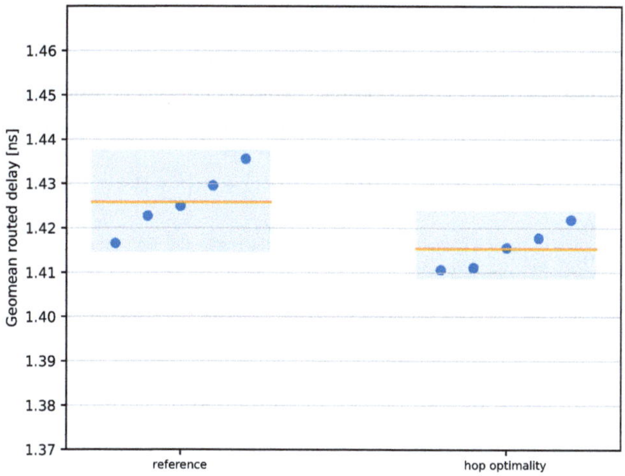

Fig. 6.15 Influence of hop-distance optimality on critical path delay

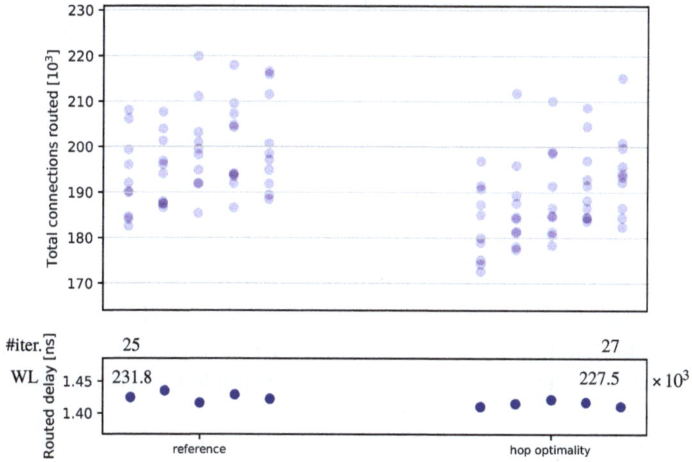

Fig. 6.16 Influence of hop-distance optimality on routability

occur when sharing of a large number of inputs is enforced, where solution times sometime also reach minutes.

In Table 6.3, we give average solution times and the number of variables and constraints needed to encode various types of regularity introduced in previous sections. All of the formulas are upper bounds. Given that modern architectures, faced with numerous technological limitations, have highly repetitive interconnect (e.g., *planes* and *lanes* in Agilex [Chr20]), the number of multiplexers that need to participate in the ILP problems is unlikely to ever become very large. Thus, we expect the proposed method to be scalable even for regularity types for which the

Table 6.3 ILP complexity

	Variables		Constraints		
Reg. type	General	Concrete	General	Concrete	Avg. sol. t [s]
# fanins (fanouts)	$(M_\varphi + 1)(\lvert V\rvert + 1)$	336	$(M_\varphi + 1)(\lvert V\rvert + 1) + 2\lvert V\rvert$	368	7.18
Input sharing	$\frac{3}{2}\lvert V\rvert^3 + \lvert V\rvert^2$	3174	$\frac{9}{2}\lvert V\rvert^3 + \lvert V\rvert^2 + \lvert V\rvert$	9418	40.66
All turns	0	0	$3\lvert V\rvert$	48	1.32
Internal (external) symmetry	0	0	$\frac{3}{2}\lvert V\rvert^2$	282	2.68
Full symmetry	0	0	$3\lvert V\rvert^2$	564	2.67
Hop-distance optimality	$(4M_H + 1)(4M_V + 1)2(M_H + M_V)\lvert V\rvert$	92,469	$(4M_H + 1)(4M_V + 1)(2(M_H + M_V)\lvert V\rvert^2 + 1)$	358,846	161.40

encoding size increases cubically with multiplexer count. Moreover, ILP problem size does not depend on the size of the circuits used in the exploration, unlike avalanche routing, which is at present the runtime bottleneck of the exploration algorithm.

6.13 Conclusions

We have shown that various types of regularity observed in switch-blocks of commercial FPGAs do not pose a significant limitation on their performance. In particular, the best performing architecture where all wires are forced to have the same fanout and fanin size ($\Lambda = 0.5$ in Fig. 6.10) has only ~1.5% worse routed critical path delay than the best architecture constructed when the router is free to choose switch types according to its needs ("∞" in Fig. 6.4). On the other hand, regularity in fanout and fanin sizes greatly increases the routability of the architectures on circuits that significantly differ from the ones used in switch-pattern construction (Fig. 6.7). This is certainly important for FPGA vendors, given the long lifetime of the products and difficulties in predicting how much the target designs will change over that period.

Regularity of the switch-pattern may allow layout optimizations that are not captured by our current area and delay models. Hence, the measured loss of performance due to regularization could be even smaller in a commercial setting. For example, we demonstrated that sharing inputs between pairs of multiplexers does not have any negative impact on routability (Fig. 6.9) and that it even provides a slight advantage in terms of performance when measured using conservative models (Fig. 6.8). On the other hand, several authors have mentioned other reasons why this technique can be beneficial [You98, Chr20], which, when combined with the above result, may more than suffice for the FPGA architect to seek only switch-patterns where all multiplexers share some inputs.

Perhaps most importantly, we have demonstrated that automated exploration methods can be used to construct competitive switch-block architectures that respect arbitrary constraints, ranging from fairly simple ones motivated by layout considerations, such as those presented in Sect. 6.6, to fairly complex ones motivated by design ideas that may enable more efficient FPGA CAD, presented in Sect. 6.11. Given that the algorithm that we proposed is generic, our hope is that it will be useful for both industry and academia in addressing the challenges that designing future FPGAs may bring.

The source code used to produce the results of the study presented in this chapter is available at https://github.com/EPFL-LAP/fpga23-regularity.

References

[And06] Anderson, J. H., & Najm, F. N. (2006). Active leakage power optimization for FPGAs. *IEEE Transactions on Computer-Aided Design of Integrated Circuits and Systems, 25*(3), 423–437.

[Bet99] Betz, V., Rose, J., & Marquardt, A. (1999). *Architecture and CAD for deep-submicron FPGAs*. Kluwer Academic Publishers. ISBN: 0-7923-8460-1.

[Chr20] Chromczak, J., Wheeler, M., Chiasson, C., How, D., Langhammer, M., Vanderhoek, T., et al. (2020). Architectural enhancements in Intel®Agilex™FPGAs. In *Proceedings of the 2020 ACM/SIGDA International Symposium on Field-Programmable Gate Arrays*. Monterey, CA, USA (pp. 140–149).

[Gue15] Gueye, S., Michelon, P. (2015). A linear formulation with $O(n^2)$ variables for quadratic assignment problems with Manhattan distance matrices. *EURO Journal on Computational Optimization, 3*(2), 79–110. ISSN: 2192-4406.

[Law63] Lawler, E. L. (1963). The quadratic assignment problem. *Management Science, 9*(4), 586–599.

[Lem00] Lemieux, G., Leventis, P., & Lewis, D. (2000). Generating highly-routable sparse crossbars for PLDs. In *Proceedings of the 2000 ACM/SIGDA Eighth International Symposium on Field Programmable Gate Arrays*. Monterey, CA, USA (pp. 155–164).

[Lew03] Lewis, D., Betz, V., Jefferson, D., Lee, A., Lane, C., Leventis, P., et al. (2003). The stratix routing and logic architecture. In *Proceedings of the 2003 ACM/SIGDA Eleventh International Symposium on Field Programmable Gate Arrays*. Monterey, CA (pp. 12–20).

[Lew13] Lewis, D., Cashman, D., Chan, M., Chromczak, J., Lai, G., Lee, A., et al. (2013). Architectural enhancements in Stratix V™. In *Proceedings of the ACM/SIGDA International Symposium on Field Programmable Gate Arrays*. Monterey, CA, USA (pp. 147–156).

[Lin10] Lin, M., Wawrzynek, J., & Gamal, A. E. (2010). Exploring FPGA routing architecture stochastically. *IEEE Transactions on Computer-Aided Design of Integrated Circuits and Systems, 29*(10), 1509–1522.

[Mar00] Marquardt, A., Betz, V., & Rose, J. (2000). Timing-driven placement for FPGAs. In *Proceedings of the 2000 ACM/SIGDA Eighth International Symposium on Field Programmable Gate Arrays*. Monterey, CA, USA (pp. 203–213).

[Mih13] Mihal, A., & Teig, S. (2013). A constraint satisfaction approach for programmable logic detailed placement. In *Proceedings of the 16th International Conference on Theory and Applications of Satisfiability Testing*, Helsinki (pp. 208–223).

[Mur20] Murray, K. E., Petelin, O., Zhong, S., Wang, J. M., Eldafrawy, M., Legault, J.-P., et al. (2020) VTR 8: high-performance CAD and customizable FPGA architecture modelling. *ACM Transactions on Reconfigurable Technology and Systems (TRETS), 13*(2), 9:1–60.

[Nik21a] Nikolić, S., & Ienne, P. (2021). Turning pathfinder upside-down: Exploring FPGA switch-blocks by negotiating switch presence. In *31st International Conference on Field-Programmable Logic and Applications*. Virtual Event (pp. 225–233).

[Nik23] Nikolić, S., & Ienne, P. (2023). Regularity matters: Designing practical FPGA switch-blocks. In *Proceedings of the 2023 ACM/SIGDA International Symposium on Field-Programmable Gate Arrays*. Monterey, CA, USA.

[Pap94] Papadimitriou, C. H. (1994). *Computational complexity*. Addison-Wesley. ISBN: 9780201530827.

[Pet21] Petersen, M. B., Nikolić, S., & Stojilović, M. (2021). NetCracker: A peek into the routing architecture of Xilinx 7-series FPGAs. In *Proceedings of the 2021 ACM/SIGDA International Symposium on Field Programmable Gate Arrays*. Virtual Event (pp. 11–22).

[Qia21] Qian, J., Shen, Y., Shi, K., Zhou, H., & Wang, L. (2021). General routing architecture modelling and exploration for modern FPGAs. In *Proceedings of the 2021 International Conference on Field-Programmable Technology*. Auckland (pp. 1–9).

[Ros91] Rose, J., & Brown, S. (1991). Flexibility of interconnection structures for field-programmable gate arrays. *IEEE Journal of Solid-State Circuits, 26*(3), 277–282.

[Rub11] Rubin, R., & DeHon, A. (2011). Timing-driven pathfinder pathology and remediation: Quantifying and reducing delay noise in VPR-pathfinder. In *Proceedings of the ACM/SIGDA 19th International Symposium on Field Programmable Gate Arrays*. Monterey, CA, USA (pp. 173–176).

[San23] Sangiovanni-Vincentelli, A., Liang, Z., Zhou, Z., & Zhang, J. (2023). Automated design of chiplets. In *Proceedings of the 2023 International Symposium on Physical Design*. Virtual Event (pp. 1–8).

[Str99] Stroobandt, D., Depreitere, J., & Campenhout, J. V. (1999). Generating new benchmark designs using a multi-terminal net model. *Integration, 27*(2), 113–129.

[Wil13] Williams, H. P. (2013). *Model building in mathematical programming* (5th ed.). Wiley.

[Wol23a] Wolf, C., & Lasser, M. (2023). *Project IceStorm*. http://bygone.clairexen.net/icestorm/.

[You98] Young, S. P. (1998). Six-input multiplexer with two gate levels and three memory cells. US 5744995. Patent assignee: Xilinx Inc.

[Zha22] Zha, Y., & Li, J. (2022). Revisiting PathFinder routing algorithm. In *Proceedings of the 2022 ACM/SIGDA International Symposium on Field-Programmable Gate Arrays*. Virtual Event (pp. 24–34).

Chapter 7
Fixed-Connectivity Pattern Design

Contents

In Chap. 4, we have seen that in order to achieve high performance of a programmable interconnect architecture, it is necessary to reduce the capacitive load on wires as much as possible. In the limit, the greatest reduction is obtained

© The Author(s), under exclusive license to Springer Nature Switzerland AG 2025
S. Nikolić, *Modern Programmable Interconnect Design*,
https://doi.org/10.1007/978-3-031-80629-2_7

when a wire is used exclusively to connect a single pair of LUTs. Not only does this reduce capacitive load on the wire to the minimum, but it also allows a signal to avoid passing through multiple levels of stored-select multiplexers, that would be necessary if the direct connection between LUTs did not exist. Given that multiplexers have been a predominant contributor to the delay of paths through programmable interconnect before delays of highly resistive metal started to dominate those of transistors, it should not come as a surprise that direct connections between LUTs, seeking to avoid this penalty, have been exploited ever since the XC2064 [Xil93]. However, patterns of such fixed connectivity that could be observed in commercial architectures were rather simple, usually not going much beyond cascades. In this chapter, largely based on our paper published at the 2020 ACM/SIGDA International Symposium on Field-Programmable Gate Arrays under the title "Straight to the Point: Intra- and Intercluster LUT Connections to Mitigate the Delay of Programmable Routing" [Nik20], we present a general algorithm for automated design of tileable fixed-connectivity patterns of arbitrary complexity, demonstrating that by introducing a relatively small number of direct connections between LUTs, significant improvements of critical path delay can be obtained. Unfortunately, since this work predated the development of the physical modeling flow presented in Chap. 4 and was implemented in VTR-7 [Luu14a] (the then latest stable version), which has certain limitations with respect to describing routing-resource graphs representative of modern plane-based architectures [Chr20], we limit this study to a mature 40nm planar technology, leaving its extension to a more advanced node for future work. Nevertheless, as stated above, fixed-connectivity patterns are relevant in any technology, since they allow timing-critical signals to avoid slow routing multiplexers and excessively loaded wires.

7.1 Straight to the Point

To illustrate the appeal of direct wires, let us take a look at Fig. 7.1a, depicting the shortest possible path between two LUTs in different clusters through programmable interconnect in a typical island-style FPGA. At the very least, the signal has to pass through one multiplexer of the switch-block, one multiplexer of the connection block, and one multiplexer of the crossbar. Needless to say, these multiplexers incur a significant delay, even in technologies in which it is the metal delay that is dominant. In such technologies in particular, the fact that wires that are part of the programmable interconnect architecture drive a large number of stored-select multiplexers has a significant impact on their overall delay.

Let us for the moment imagine that, most of the time, signals on the critical path of the implemented circuit will connect LUTs at the same relative location on the FPGA grid. If this assumption were to hold, we could resolve the problem of both the large number of slow stored-select multiplexers on the critical path, and the high capacitive load on the shared programmable interconnect wires by simply hardening a direct connection between the LUTs at the two designated locations.

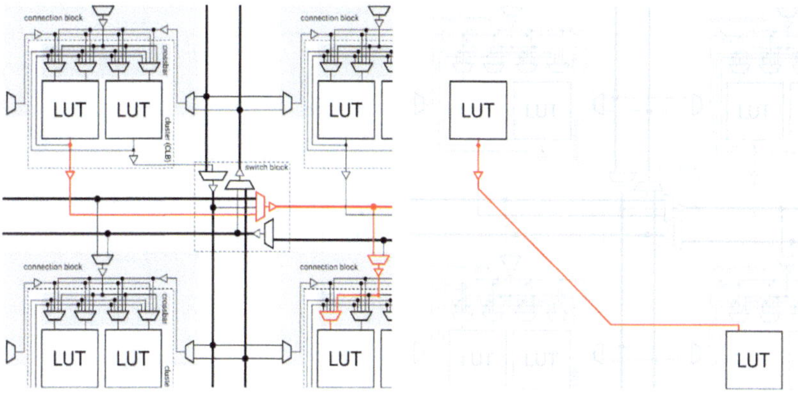

(a) No direct connections.

(b) With direct connections.

Fig. 7.1 Potential benefits of introducing direct connections between LUTs. The figure shows a minimum length path traversed by a signal generated by one LUT and consumed by another, residing in a neighboring cluster. In (**a**), the connection is in an architecture without direct connections between LUTs and, in (**b**), in an architecture with direct connections between LUTs

This is depicted in Fig. 7.1b. Not only does the direct connection avoid all stored-select multiplexers, apart from (depending on implementation; see Fig. 7.2) maybe the one driving an input pin of the target LUT, but its capacitive load is reduced to that of only one LUT or stored-select multiplexer input.

How successful hardening of direct connections between LUTs will be in practice depends entirely on the extent to which the assumption that these direct connections bridge locations that typically form the endpoints of timing-critical signals actually holds. Ensuring that a particular pattern of direct connections is successful at optimizing the critical path delay of a typical user circuit hence depends on 1) constructing the pattern so that it captures the most common locations of endpoints of critical signals in placed circuits and 2) altering the placement of the circuit so that it maximizes the extent to which the direct connections are used by the timing-critical signals. In this chapter we provide an algorithm for solving the first, exploratory, task. Optimizing exploitation of the constructed patterns by providing adequate algorithmic support for solving the second task will be the topic of the next chapter.

7.2 Related Work

As we have already mentioned, the idea of using direct connections between LUTs is nothing new. For example, XC2064 allowed one of the two outputs of the fracturable 4-LUT to connect directly to its right neighbor LUT, whereas the other output could connect to the LUTs above and below [Xil93]. Many other early FPGAs featured hardened direct connections between LUTs. Among them were the Xilinx XC4000,

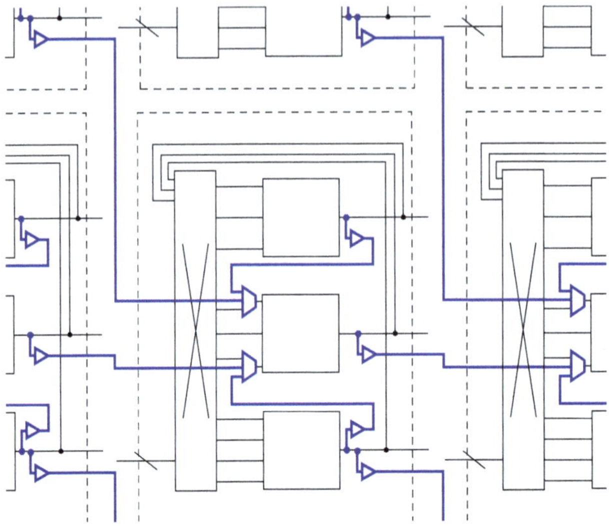

Fig. 7.2 Direct connections considered in this work. Blue, thicker lines show direct connections between LUTs that can be both within the same cluster and in different clusters. Each direct connection is driving the particular LUT input through a multiplexer, which enables selecting the driver from multiple direct connections, as well as the programmable routing structure. The purpose of this style of connections is to retain all of the original flexibility of the architecture. In modeling timing in this work, we assume that decoupling multiplexers shown in blue are explicitly introduced. Alternatively, direct connections can be added to the existing crossbar multiplexer driving the corresponding LUT input, thus trading off some of the direct connection performance by passing it through a larger multiplexer, for reduced impact when the direct connection is not used, by removing the additional decoupling multiplexer from the programmable path

which contained a two-level binary tree pattern of direct connections, *UTFPGA-1* from the University of Toronto, with a chain of three LUTs [Cho91], and the *Triptych* [Bor95] from the University of Washington, with a pattern mimicking the reconvergent fanouts that commonly occur in digital circuits. However, all these architectures appeared before logic clusters were widely adopted. As described in Sect. 3.3, logic clusters provided a single-level multiplexing structure and could essentially be considered an implementation of a direct connection pattern corresponding to a complete graph of N LUTs. The flexibility of this pattern diminished the prior interest in sparser and more efficient patterns, reducing them mostly to simple cascades, for both passing special carry-style signals [Lew03] and general LUT-generated ones [Hut02, Par11, Gai19]. However, it is important to remember that clusters are themselves only a rough, if very effective, approximation of inherent circuit connectivity that older architectures like the Triptych attempted to capture by different direct connection patterns. As illustrated in Fig. 7.1, it is quite reasonable to expect that additional gains can be obtained by combining the approximation of variable connectivity density through the use of logic clusters with approximation of recurring sparse connectivity patterns, by hardening specific LUT-to-LUT connections regardless of whether they are contained within the cluster or cross its boundaries. To the best of our knowledge, there were no attempts to bridge this divide with something bolder than a simple chain. It is the goal of this chapter to rectify that and provide a way to design appropriate fixed-connectivity patterns automatically.

Some FPGAs, such as Stratix 10 [Int20], contain direct connections between neighboring *clusters* and not individual LUTs. This could be considered a straightforward generalization of the nearest-neighbor connections of XC2064 to cluster-based architectures. Yet, these connections are broadcast to all LUTs within the cluster and hence both suffer from a high capacitive load and increase the size of a large number of multiplexers. Consequently, they have been removed from the latest 7nm Agilex architecture [Int23a]. Readers interested in the impact of cluster-level nearest-neighbor interconnect on performance of FPGAs in older technologies in which broadcasts were not as expensive can refer to the work of Roopchansingh and Rose [Roo02].

7.3 Optional Direct Connections

In this chapter we attempt to assess the value of introducing direct connections between LUTs both within the same cluster and in different clusters, with respect to critical path delay reduction. We assume that each of the direct connections can be optionally selected by configuring the appropriate multiplexer, as illustrated in Fig. 7.2. This enables us to add direct connections on top of any standard architecture, without reducing its routing flexibility, and thus always makes it possible to use the same implementation of a particular circuit as the one found for the original architecture. Essentially, this limits the possible damage to the critical path delay caused by the introduction of the direct connections to the delays of the added multiplexers—any improvement has only this small penalty to overcome. Also, we thus retain the possibility to use the existing CAD algorithms (those in the VTR framework [Luu14], in particular) without any modification. This is certainly suboptimal, but we wish to focus on the exploration of the design space: We aim at answering the question "What pattern of direct connections serves best the critical path delay reduction?" Development of a dedicated placer is left to the next chapter.

7.3.1 And What About Fully Hardened Direct Connections?

Such decoupled direct connections have already been used in real commercial products such as XC4000 [Xil98] and Versal [Gai19], as well as in research publications by, e.g., Feng et al. [Fen18] and Tang et al. [Tan14]. The alternative is to restrict a particular LUT input to only be driven by one particular other LUT. This avoids the delay penalty of introducing another multiplexer at the LUT input, or increasing the size of the existing one. However, based on the experience from our prior work on automated exploration of such architectures [Nik19], developing CAD tools that can successfully overcome the restrictive nature of these fully hardened connections is a great challenge. Most likely, in order for this endeavor to be successful, it has to be perceived already from the synthesis and technology mapping stage of the CAD flow, as has been done by Feng et al. [Fen18]. Since advanced mapping algorithms required for this are highly computationally intensive [Ray12], the size of patterns that can be successfully processed is very limited. Moreover, it is difficult to envision how these algorithms could be applied to intercluster connections, since at the time of technology mapping, it is typically not yet known which LUTs will be part of the same cluster; making assumptions on this before placement and trying to retain decisions based on these assumptions have, on the other hand, been shown to lead to inferior performance of implemented designs [Che07]. As we will demonstrate in this chapter, replacing the intercluster routing paths with direct connections is precisely what brings the highest benefit, because of the larger number of stored-select multiplexers that can be avoided, as well as by avoiding the penalty of entering the cluster through highly loaded thin-metal wires of the connection block (see Chap. 4). Note that in case of intracluster connections, only a single multiplexing level can be avoided. Furthermore, Fig. 2.17 indicates that in commercial designs, intracluster connections are rarely critical.

7.3.2 *Endpoint Alignment*

To benefit from the tight bound on negative improvement offered by optionally used direct connections, in this exploratory work, we always keep the same packing and placement as produced for the underlying architecture. We only use the direct connections opportunistically, where the possibility to do this is naturally created during the placement process. Apart from the placement of clusters in the programmable fabric, the number of such opportunities is highly dependent on the local placement of LUTs within the clusters—and thus potentially underwhelming, given that often, and especially in VPR, intracluster placement is entirely arbitrary [Luu14a]. Figure 7.3a shows in red the signals of a particular benchmark circuit implemented on one particular architecture that were successfully converted to use direct connections: as one could fear, a negligible number. Permuting the LUTs inside the clusters in an attempt to maximize the usage of direct connections results in the situation of Fig. 7.3b. For the reasons apparent from these two figures, we slightly depart from a purely opportunistic approach by performing explicit LUT permutation between the placement and the routing stage. Ideally, the routing algorithm would dynamically permute the LUTs where appropriate [Lew03], but, due to the current limitations of academic tools [Luu14a], we are forced to resort to explicit permutation.

The previous paragraph anticipates our only modifications to an ordinary FPGA CAD flow, more precisely described in Sect. 7.6.3. More aggressive connection

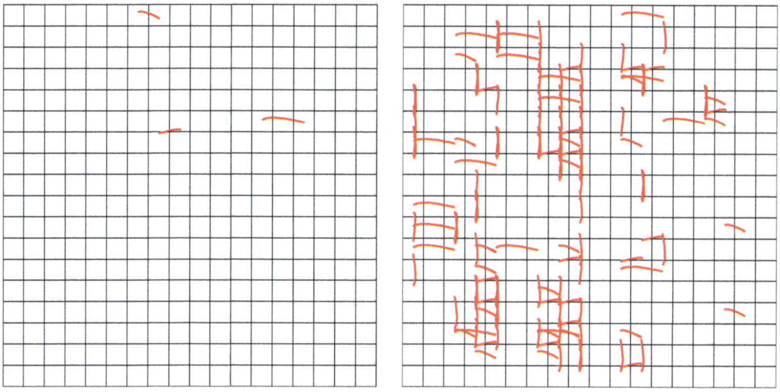

(a) Before permutation. (b) After permutation.

Fig. 7.3 Influence of LUT permutation on usability of direct connections. The same placement of the *sha* benchmark from the *VTR* set is shown before (**a**) and after (**b**) permuting LUTs inside the clusters. Each cell is a location on the FPGA. Connections implemented as direct are shown in red

endpoint alignment, by replacing LUTs across clusters, will be covered in the next chapter. Our main contribution here, with this minimalist approach to exploiting direct connections, is to systematically explore the large space of possible tileable direct connection patterns.

7.4 The Space of Tileable Fixed-Connectivity Patterns

As mentioned in Chap. 3, the defining feature of island-style FPGAs is that they are composed by abutting a large number of copies of a single tile (or a handful of different tiles). Fixed-connectivity patterns must respect this tileability as well. In Sect. 3.6, we introduced the notion of periodic graphs [Hof94] as a unified way of representing tileable architectures; we also mentioned that they are particularly useful for representing fixed-connectivity patterns.

When we only want to augment an existing programmable interconnect architecture with a collection of direct connections between LUTs, the node set of the static graph S which we have to construct reduces to the set of LUTs in a cluster. For instance, an island-style FPGA without any direct connections between LUTs would be represented by an edgeless graph. On the other hand, an FPGA with a chain connecting LUTs in the same column would be represented by the static graph of Fig. 7.4a, for the case of a four-LUT cluster. Another example of a slightly more complex pattern—the one for which the mapping of Fig. 7.3b was produced—is shown in Fig. 7.4b. Here we can see, for example, that the edge $(u, v, \mathbf{w}) = (5, 4, (2, 0))$ represents the connection between the fifth LUT of each

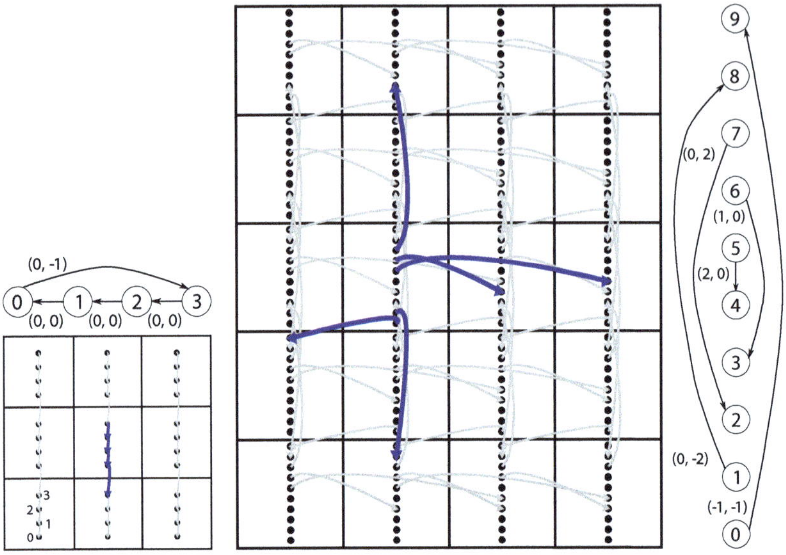

(a) (b)

Fig. 7.4 Using static graphs to describe patterns of direct connections. Each square represents a cluster and each point represents an LUT. Outgoing edges of one replica of the node set (one cluster) are emphasized in blue. (**a**) Columnar chain. (**b**) Pattern used in Fig. 7.3b

cluster and the fourth LUT of the second cluster to the right, within the same row of the fabric.

The architectural search space considered in this chapter coincides with the space of all patterns representable as a static graph on N nodes with M edges, and edge weight bounded by a Chebyshev length of w (that is, absolute values of both weight components are $\leq w$). As before, N is the cluster size, while M and w are parameters. To give some numerical sense of the size of this design space, we could consider the following. There are N^2 ways to choose the endpoints of each edge and $2w + 1$ possible values for each component of its weight. Hence, we can choose the M edges from $((2w + 1)\,N)^2$ possible ones. For $(N, M, w) = (10, 20, 4)$, which are not particularly large numbers, given the cluster sizes and channel wirelengths of modern FPGAs, this amounts to $\sim 10^{59}$. Of course, this is a loose bound because it does not account for isomorphisms; the intention is only to give some rough sense of the scale of the problem.

7.4.1 *Fully Specifying an Architecture*

The reader may notice that in Fig. 3.12a, multiplexers driving the LUT inputs are also included in the static graph. To fully specify the fixed-connectivity pattern, the

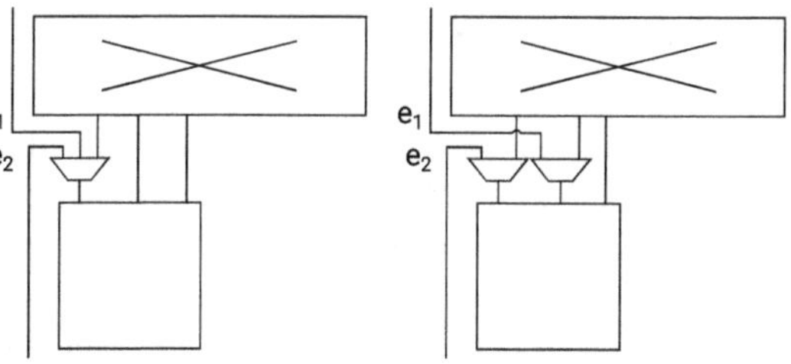

Fig. 7.5 Multiplexer placement trade-offs. In the architecture on the left, connections e_1 and e_2 are sharing a common multiplexer and, thus, cannot be used simultaneously. This is addressed in the right architecture where both direct connections can be used and are delayed by faster 2:1 multiplexers. However, another input of the LUT is now driven by an additional multiplexer, increasing the probability that the critical path delay would increase when the direct connection is not used

direct connection decoupling multiplexers must be represented in the static graph as well. Nevertheless, to prevent further search space size increase, we retain the level of detail of Fig. 7.4, distributing direct connections among the input pins of the target LUT in a round-robin fashion. This roughly maximizes the number of direct connections that can be simultaneously used to drive inputs of any LUT and keeps the LUT input delay increase due to the added decoupling multiplexers roughly balanced. An alternative approach is shown in the left part of Fig. 7.5.

7.5 Searching a Large Design Space

In this section we describe the reasoning about our approach to searching the design space, as well as the details of the search algorithm itself. The greatest challenge is, of course, navigating the large size of the search space which, much like in the case of switch-pattern design presented in Chap. 5, renders the classical brute-force approach [Zgh17] prohibitive. To explore spaces of comparable size, in Chap. 5, we relied on extracting valuable information from the routing process, instead of discarding it by perceiving the router as a mere black box that measures performance. Here, however, the direct connections are intended to replace entire point-to-point paths, meaning that all the required information—that is, the locations of the various connection endpoints—is available from placement; this removes the need to observe the behavior of the router. By formulating the design task as a covering problem and leveraging a well-known approximation result, we are able to solve it efficiently. This is similar to optimizing a tractable surrogate, chosen by analyzing the nature of the problem at hand, that Lemieux et al. used to automatically design highly routable sparse crossbars [Lem00].

7.5.1 Our Search Space: A Naive View

Given an FPGA architecture, our task is to select some M edges that will enter the edge set of the static graph describing the pattern of direct connections that we intend to augment it with. Let us, for the moment, disregard the fact that our primary goal in doing all this is to reduce the critical path delay. Then, the value of adding a new edge to the edge set is in increasing the number of circuit connections covered by the direct connections. Each potential static graph edge can be assigned a set of circuit connections that it can cover. This set is dependent on the topology of the circuit to be mapped on the FPGA and its particular placement. Because we allow arbitrary LUT permutations within the cluster, the set of connections that can potentially be covered by an edge of the static graph is the set of all connections between clusters whose relative position corresponds to the weight of the edge, irrespective of the precise location of the endpoint LUTs in their respective clusters. In Fig. 7.6a, these sets are represented in gray, for the edges $e_1 = (1, 9, (0, -1))$, $e_2 = (2, 8, (0, -1))$, and $e_3 = (8, 3, (0, -3))$, and a particular placement of the *sha* VTR benchmark. Because edges e_1 and e_2 have the same weight, the sets of connections they can cover are identical, while the set of edge e_3 is disjoint from

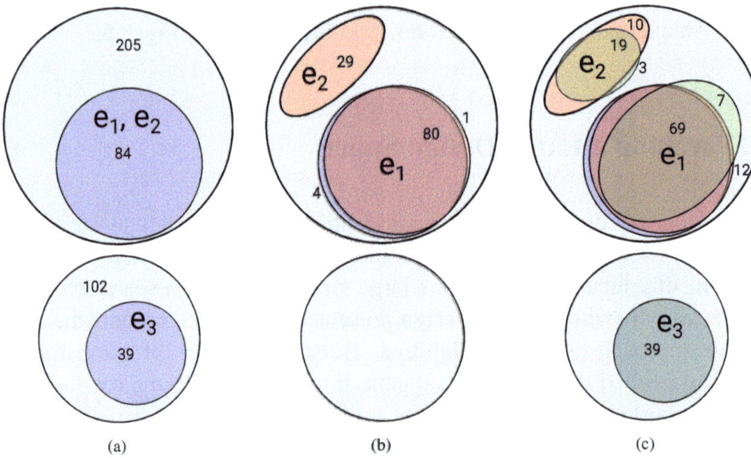

Fig. 7.6 Analogy between our best pattern search and the maximum coverage problem [Vaz04]. Sets of connections of a particular placement of the *sha* benchmark coverable by the edges $e_1 = (1, 9, (0, -1))$, $e_2 = (2, 8, (0, -1))$, and $e_3 = (8, 3, (0, -3))$ are shown. The gray sets include all coverable connections, while the blue ones include the edges covered by the pattern composed of the single respective edge, after a particular permutation of LUTs. Connections covered by the patterns composed of e_1 and e_2, and all three edges are shown in red and green, respectively. Coverage achieved by the combinations of multiple edges is larger than each of their individual coverages, even if they overlap entirely (e.g., e_1 and e_2), but smaller than their sum, even if the individual coverage sets are disjoint (e.g., e_2 and e_3)

them, due to its different weights. We could attempt to choose the M edges so as to maximize the union of the sets of connections they can cover. Because the gray sets of Fig. 7.6a are either identical or disjoint, we could trivially achieve this by greedily choosing edges with disjoint sets. Note, however, that this would be misleading, as it would appear that choosing an edge with the weight equal to the weight of some edge already in the static graph would bring no benefit, due to their sets of coverable connections coinciding. Intuitively, of course, we would expect improvement in some cases, as there could be more than one connection between any two clusters, whereas a single static graph edge of the appropriate weight could cover only one of them. Likewise, it would appear that the coverage of edges with different weights is entirely additive, which is not really the case, as we will soon see.

7.5.2 *Our Search Space: One Step at a Time*

A more realistic view of the extent to which a particular edge may contribute to covering the connections of the circuit to be mapped can be obtained by measuring the coverage achieved by the pattern composed only of that edge. To perform such a measurement, we already need to consider LUT permutation. When the pattern is composed of any single one of these edges, or more generally, when the total degree of the static graph is bounded by one (that is, no two edges share a node and there are no loops), the set of covered edges constitutes a matching [Bon08]. We search for maximal matchings with the following simple greedy algorithm:

1. Mark position of all LUTs as *free*.
2. Sort the circuit connections in decreasing length.
3. For each connection (u, v), between LUTs in clusters at an offset $\mathbf{w}$, if there is an edge $(u_p, v_p, \mathbf{w})$ in the static graph, and u and v are either free or fixed to u_p and v_p, respectively, cover (u, v) and fix positions of u and v to u_p and v_p, respectively.

After applying this algorithm to each of the three single-edge patterns, we obtain the blue sets of Fig. 7.6a. Whereas the gray sets corresponded to the unions of all sets of connections covered by the pattern under all possible permutations, the blue ones are the sets covered under one particular permutation. Hence, for instance, if the pattern includes only a single edge of weight $(0, -1)$, a subset of the 205 connections with 84 elements can be covered. Note that the covered sets of e_1 and e_2 again coincide, while that of e_3 is again disjoint from them. Apart from the numerical changes of the set cardinality, we are back to where we begun—we have the same coverage model that appears to be trivial to solve, but counters our intuition.

7.5.3 Our Search Space: Combining Steps

So let us combine e_1 and e_2 into one common pattern. The degree of this pattern is still bounded by one, so our algorithm still works, and the coverages it produces are overlaid in red, in Fig. 7.6b. Now the situation already starts becoming more intuitive: The sets of connections actually covered by e_1 and e_2 are disjoint, as we can cover each connection only once. The combined coverage is now larger than either of the individual coverages (that were in fact the same), so there is indeed benefit from selecting a second edge of the same weight, as we intuitively predicted in the beginning. However, the combined coverage is smaller than the sum of the two individual coverages, indicating that coverage is really not additive.

What happens if we add e_3 to the pattern as well? Its coverable connection set (gray) was disjoint from the others, so will the combined coverage simply be the sum of its coverage and that of e_1 and e_2 combined? The new coverage sets are overlaid in green, in Fig. 7.6c. Both the sets of e_1 and e_2 shrunk with the introduction of e_3. This is because e_3, which had precedence due to sorting of circuit connections based on length, fixed the positions of its endpoints in such a manner that e_2 could no longer cover some of the connections it covered previously. In turn, e_2 took over some of the connections from the intersection with e_1, causing a change in its set as well. Hence, even the coverage sets that originally appeared to be disjoint can have "induced" intersections, caused by conflicting permutation requirements.

7.5.4 Our Search Problem: An Analogy

The last subsections give us a taste of the challenges that our search problem faces. *Maximum Coverage* is a well-known intractable problem where the goal is to select a fixed number of sets out of several sets that may have some elements in common; in doing so, one wants to maximize the cardinality of the union of the selected sets [Vaz04]. A standard greedy approach to solving the problem is known to give the best achievable polynomial-time approximation, unless $\mathbf{P} = \mathbf{NP}$ [Fei98]. Without any rigorous analysis of the relations between the problem of finding the optimum set of direct connections and maximum coverage, relying on the intuition developed by the observations in the previous example instead, we adopt the greedy approach to solving our search problem.

7.5.5 The Greedy Algorithm

The pseudocode of our search algorithm is shown in Algorithm 7.1. The algorithm starts from an edgeless graph on N nodes and performs M rounds of listing all unique expansions of the previous best-scoring pattern and choosing the new best one among them. During expansion listing, edges are first added to the static graph without weight assignment. All N^2 such expansions are generated and the resulting

Algorithm 7.1 Greedy pattern search

```
function SEARCH(N, M, w, cA, cB, cC)
    g_best = ([0, N), { })
    for m in [0, M) do
        G = GET_ALL_UNIQUE_EXPANSIONS(g_best, w)
        FILTERA(G, cA)
        FILTERB(G, cB)
        FILTERC(G, cC)
        g_best = PICK_BEST(G)
    return g_best
function FILTER(G, c)                                  ▷ A prototypical filter
    for g in G do
        for b in benchmarks do
            scores[b][g] = COMPUTE_SCORE(g, b)
        for b in benchmarks do
            SORT(scores[b])
            for g in G do
                ranks[g] += INDEX(scores[b], g)
    SORT(ranks)
    return ranks[0:c]
```

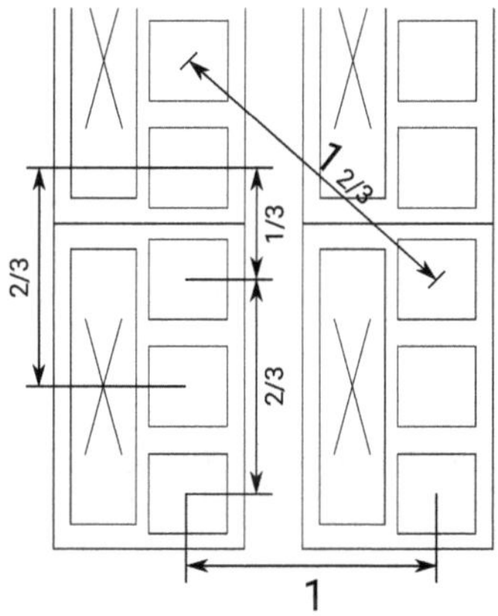

Fig. 7.7 Connection length calculation. Several examples of connection length calculation are shown for a cluster size of three with LUTs vertically stacked. The distance between two vertically adjacent LUTs is equal to $1/N \times L_T$, where N is the cluster size and L_T the height of the tile. The distance between two horizontally adjacent LUTs is equal to L_T. N and L_T are 3 and 1 in this example, respectively. Note that, contrary to the more precise model of Chap. 4, this assumes a 1:1 aspect ratio of the tile. Although not realistic for commercial architectures [Lew03], this assumption was necessary to be in line with the identical horizontal and vertical routing channel constraint of VTR 7 [Luu14]

graphs are split into classes of isomorphic ones. The next stage is weight assignment, where, for each tried weight, a class representative that results in the minimum length of the added connection is chosen. This is based on the assumption that the LUTs are stacked vertically, like in the Stratix FPGAs [Lew13] (Fig. 7.7).

The best pattern is chosen based on the postrouting critical path delay, as this is what we wish to optimize. Because there are thousands of new patterns appearing in each search iteration, running the complete CAD flow is infeasible, and we need fast predictors to filter out most of the weak candidates. A prototypical filter is represented by the function *filter* of Algorithm 7.1. For each circuit in the benchmark set and each pattern, a score is computed. Patterns are then ranked for each of the circuits and c of them (where c is a parameter) with the minimum sum of ranks are kept for the next filtering stage. Details of the filters are discussed below. The first two are designed to be fast and based only on coverage. They also serve the purpose of anticipating the accumulated effect of adding multiple edges to the pattern that may not be immediately reflected on critical path delay. The final filter is considerably slower, but also more accurate and tries to predict the critical path delay itself.

7.5.6 Pruning the Candidates: The First Filter

The first filter groups the pattern edges by their weights and treats the groups completely independently, attempting to quickly assess how much circuit connection coverage the chosen set of weights can achieve. For a given pattern P, edge weight $\mathbf{w}$, and a cluster C of the placed circuit, it works as follows:

1. For each LUT u in P, count edges $(u, v, \mathbf{w})$ and store these counts in $\mathbf{s_p}$.
2. For each LUT u' in C, count the number of connections starting at u' and ending in a cluster at an offset equal to $\mathbf{w}$ and store these counts in $\mathbf{s_c}$.
3. Sort the vectors $\mathbf{s_p}$ and $\mathbf{s_c}$ and compute the score $S(P, C, \mathbf{w})$ as $\sum_i min(\mathbf{s_p}(i), \mathbf{s_c}(i))$.

In the example of Fig. 7.8, for $\mathbf{w} = (1, 0)$, vectors $\mathbf{s_p}$ and $\mathbf{s_c}$ are $(2, 0, 0)$ and $(1, 1, 0)$, respectively. Hence, $S(P, C, (1, 0))$ is 1 in this case. The score $S(P, C)$

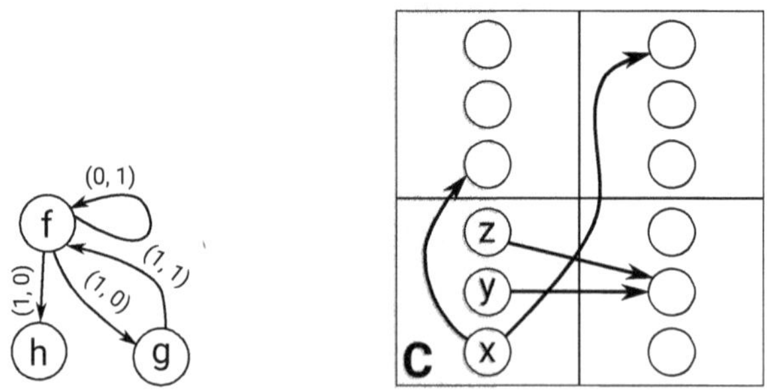

Fig. 7.8 An example to demonstrate our coverage-based filters. The static graph on the left describes a pattern, while the fabric on the right shows a cluster, C, of a packed and placed circuit, along with three of its immediate neighbors. Only circuit connections originating in cluster C are shown

is obtained simply as $\sum_{\mathbf{w}} S(P, C, \mathbf{w})$, while the final score $S(P)$ is obtained by summing $S(P, C)$ for all C. In the running example, all the remaining count vectors are $(1, 0, 0)$ so the score $S(P, C)$ is 3. A dual version of the filter considering incoming connections is applied in the same manner, and the two scores are added for each pattern.

This is a very simple filter that completely ignores the mutual influence of various pattern edges through conflicting permutation requirements, as well as distribution of edges of different weights among LUTs, therefore largely overestimating the possible coverage. It has one important virtue, however—it is very fast to compute. Moreover, because we are keeping the same packing and placement for a given benchmark and random seed pair, we can precompute the score for each $(\mathbf{w}, \mathbf{s_p})$ pair on each particular placement. Despite its simplicity, this filter is effective in reducing the number of possible expansions from several thousands, with many clearly inferior to others and many essentially equivalent, down to a figure manageable by the further stages of the algorithm.

7.5.7 Pruning the Candidates: The Second Filter

The second filter attempts to fix some of the problems of the first one. The first filter essentially assumes that the static graph edges can be swapped between different LUTs dynamically, to maximize the number of covered circuit connections incident to the given cluster. Here, the static graph edges are really tied to their endpoint LUTs. An optimal permutation of the LUTs in a given cluster is computed, considering that each LUT is one endpoint of its incident connections, and that the other endpoints can be freely permuted, so as to make the connections truly covered. This assumption means that the effects of conflicting permutation requirements are still largely neglected, but also that the scores can still be quickly computed. Computation of the score $S(P, C)$ proceeds as follows:

1. For each LUT u in P, form the multiset of the incoming edge weights $m_i(u)$ and the multiset of the outgoing edge weights $m_o(u)$.
2. Perform the same for each LUT u' in C.
3. For each u in P and each u' in C, compute the score $s(u, u')$ as $|m_i(u) \cap m_i(u')| + |m_o(u) \cap m_o(u')|$.
4. Populate the score matrix M_s with rows corresponding to all u in P and columns to all u' in C by computing the appropriate scores $s(u, u')$.
5. Solve the assignment problem [Mun57] on M_s to obtain the score $S(P, C)$.

The final score $S(P)$ is again computed as $\sum_C S(P, C)$. In the running example, $m_i(f) = \{(0, -1), (-1, -1)\}$, $m_o(f) = \{(0, 1), (1, 0), (1, 0)\}$, while $m_i(x) = \{\}$ and $m_o(x) = \{(0, 1), (1, 1)\}$. Hence, $s(f, x) = 1$. The complete score matrix is

$$M_s = \begin{matrix} & x & y & z \\ f & 1 & 1 & 1 \\ g & 1 & 0 & 0 \\ h & 0 & 0 & 0 \end{matrix}$$

with the entries participating in an optimum assignment highlighted. $S(P, C)$ is now equal to 2, which is less than 3 reported by the first filter, because the distribution of edges with different weights among the LUTs is now taken into account.

7.5.8 *Pruning the Candidates: The Last Filter*

So far, the filters had only focused on assessing coverage, without any relation to optimizing delay. We note that this could have been different if each edge was weighted by its postplacement criticality, when computing coverage. However, in this study, the third filter is the first point when timing directly enters the edge selection process. It relies on optimal permutation of a small subset of (near) critical nodes, using *Integer Linear Programming* (ILP), to obtain a critical path delay prediction, based on the postplacement assessment and the improvement achieved after LUT permutation. The details of choosing which nodes to permute are described in Sect. 7.6.3, where the process is referred to as *solving the critical core*.

Finally, because intracluster connections occur about as often as all intercluster connections combined, the first two filters would tend to select mostly zero-weighted edges. To prevent this, we split the search into two stages—first for nonzero-weighted edges, followed by only zero-weighted edges, with the previous set kept fixed. We will come back to the problem of intracluster connections in Sect. 7.7.5.

7.6 Experimental Setup

The evaluation flow that we use in this chapter is based on VTR-7 [Luu14] and depicted in the flowchart of Fig. 7.9. For comparison, the underlying architecture is passed through the traditional CAD flow of VTR, before being augmented by direct connections. The implementation flow of the pattern-enhanced architectures starts by reading the underlying architecture description and the description of the pattern and generating a modified architecture that incorporates the pattern into the underlying one. The packing and placement files produced in the reference flow are passed to the LUT permutation algorithm along with all the postplacement timing information, the benchmark circuit netlist, and the modified architecture model. After LUT permutation, those circuit connections that are implementable as

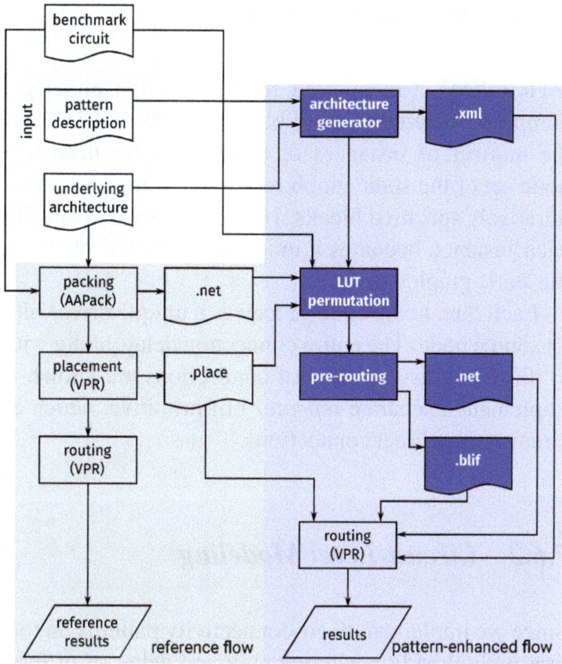

Fig. 7.9 Pattern evaluation flow. Stages that are not part of the usual VTR flow are shown in blue

direct are prerouted, and the modified packing and circuit netlist files are generated. These two files are then passed to the router along with the unmodified original cluster-level placement, to obtain the final implementation of the circuit in the new architecture.

To reduce the measurement noise, all benchmark circuits are placed with five different placement seeds; reported results are the median values of the five experiments. To further minimize noise, we implement the *delay targeted routing* algorithm of Rubin and DeHon [Rub11].

In all experiments, the 40 nm *k6_N10_mem32K_40 nm* architecture from the VTR project is used as the underlying (reference) architecture. In the following sections, we describe how this architecture is transformed to include the specified direct connections. To appropriately model the added circuitry, we perform SPICE simulations in the closest technology node we had access to—*45 nm PTM HP* [Zha07]—and scale the delays to match those reported in the underlying architecture. The details on modeling and the stages of the pattern-enhanced architecture CAD flow where it departs from the reference one are also discussed in the following sections.

7.6.1 Architecture Generation

VTR's XML format used to describe the underlying architecture strives to be compact by defining each block type (e.g., LUT) only once and then specifying the number of instances of that block. We need to find a mapping between the node set of the static graph representing the pattern of direct connections and these iteratively specified blocks. Hence, the original specifications are "unrolled" so that each instance becomes a uniquely identified block, corresponding to one node of the static graph.

Each direct connection receives a unique driver block that is tied to the output of its source node. The entire connection delay, along with that of the driver, is assigned to this block as the loads of connections are known in advance. The driver blocks implement a separate *repeater* blif primitive, which enables simple prerouting of a circuit using direct connections.

7.6.2 Circuit-Level Modeling

Since we implement fixed-connectivity patterns on top of an existing VTR architecture with already annotated area and delay of different blocks, we need to model the delays of the direct connections to a comparable precision. We do this by representing each direct connection as a sequence of Π-type RC stages [Bet99]. The per-tile-length RC parameters are those of the channel wires of the underlying architecture. Each direct connection is modeled by $\lfloor L/N \rfloor$ full-tile stages plus one stage with RC parameters scaled by $L/N - \lfloor L/N \rfloor$, where L is the length of the connection calculated as shown in Fig. 7.7 and N the size of the cluster ($N = 10$ in all experiments presented here). The drive strength of the first driver stage is set to that of the first stage of the global wire driver of the underlying architecture, while the second stage drive strength is swept in both directions, starting from that of the second stage of the global wire driver. Once the delay stops decreasing, the sweep is terminated and the last considered size is taken. When that is advantageous (for short wires), we remove the second stage of the driver. The SPICE measurements on direct connections are made using the circuit shown at the bottom of Fig. 7.10. The model of a single global wire of the underlying architecture is shown at the top of the same figure. The slightly reduced load applied to the direct connection (i.e., 3/3 inverter instead of a 5/5 one) reflects the fact that it drives only a single LUT input. We multiply all measured delays by the ratio between the delay of the reference wire reported in the underlying architecture and the corresponding delay obtained in our technology through SPICE simulation. The final delays are reported in Fig. 7.12a.

Multiplexers are modeled as two-level structures [Chi13]. The pass transistor and first-stage driver sizes are taken from the switch-block of the underlying architecture. The second-stage drivers NMOS and PMOS are assumed to be 3 and

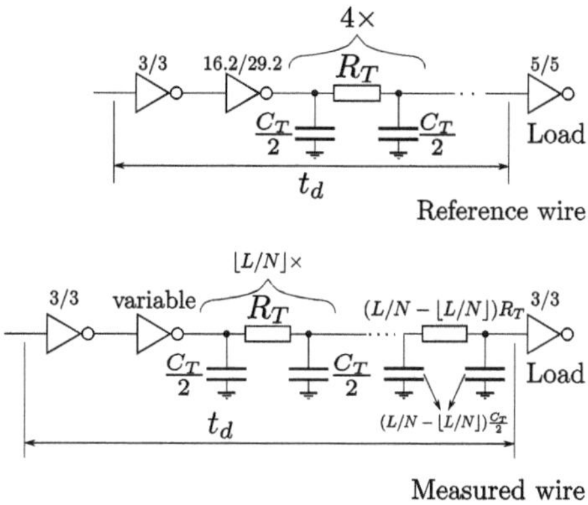

Fig. 7.10 Experimental setup for direct connection delay measurement

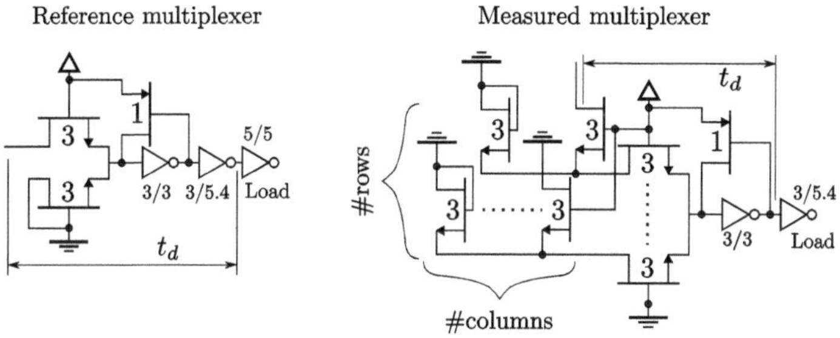

Fig. 7.11 Experimental setup for multiplexer delay measurement

5.4× wider than the minimum width transistor, respectively. We use the second stage as the load, assuming that a single-stage driver is sufficient, given that the multiplexer is driving only one LUT input. Figure 7.11 shows the circuit used for measuring the multiplexer delay (on the right) and the circuit of a 2:1 one-level multiplexer with a two-stage driver used for reference delay measurement (on the left). We scale all the delays by the ratio of the delay of the *BLE* [Bet99] output multiplexer reported in the underlying architecture and the delay measured on the reference circuit. The resulting delays are reported in Fig. 7.12b.

7.6.3 LUT Permutation

As mentioned in Sect. 7.3, the initial placement of LUTs within the clusters, oblivious to direct connections, does not offer sufficient opportunity to use them. To overcome this issue, we employ the LUT permutation algorithm the pseudocode of which is shown in Algorithm 7.2. First, we find the circuit connections that might possibly be covered by a direct connection—that is, all connections between clusters whose relative distance equals the weight of some edge of the pattern (function *find_candidate_cons*). All other connections are irrelevant for the permutation

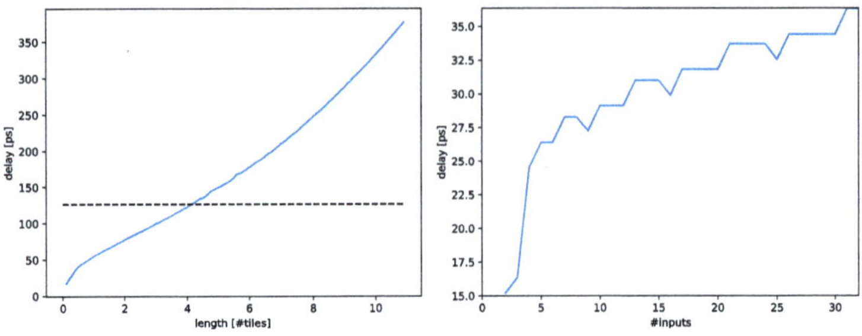

(a) (b)

Fig. 7.12 Direct connection and multiplexer delays. The dashed line corresponds to the delay of the reference four-tile-long global-routing channel wire of the underlying architecture. (**a**) Direct connection delays. (**b**) Multiplexer delays

Algorithm 7.2 LUT permutation

```
function PERMUTE(pattern, clusters, placement, timing_graph)
    for c in clusters do
        RANDOMLY_PERMUTE(c)
    E = FIND_CANDIDATE_CONS(clusters, placement, pattern)
    core = EXTRACT_CRITICAL_CORE(timing_graph, E, size)
    permutations, fixed_nodes = SOLVE_CORE(core)
    permutations = ANNEAL(clusters \ fixed_nodes)
    return permutations
function EXTRACT_CORE(timing_graph, E, size)
    for e in E do
        timing_graph.delay(e) = min_direct_con_delay(e)
    DO_STA(timing_graph)                      ▷ STA = Static Timing Analysis
    covered = E
    V = NODES_INCIDENT_TO(E)
    while |V| > size do
        e = LEAST_CRITICAL(covered, timing_graph)
        covered.remove(e)
        timing_graph.delay(e) = programmable_delay(e)
        DO_INCREMENTAL_STA(timing_graph)
        UPDATE(V)
    return V
```

process. Our primary goal in permuting the LUTs is to cover as many critical circuit connections as possible by the fast direct connections of the pattern. Since we are making decisions after placement, the timing predictions are already quite accurate and the number of (close to) critical nodes is usually fairly limited. Hence, we can extract the critical portion of the circuit—the *critical core*—and optimize its coverage exactly. For this, we proceed as follows: Initially, we assume that all potentially coverable connections we just identified are part of the critical core. We then reduce the core size greedily, by removing the edge of the maximum slack, until the core size is less than a given parameter (function *extract_core*). Finally,

ILP is used to find a legal permutation of the clusters containing the core nodes so that the critical path delay is minimized. In order to keep the runtime reasonable, a timeout of one minute is imposed on the solver, but, in practice, the optimum is often reached much faster. Because the core contains only a small fraction of the entire circuit, the resulting permutation does not maximize the overall usage of direct connections, which is important for reducing pressure on general routing. To achieve that, we conclude by performing a low temperature simulated annealing, with a setup similar to that of VPR's non-timing-driven placement algorithm. A concrete example of the difference created by LUT permutation with respect to the possibility of using direct connections was shown in Fig. 7.3b.

7.6.4 Prerouting

After completing LUT permutation, all circuit connections that match an edge of the static graph are implemented using the appropriate direct connection. This necessitates LUT input permutation. When several connections share a multiplexer, the one with the least postplacement slack is selected, while the others are left to be implemented using general routing. The packed netlist is modified accordingly, as is the benchmark circuit netlist, by instantiating *repeater* subcircuits between sources and targets of the prerouted connections. Thus, instead of truly prerouting the circuit, we leave VPR with only one choice for routing each prerouted connection.

7.6.5 Further Assumptions and Limitations

To preserve the legality of the circuit implementation after explicit permutation of LUTs, the routing algorithm must no longer consider all LUT outputs to be equivalent. Due to VPR's current lack of support for selectively disabling the permutation of outputs [Luu14a], this calls for preventing any route-time output permutation whatsoever. Allowing LUT output permutation by the router for the reference architecture alone would be too unfair to the pattern-enhanced architectures. On the other hand, disabling permutations altogether would show unrealistically large benefits of direct connections, because there is clearly a trade-off between choosing the permutation that optimizes direct connection utilization and the one that optimizes congestion in programmable routing. Missing a better alternative at the moment, we generate 30 different random permutations of each packing for the reference architecture, prohibit any further output permutation by the router, and choose the permutation resulting in the median critical path delay as the representative one. Once we establish that fixed-connectivity patterns are indeed effective for optimizing critical path delay of the implemented circuits through pattern exploration in this chapter, we will extend VPR's capabilities to include selective permutability in Chap. 8, which deals with dedicated CAD support for such patterns.

Since our permutation algorithm is currently unable to determine local routability, we require that the underlying architecture has a fully populated crossbar and that all LUT inputs are permutable. This means that we cannot consider fracturable LUTs for the time being. Similarly, because the algorithm is not aware of a priori fixed LUT positions in certain clusters, we currently do not support carry chains; multipliers and memories of the underlying architecture remain fully supported. These limitations could cause direct connections to appear slightly more appealing than they would have, had the architectures more representative of the current state of the art been used; still, we do not believe that this changes our final conclusions.

The search space is limited to patterns on 10 nodes, which conforms to the cluster size of the underlying architecture, and up to 20 connections of Chebyshev length ≤ 4 (that is, $(N, M, w) = (10, 20, 4)$). The search algorithm filters are set to pass 100, 10, and 3 expansions, respectively. The size of the critical core is set to 100 nodes, and *glpk* version 4.64 [Mak19] is used to solve it. In all experiments, routing channel width is set to 300. A subset of VTR benchmarks is used for all experiments, with the four largest excluded due to prohibitive runtime requirements.

7.7 Experimental Results

We now discuss the outcome of the search described in the preceding sections.

7.7.1 Intercluster Connections: Convergence

As mentioned in Sect. 7.5.6, the much larger number of intracluster connections appearing in an average circuit would overwhelm the coverage-based filters, requiring the search to be split into two phases. The first phase considers only intercluster connections, which have higher potential to reduce delay when used successfully. Figure 7.13 shows the evolution of the relative change of the postrouting critical path delay over the iterations of this first phase of pattern search. The pattern itself is shown in Fig. 7.14, with edges labeled as *weight/iteration*, where *iteration* is the iteration of the search algorithm when the particular edge was added.

Most likely due to the noise of the router [Rub11], Fig. 7.13 is not perfectly monotonic. This should not be surprising, since we have already seen a similar situation in Fig. 4.1, adopted from Ahmed [Ahm01]. For reference, it is interesting to also inspect Fig. 7.15, which shows the evolution of the critical path delay change after critical core solving (third filter of the search algorithm), as predicted from VPR's postplacement data. This curve is not monotonic either, due to the heuristic nature of critical core extraction and the limited time given to the ILP solver, but it is much smoother than the postrouting curve of Fig. 7.13. This indicates that, as

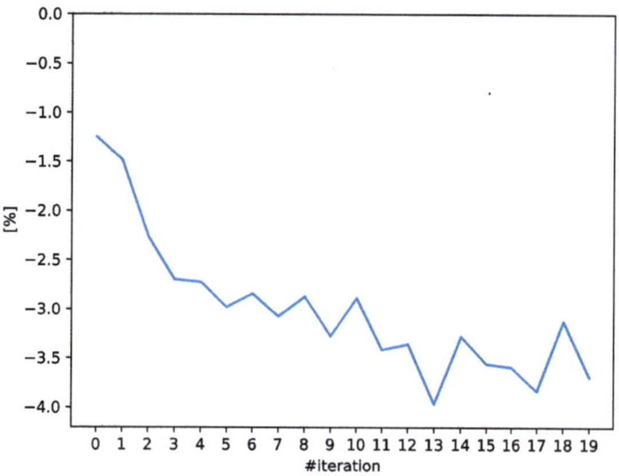

Fig. 7.13 Evolution of the relative change of the geomean postrouting critical path delay. Despite some unsurprising noise, the overall trend seems clearly toward a monotone improvement. Iteration counter is zero-based, meaning that the first direct connection was added to the pattern at $x = 0$

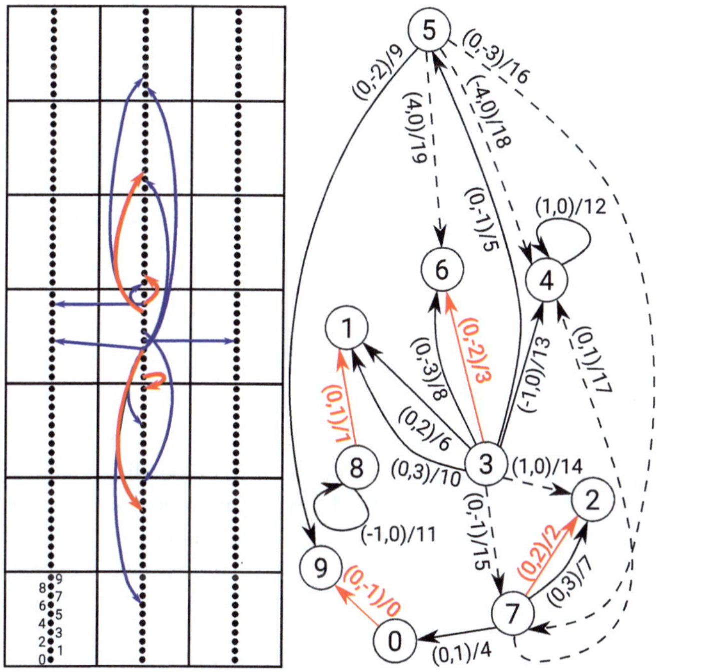

Fig. 7.14 Pattern obtained through the search process. Dashed edges are added after the iteration that results in the maximum achieved delay improvement (iteration #13) and are thus not considered further in our analysis. Connections in the fabric depiction on the left correspond to the edges of this best found pattern originating in one particular cluster. The four red edges that were the first ones to be added to the pattern are responsible for 68% of the total achieved gain

noisy as it is, the evolution of Fig. 7.13 is not a result of chance but truly driven by a sound, albeit imperfect, predictor. Note that during the search, the benchmarks *bgm*, *LU8PEEng*, *stereovision0*, and *stereovision2* were temporarily removed to

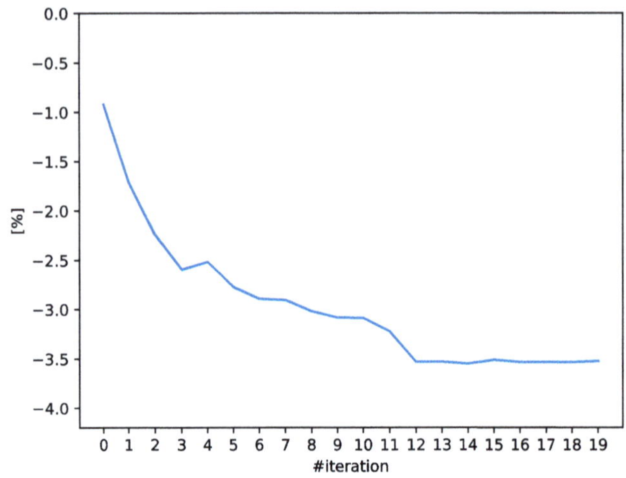

Fig. 7.15 Evolution of the relative change of the geomean postplacement critical path delay prediction after critical core solving

reduce the runtime, while *mkPktMerge* and *stereovision1* were removed to enhance stability. Critical paths of these latter benchmarks contain no connections between LUTs and are thus not directly improvable by the considered patterns.

7.7.2 Intercluster Connections: Delay Impact

Because our ultimate goal is to minimize the postrouting critical path delay, we take the pattern at iteration #13 (with 14 edges), corresponding to the minimum of the curve of Fig. 7.13, as the final one. Its edges are represented by solid lines in Fig. 7.14. The per-benchmark delay changes for this pattern are shown in Fig. 7.16. It is interesting to note the significant differences in the amount of improvement that the obtained pattern brings to different circuits. In the future, it could be useful to try to understand what circuit characteristics are causing these differences, perhaps by applying techniques similar to those developed by Hutton [Hut97]. That information could then be used to design different patterns of direct connections for different classes of circuits, in the spirit of what Betz and Rose proposed [Bet95]. The pattern exploration algorithm presented in this chapter could be readily used for that, by simply appropriately changing the set of circuits used during exploration to contain only those from the target class. Due to the decoupling multiplexers through which the direct connections drive the target LUT pins, all other circuits would still be effectively supported, with the direct connections merely being less useful for them. The delay penalty which the circuits not belonging to the target class would have to pay is limited to the delay of the decoupling multiplexers themselves. Due to sparsity of the direct connection patterns, which is necessary to maintain their speed, this penalty is typically very small, as we will see shortly.

The postplacement predictions of Fig. 7.15 suggest that the obtained improvements did not originate primarily from noise, but implementing connections as direct. Nevertheless, it is important to know how much of this improvement comes from fixing a different LUT permutation for the pattern-enhanced than for the reference architecture. To measure that, we fix the permutations that led to the improvements of Fig. 7.16 and report the postrouting delay change without using the direct connections in Fig. 7.17. As we can see, part of the observed improvement indeed comes from the chosen permutation. This explains, for instance, why the circuits *mkPktMerge* and *stereovision1* that do not even have connections between LUTs on their critical paths appear as improved. Moreover, the magnitude of the improvement contributed by the chosen permutation corresponds well with the mismatch between the observed postrouting delay of Fig. 7.13 and the postplacement delay prediction of Fig. 7.15. Despite this noise, the main source of improvement clearly does not lie in different permutation fixing. We can conclude that the actual improvement due to the introduction of direct connections is 2.77% instead of the 3.26% reported in Fig. 7.16, with the 0.49% difference contributed by permutation fixing.

In this chapter we have taken a specific avenue in adding direct connections. We have introduced it in Sect. 7.3 with the visual help of Fig. 7.2: We used direct connections connected through multiplexers at the input of the LUTs; the rationale is that this would bring a minimal penalty to the normal routing process whenever direct connections are not helpful. Figure 7.18 shows the relative postrouting delay changes for the benchmark circuits implemented in the pattern-enhanced architecture with unmodified packing of the reference (i.e., without any LUT

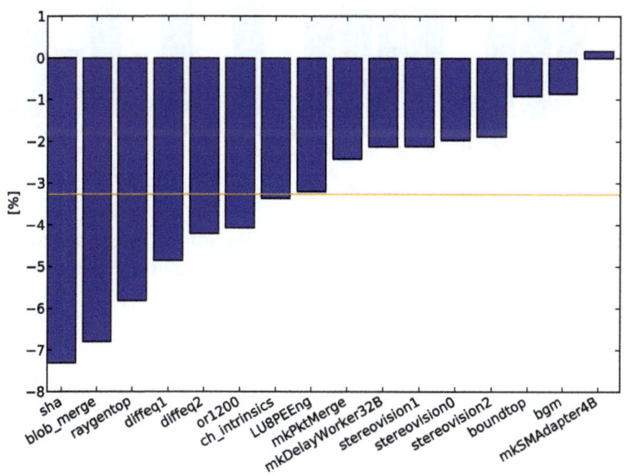

Fig. 7.16 Relative change of the postrouting critical path delay per benchmark. The orange line shows the decrease in the geomean critical path delay. Practically, no circuit is worsened by the presence of the direct connections and the benefit is up to 7%

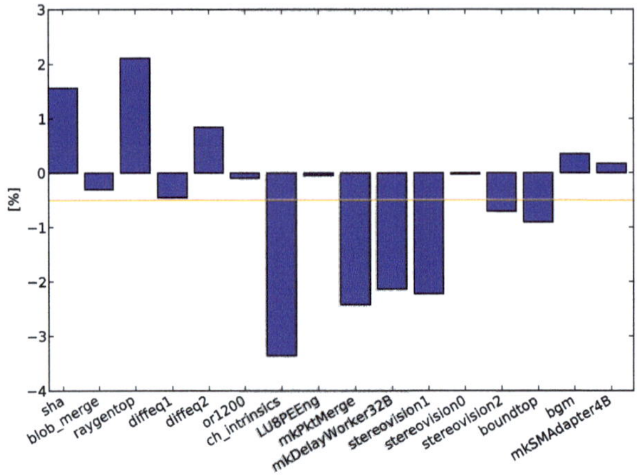

Fig. 7.17 Influence of the permutation on delay changes. This graph shows the delay changes resulting from fixing the permutations that led to improvements of Fig. 7.16, but not using any of the direct connections. The plot suggests that a part of the previously observed improvement is caused by the different permutations. The magnitude roughly corresponds to the difference between postplacement and postrouting delays (Figs. 7.15 and 7.13)

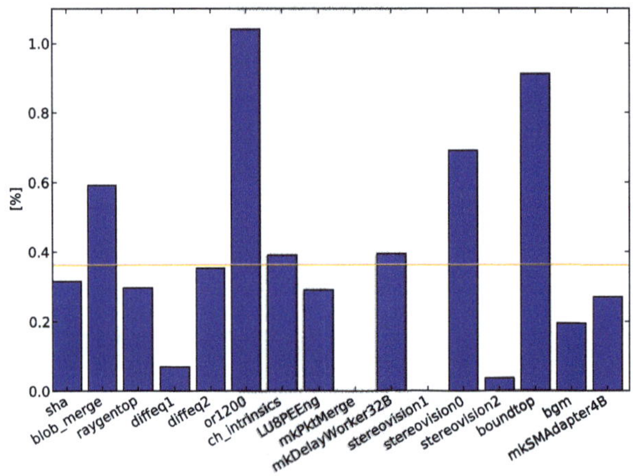

Fig. 7.18 Maximum delay overhead. The graph shows the impact of the added multiplexer delay on the circuit implementation found for the underlying architecture. We can always opt for these implementations

permutation) and no use of direct connections. As we can see, the penalty is indeed pretty insubstantial. Had the decoupling multiplexers been merged into those of the crossbar, the penalty would likely have been even smaller, at the expense of a slightly reduced effectiveness of direct connections and perhaps making layout more difficult (see Chap. 6).

7.7.3 *Intercluster Connections: The Pattern*

Let us comment briefly on the found pattern itself. It is interesting that most of the selected edges are vertical. This results from both the placement of circuits and the search algorithm favoring vertical edges in case of ties. The latter decision is due to their potentially shorter length (see Fig. 7.7). To illustrate the distribution of postplacement intercluster connections, irrespective of any direct connection, we show in Fig. 7.19 a heat map of the average fraction of signals going to each location in the neighborhood of a cluster. For this we use the *sha* benchmark that appears fairly representative of the whole set. The map indicates, for instance, that 6.2% of the signals leaving any cluster, on average, connect to an LUT in the cluster just above it. As we can see, there are indeed more connections that are vertical than those that are horizontal, while the diagonal ones are even rarer. Another interesting observation is that the heat map is fairly symmetrical, which could explain why pairs of opposed vectors often occur as consecutive pairs of edge weights chosen by the search algorithm. One possible reason for the larger number of vertical connections could be the columnar architecture of modern FPGAs (see Fig. 2.15). We note that, unlike in Chaps. 4, 5, and 6, the aspect ratio of the tile is here assumed to be 1, the device is logically square, and capacities and segmentations of the vertical and the horizontal channels are identical. The observed directional bias of the placed circuits might point to a need to revisit the conclusions about the directional bias in architectures composed of square tiles that were previously made by Betz and Rose [Bet96], in the context of modern FPGAs that incorporate columns of hardened blocks. We discussed some aspects of device and tile aspect ratios in Setion 4.7.3.1.

Finally, the area overhead of the added connections is equivalent to 612 minimum width transistors, with 374 contributed by the drivers and 238 by the multiplexers. This represents a mere 1.13% increase of the cluster area of the underlying architecture (not counting the area of the switch-blocks and the connection blocks).

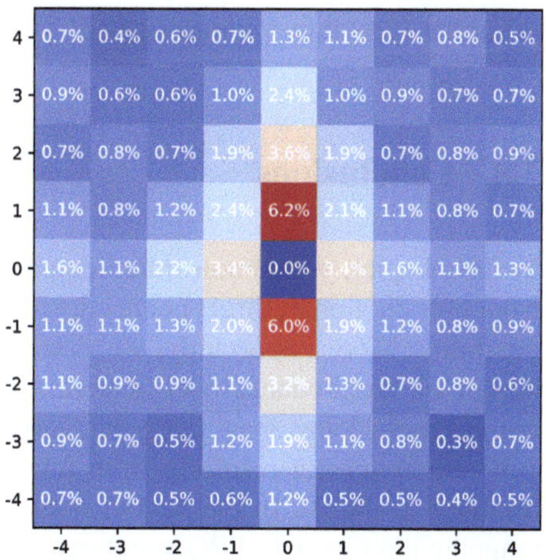

Fig. 7.19 Distribution of target-cluster locations of intercluster connections bounded by $w = 4$. The x-axis corresponds to the horizontal offset and the y-axis to the vertical offset. The data comes from the *sha* benchmark. Dominant frequency of vertical connections and rough symmetry of the heat map help explain the choices made by the search algorithm

7.7.4 Intercluster Connections: A Trade-Off

We may note that 68% of the entire achieved gain came from the first four edges added to the pattern. They are shown in red in Fig. 7.14. By sacrificing some of the performance gain, the required minimum width transistor investment reduces to 147 or merely 0.27% of the cluster area. Not only is the pattern formed by these four edges perfectly symmetrical in terms of edge weights, but the edges were also chosen at each step to be the shortest ones with the given weight, while maintaining the node degree bound of one. If we recall the discussion of Sect. 7.5.2, this feature greatly simplifies the problem of permutation. Hence, it is probably not only the good selection of edge weights, but also the possibility for an imperfect mapping algorithm to actually use them that makes this pattern successful.

7.7.5 Intracluster Connections

As mentioned before, starting from intercluster connections seemed more reasonable both in terms of easier filtering of best connection candidates and in terms of optimization potentials. In fact, contribution of the intracluster connections to the geomean critical path delay of the benchmark set used during the search is a mere 2.84%, compared to 33.6% for intercluster connections. This confirms the expectations derived from the absence of local connections from the typical critical path of Fig. 2.17. Despite the drastically smaller margin for improvement, we did an experiment to add local direct connections to the best previous pattern, by including in the ILP formulation only those connections that have both endpoints in the critical core after it has already been extracted: This way, the solver can still return the previous solution but can also opt for covering a local connection instead, if beneficial. As we can see from Fig. 7.20, there is little improvement in critical path delay, without any visible trend toward this improvement increasing with addition of further local direct connections.

7.8 Conclusions

In this chapter, we demonstrated empirically that FPGA performance can benefit from introduction of direct connections between LUTs, without any compromise in flexibility. More importantly, we developed an efficient algorithm for automated design of fixed-connectivity patterns. The 2.77% average improvement of the critical path delay may not seem large at first, but it is comparable with the $\leq$3% improvement resulting from introduction of time-borrowing capabilities to Stratix V [Lew13]. In the next chapter, we will develop dedicated CAD tools for mapping circuits onto FPGA architectures with direct connections between LUTs,

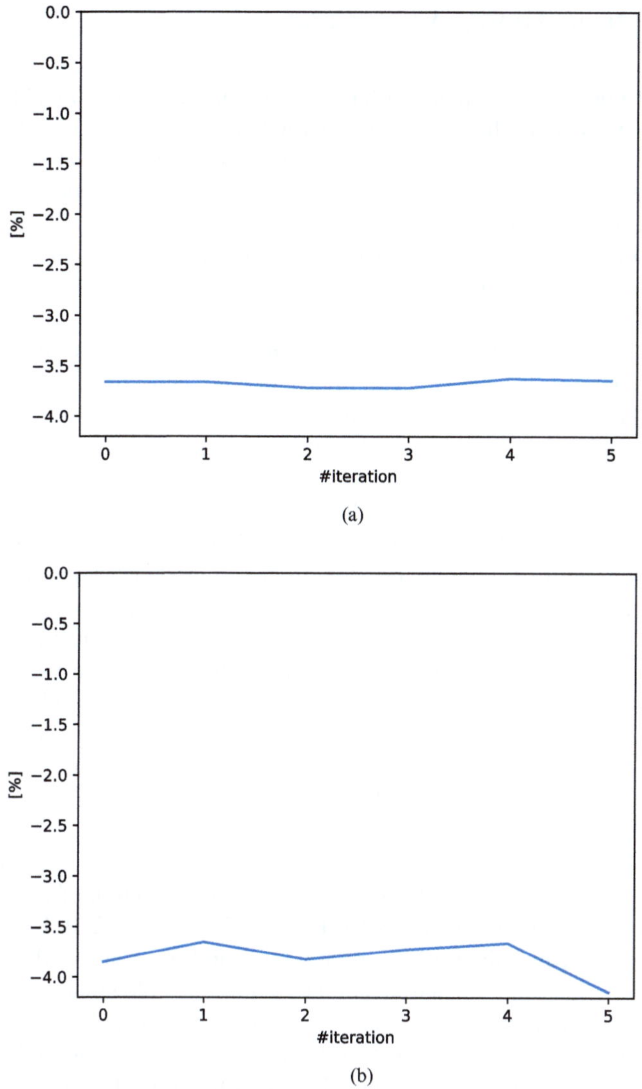

Fig. 7.20 Relative change of the geomean critical path delay while adding local connections to the pattern of Fig. 7.14. (**a**) Postplacement. (**b**) Postrouting

to further increase their effectiveness. Exploring the use of direct connections for replacing portions of programmable interconnect in order to improve area efficiency (and hence in turn delay through reduction in wirelength) of reconfigurable fabrics, rather than just augment the programmable interconnect with direct connections to improve its performance, is an interesting avenue that we plan to pursue in future work.

7.9 A Note on Timing Assumptions

Since direct connections replace different paths through programmable interconnect, conclusions about their effectiveness fundamentally depend on the delays of the resources which they replace. This, in turn, depends on fabrication technology. For example, it may happen that the utility of intracluster direct connections would be much higher in scaled technologies, as they could provide a low-load alternative to crossbar wires, especially if a relatively large cluster size of previous technologies is retained (see Chap. 4). Unfortunately, due to some logistical details related to the version of VTR that this study was based on, at the moment we are not able to repeat the experiments presented here in a newer technology. Nevertheless, the exploration algorithm which we developed is entirely technology-agnostic and can be used to this end in the future, without any modification.

A more fundamental problem with timing assumptions is related to handling differences in LUT input delays. Namely, delays of different inputs of a 6-LUT in Stratix IV, which formed the basis for the underlying architecture used in this study, are [Luu14]: 82, 173, 261, 263, 398, and 397 ps. However, since VTR, as of version 8, cannot route until LUT inputs in a timing-driven mode (rather projecting sinks to cluster inputs during packing) [Luu14, Mur20], all inputs are reported to have the average delay of 261 ps [Luu14]. This means that the present study may have slightly overestimated the utility of direct connections, since their use is perceived as beneficial whenever it is possible. If LUT input delays are faithfully represented, however, if, e.g., a direct connection is driving the input with the delay of 398 ps and the router can reach the input with the delay of 82 ps through programmable interconnect, delay savings caused by replacing the programmable path with the direct connection will be reduced by $398 - 82 = 316$ ps—a very large number in comparison with the 126 ps delay of a length-4 channel wire in this architecture. Of course, this input-rotation penalty is exaggerated for the pattern of Fig. 7.14, since the in-degree distribution among its LUTs is $(1, 3, 2, 0, 2, 1, 2, 0, 1, 2)$. Not only does this greatly reduce the rotation penalty (even to 0 for some of the LUTs), but the low in-degrees also allow introduction of a more flexible input access structure without incurring a large area and delay cost; this would make it possible to bring the most critical signals to the fastest LUT pins through direct connections as well. Extensions to VTR that enable routing all the way to the LUT inputs have recently been introduced [Moh23] and will most likely enter the next stable version of VTR (VTR 9), which will allow the above concerns to be resolved in the future.

We should also note that some of the large delay discrepancy between the above LUT input delays are caused by high complexity of the fracturable ALM used in Stratix IV [Lew05] that would not exist in other implementations. In fact, alternatives that largely suppress even the natural differences in input delays caused by the tree structure of the LUT's multiplexer have been invented [Chi08]. Finally, and perhaps most importantly, as technology scales, LUT delays become less and less problematic in comparison with wire delays, meaning that input-rotation penalty will be more easily offset by replacing wires of programmable interconnect by low-load, minimum length direct connections.

References

[Ahm01] Ahmed, E. (2001). The effect of logic block granularity on deep-submicron FPGA performance and density. Master Thesis. Toronto: University of Toronto.

[Bet95] Betz, V., & Rose, J. (1995). Using architectural "families" to increase FPGA speed and density. In *Proceedings of the 1995 ACM Third International Symposium on Field-Programmable Gate Arrays*. Monterey, CA, USA (pp. 10–16).

[Bet96] Betz, V., & Rose, J. (1996). Directional bias and non-uniformity in FPGA global routing architectures. In *Proceedings of International Conference on Computer Aided Design*, San Jose, CA, USA (pp. 652–59).

[Bet99] Betz, V., Rose, J., & Marquardt, A. (1999). *Architecture and CAD for deep-submicron FPGAs*. Kluwer Academic Publishers. ISBN: 0-7923-8460-1.

[Bon08] Bondy, J. A., & Murty, U. S. R. (2008). *Graph theory* (1st ed.). Springer Publishing Company, Incorporated. ISBN: 1846289696.

[Bor95] Borriello, G., Ebeling, C., Hauck, S., & Burns, S. M. (1995). The triptych FPGA architecture. *IEEE Transactions on Very Large Scale Integration Systems, 3*(4), 491–501.

[Che07] Chen, D. T., Vorwerk, K., & Kennings, A. (2007). Improving timing-driven FPGA packing with physical information. In *2007 International Conference on Field Programmable Logic and Applications*. Amsterdam, The Netherlands (pp. 117–23).

[Chi08] Chirania, M. (2008). Lookup table with relatively balanced delays. US 7471,104 B1. Patent assignee: Xilinx Inc.

[Chi13] Chiasson, C. (2013). Optimization and modeling of FPGA circuitry in advanced process technology. MA thesis. University of Toronto.

[Cho91] Chow, P., Seo, S. O., Au, D., Fallah, B., Li, C., & Rose, J. (1991). A $1.2\mu m$ CMOS FPGA using cascaded logic blocks and segmented routing. In *Proceedings of the International Workshop on Field Programmable Logic and Applications*, Oxford, UK.

[Chr20] Chromczak, J., Wheeler, M., Chiasson, C., How, D., Langhammer, M., Vanderhoek, T., et al. (2020). Architectural enhancements in Intel®Agilex™FPGAs. In *Proceedings of the 2020 ACM/SIGDA International Symposium on Field-Programmable Gate Arrays*, Monterey, CA, USA (pp. 140–49).

[Fei98] Feige, U. (1998). A threshold of Ln N for approximating set cover. *Journal of the ACM, 45*(4), 634–52. ISSN: 0004-5411.

[Fen18] Feng, W., Greene, J. W., & Mishchenko, A. (2018). Improving FPGA performance with a S44 LUT structure. In *Proceedings of the 2018 ACM/SIGDA International Symposium on Field-Programmable Gate Arrays*, Monterey, CA, USA (pp. 61–66).

[Gai19] Gaide, B., Gaitonde, D., Ravishankar, C., & Bauer, T. (2019). Xilinx adaptive compute acceleration platform: Versal architecture. In *Proceedings of the 2019 ACM/SIGDA International Symposium on Field-Programmable Gate Arrays*, Seaside, CA, USA (pp. 84–93).

[Hof94] Höfting, F., & Wanke, E. (1994). Polynomial time analysis of torodial periodic graphs. In *Proceedings of the 21st International Colloquium on Automata, Languages and Programming*. Jerusalem, Israel (pp. 544–555).

[Hut02] Hutton, M., Chan, V., Kazarian, P., Maruri, V., Ngai, T., Park, J., et al. (2002). Interconnect enhancements for a high-speed PLD architecture. In *Proceedings of the 2002 ACM/SIGDA Tenth International Symposium on Field-Programmable Gate Arrays*, Monterey, CA, USA (pp. 3–10).

[Hut97] Hutton, M. (1997). Characterization and parameterized generation of digital circuits. PhD thesis. University of Toronto.

[Int20] Intel Corporation. (2020). Intel®Stratix®10 Logic Array Blocks and Adaptive Logic Modules User Guide. UG-S10LAB, ver. 2020.04.24.

[Int23a] Intel Corporation. (2023). Intel Agilex®7 Logic Array Blocks and Adaptive Logic Modules User Guide. UG-20204, ver. 2023.03.27.

[Lem00] Lemieux, G., Leventis, P., & Lewis, D. (2000). Generating highly-routable sparse crossbars for PLDs. In *Proceedings of the 2000 ACM/SIGDA Eighth International Symposium on Field Programmable Gate Arrays*, Monterey, CA, USA (pp. 155–164).

[Lew03] Lewis, D., Betz, V., Jefferson, D., Lee, A., Lane, C., Leventis, P., et al. (2003). The stratix routing and logic architecture. In *Proceedings of the 2003 ACM/SIGDA Eleventh International Symposium on Field Programmable Gate Arrays*, Monterey, CA (pp. 12–20).

[Lew05] Lewis, D., Ahmed, E., Baeckler, G., Betz, V., Bourgeault, M., Cashman, D., et al. (2005). The stratix II logic and routing architecture. In *Proceedings of the 2005 ACM/SIGDA 13th International Symposium on Field-Programmable Gate Arrays*, Monterey, CA, USA (pp. 14–20).

[Lew13] Lewis, D., Cashman, D., Chan, M., Chromczak, J., Lai, G., Lee, A., et al. (2013). Architectural enhancements in stratix V™. In *Proceedings of the ACM/SIGDA International Symposium on Field Programmable Gate Arrays*, Monterey, CA, USA (pp. 147–56).

[Luu14] Luu, J., Goeders, J., Wainberg, M., Somerville, A., Yu, T., Nasartschuk, K., et al. (2014). VTR 7.0: Next generation architecture and CAD system for FPGAs. *ACM Transactions on Reconfigurable Technology and Systems, 7*(2), 1–30. ISSN: 1936-7406.

[Luu14a] Luu, J. (2014). Architecture-aware packing and CAD infrastructure for field-programmable gate arrays. Ph.D. Thesis. Toronto: University of Toronto.

[Mak19] Makhorin, A. O. (2019). GLPK (GNU Linear Programming Kit). https://www.gnu.org/software/glpk/.

[Moh23] Mohaghegh, A., & Betz, V. (2023). Tear down the wall: Unified and efficient intra-and inter-cluster routing for FPGAs. In *Proceedings of the 33rd International Conference on Field-Programmable Logic and Applications (FPL)*. Gothenburg, Sweden (pp. 130–36).

[Mun57] Munkres, J. R. (1957). Algorithms for the assignment and transportation problems. *Journal of the Society for Industrial and Applied Mathematics, 5*(1), 32–38.

[Mur20] Murray, K. E., Petelin, O., Zhong, S., Wang, J. M., Eldafrawy, M., Legault, J.-P., et al. (2020). VTR 8: High-performance CAD and customizible FPGA architecture modelling. *ACM Transactions on Reconfigurable Technology and Systems (TRETS), 13*(2), 9:1–60.

[Nik19] Nikolić, S., Zgheib, G., & Ienne, P. (2019). Finding a needle in the haystack of hardened interconnect patterns. In *2019 29th International Conference on Field Programmable Logic and Applications (FPL)*. Barcelona, Spain (pp. 31–37).

[Nik20] Nikolić, S., Zgheib, G., & Ienne, P. (2020). Straight to the point: Intra-and intercluster LUT connections to mitigate the delay of programmable routing. In *FPGA '20: The 2020 ACM/SIGDA International Symposium on Field-Programmable Gate Arrays*, Seaside, CA, USA, February 23–25, 2020 (pp. 150–60).

[Par11] Parandeh-Afshar, H., Zgheib, G., Brisk, P., Ienne, P. (2011). Reducing the pressure on routing resources of FPGAs with generic logic chains. In *Proceedings of the ACM/SIGDA 19th International Symposium on Field*, Monterey, CA, USA (pp. 237–46).

[Ray12] Ray, S., Mishchenko, A., Eén, N., Brayton, R. K., Jang, S., & Chen, C. (2012). Mapping into LUT structures. In *2012 Design, Automation & Test in Europe Conference & Exhibition*, Dresden, Germany (pp. 1579–1584).

[Roo02] Roopchansingh, A., & Rose, J. (2002). Nearest neighbour interconnect architecture in deep submicron FPGAs. In *IEEE Custom Integrated Circuits Conference*, San Diego, CA, USA (pp. 59–62).

[Rub11] Rubin, R., & DeHon, A. (2011). Timing-driven pathfinder pathology and remediation: Quantifying and reducing delay noise in VPR-pathfinder. In *Proceedings of the ACM/SIGDA 19th International Symposium on Field Programmable Gate Arrays*, Monterey, CA, USA (pp. 173–176).

[Tan14] Tang, X., Gaillardon, P.-E., & De Micheli, G. (2014). Pattern-based FPGA logic block and clustering algorithm. In *24th International Conference on Field Programmable Logic and Applications*, Munich, Germany (pp. 1–4).

[Vaz04] Vazirani, V. V. (2004). *Approximation algorithms*. Springer. ISBN: 9783540653677.

[Xil93] Xilinx Inc. (1993). *XC2000 logic cell array families*. San Jose, CA, USA.

[Xil98] Xilinx Inc. (1998). *The programmable logicdata book.*

[Zgh17] Zgheib, G., & Ienne, P. (2017). Evaluating FPGA clusters under wide ranges of design parameters. In *Proceedings of the 27th International Conference on Field Programmable Logic and Applications*. Ghent, Belgium (pp. 1–8).

[Zha07] Zhao, W., & Cao, Y. (2007). Predictive technology model for nano-CMOS design exploration. *ACM Journal on Emerging Technologies in Computing Systems, 3*(1), 1–17.

Chapter 8
Dedicated Placement for Fixed-Connectivity Patterns

Contents

© The Author(s), under exclusive license to Springer Nature Switzerland AG 2025

S. Nikolić, *Modern Programmable Interconnect Design*,
https://doi.org/10.1007/978-3-031-80629-2_8

In the last chapter, we developed an algorithm to automatically design patterns of direct connections between LUTs by extracting information from placed circuits. In doing so, we were constructing the pattern to fit the direct-connection-oblivious mapping of circuits, with little effort invested in the opposite direction, to map the circuits to fit the constructed pattern: Our sole modification to the usual CAD flow was to permute LUTs within their respective clusters after placement, so as to align them with the endpoints of the added direct connections, which only altered the otherwise arbitrary intracluster placement produced by VPR.

There are multiple reasons why basing early exploration on existing standard CAD tools is beneficial, one of them being purely practical. Namely, a problem in exploring such architectures is that there could be two different causes for failing to achieve the anticipated effect of the additional connections. One could, for instance, expect that a cascade of LUTs is reasonably useful for reducing the critical path delay of a typical circuit. Failure to observe any benefit could lead to a guess that the CAD tools do not provide adequate support for such cascades. However, before dedicating effort to envisioning new algorithms, it would be useful to know that the problem does not lie in the simplicity of the cascade itself because, for instance, it cannot cover multiple fanouts or fanins. Unfortunately, without an optimal algorithm for putting the cascades to use, one cannot be sure which of these two potential sources led to the unexpected result. In other words, a lack of good algorithms makes it hard to assess the quality of architectures, while a lack of a good architecture makes it hard to assess the quality of algorithms, unless they are proven to be optimal.

By demonstrating in the last chapter that fixed-connectivity patterns are indeed effective for critical path delay optimization, we have broken this cycle of uncertainty: The constructed patterns can only be rendered still more effective if coupled with dedicated CAD tools. In this chapter, largely based on our paper previously published at the 30th International Conference on Field-Programmable Logic and Applications in 2020, under the title "Timing-Driven Placement for FPGA Architectures with Dedicated Routing Paths" [Nik20a], and in particular its journal extension which appeared in 2022 in ACM Transactions on Reconfigurable Technology and Systems, under the title "Detailed Placement for Dedicated LUT-Level FPGA Interconnect" [Nik22], we develop a dedicated detailed placer that uses ILP to move LUTs across clusters, in order to better align them with the direct connections of the FPGA.

8.1 Quantifying Expectations

Before diving into details, it is useful to quantify our hopes. In the previous chapter, we have observed a $\sim 3\%$ improvement of the average critical path delay of a subset of VTR circuits. If the delays of all connections between LUTs were reduced to the average of the delays of the direct connections in the best found architecture

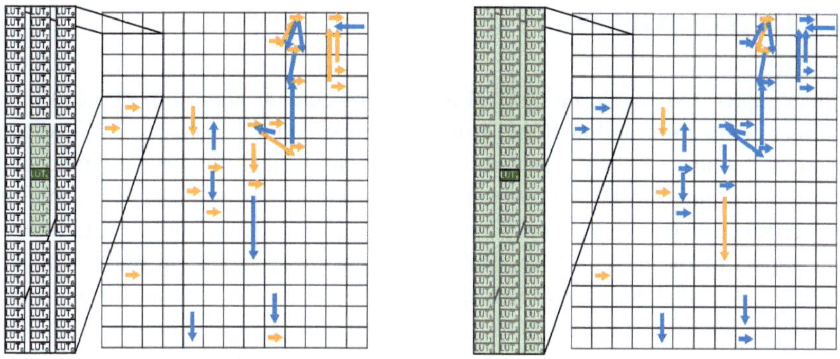

Fig. 8.1 Influence of movement freedom on delay minimization. Arrows depict a subset of connections of the *or1200* circuit, placed by standard VPR, on the architecture of Fig. 8.2. The blue arrows mark the connections that are successfully implemented as direct after appropriately moving their endpoint LUTs, while those that remain programmable are shown in orange. In the left figure, LUTs can only move within their respective cluster (indicated by the light green region depicting the allowed positions for the central LUT shown in dark green), while in the right, they can also move to the adjacent clusters, resulting in a 900 ps smaller delay

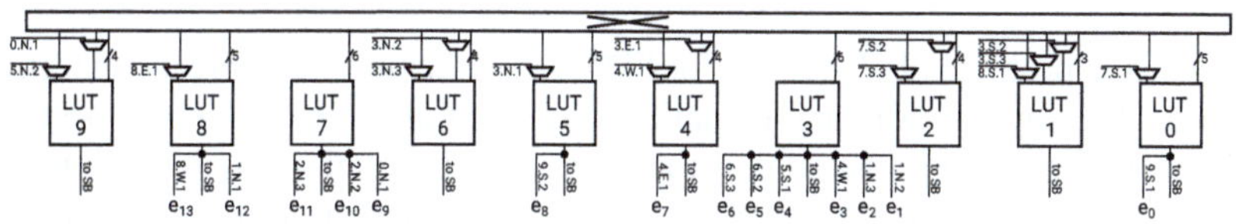

Fig. 8.2 Target architecture corresponding to solid edges of the static graph of Fig. 7.14. The other endpoint of each direct connection is labeled as $L.O.D$, where $L \in (0, 9)$ is the index of the LUT in its cluster, $O \in \{N, S, E, W\}$ is the connection orientation with respect to the shown cluster, and D is the distance to the other cluster. Diagonal connections are also supported by the algorithm but are not present in this architecture; hence, we use this simpler notation

of the last chapter, the improvement would rise to about 19%. This is clearly not achievable, but it shows that there is likely a fairly big margin for improvement.

A more illustrative example is given in Fig. 8.1. Each cell represents one 60-input cluster of ten 6-LUTs. The architecture also contains a number of direct connections between individual LUTs, shown in Fig. 8.2. The arrows show a subset of connections of the *or1200* circuit placed on this architecture by standard VPR [Luu14], oblivious of existence of the direct connections, with a resulting postplacement delay of 13 ns. Moving the LUTs within their respective clusters, in an attempt to improve this delay by aligning the connections depicted by arrows with the direct connections of the architecture, produces the figure on the left. The blue connections are the ones successfully aligned, resulting in a delay of 12.57 ns. Allowing the LUTs to move to the adjacent clusters as well produces the situation on the right, with the delay of 11.67 ns. These numbers were produced by actual optimization, as described in the following sections, and clearly demonstrate the benefits of moving LUTs across clusters.

8.2 Target Architectures

In this chapter, we develop a general detailed placement algorithm that can target all direct-connection architectures from the space defined in Sect. 7.4. The most important feature of these architectures, without which the detailed placement approach presented in the subsequent sections would not be possible, is that any direct connection can be used only optionally, owing to the decoupling multiplexers at the inputs of the target LUTs. This guarantees that every legal implementation of a circuit on the underlying FPGA architecture without the dedicated interconnect will also be legal for the architecture augmented with the direct connections. Even though the algorithm supports any architecture from the aforementioned space, throughout this chapter, most experiments will use the best found architecture from the previous chapter, the schematic of which is repeated in Fig. 8.2. Whenever another architecture is used, this will be stated explicitly.

8.3 General Approach

We tackle the problem of placement for FPGA architectures with direct connections between LUTs by constructing a detailed placement algorithm which (1) selects a minimal subset of LUTs that allows the desired critical path delay reduction to be obtained by implementing some of the connections incident to the selected LUTs as direct and (2) then solves an ILP to determine the new positions of the selected LUTs such that the critical path delay is actually improved. Technical details of these two steps are explained in Sects. 8.5 and 8.6, respectively, while their relation to the existing work on detailed placement algorithms is presented in Sect. 8.4.

In this section, we attempt to give a higher level view of the important decisions that formed our approach to solving the problem. Notably, we answer the question of why a *placement* algorithm is imperative in the first place, why it is necessary to move individual LUTs, and why we opted for a detailed placer that acts upon an already constructed placement that neglects the existence of the direct connections. The position of the proposed algorithm in the overall CAD flow is shown in Fig. 8.3.

This section also includes a discussion of why arguably the most obvious solution to the placement problem from an academic standpoint—simulated annealing—is not particularly well suited to the situation when the objective is targeted implementation of critical connections of the circuit by direct connections of the FPGA.

Fig. 8.3 Position of the proposed detailed placement algorithm in a typical FPGA CAD flow

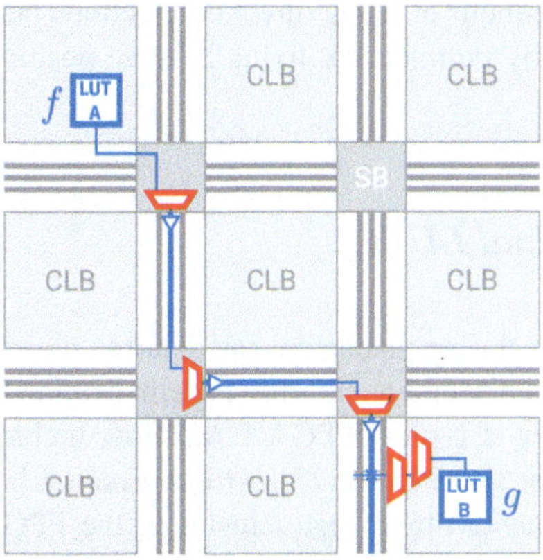

(a) Implementing a connection using programmable interconnect. Determined during routing.

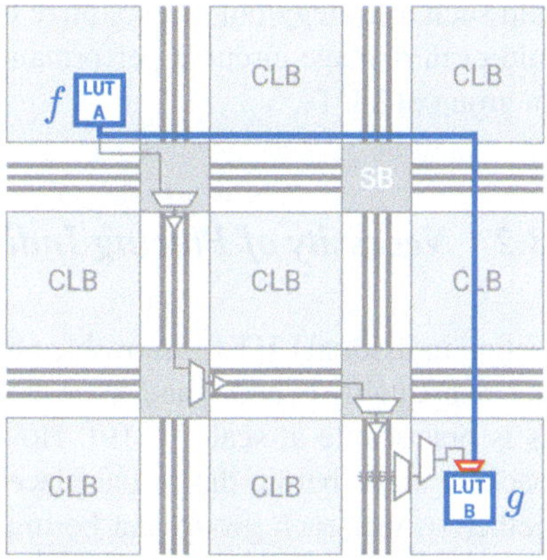

(b) Implementing a connection using a direct connection of the FPGA. Must be determined during placement.

Fig. 8.4 Importance of placement for using direct connections. Which wires and multiplexers will implement a connection of the circuit using programmable interconnect (a) can be determined at route time. However, whether it is possible to use a particular direct connection of the FPGA instead is fully determined by the placement of the two endpoint LUTs of the circuit's connection (b)

8.3.1 Is This Not a Routing Problem?

The purpose of this chapter is to develop adequate CAD support for FPGA architectures equipped with optional direct connections between LUTs, so that these fast dedicated connections may be used to implement the most critical connections of the user's circuit and increase its performance. In a standard FPGA CAD flow [Bet99] (Fig. 8.3), it is typically the router which determines the exact path through the programmable interconnect that will implement a particular connection of the circuit, once its endpoints have been fixed during placement. This is the case illustrated in Fig. 8.4a.

Given that our goal is to determine if some connections of the circuit can be profitably routed by the direct connections of the FPGA, a question could be raised if it is actually a routing and not a placement algorithm that is required. Similarly to the carry chains [Luu14b], the direct connections of the FPGA have uniquely defined endpoints. Hence, if a circuit connection (A, B) is to be implemented using a direct connection between points f and g, A (respectively, B) must be aligned with f (respectively, g) during placement; otherwise there will be no way of accessing this particular direct connection. This is illustrated in Fig. 8.4b.

Unlike the case of carry chains, timing criticality of a set of connections between LUTs cannot be readily ascertained during synthesis. At the same time, the number of possible topologies that the direct connections between LUTs can support vastly surpasses the columnar cascade of the carry chains. For these reasons, strategies such as a priori locking blocks together and moving them in unison during placement [Luu14b] would be too constraining for the problem at hand; not only could such a strategy fail to maximize the benefit of using direct connections but it could even damage circuit's performance, by prematurely fixing relative positions of a group of LUTs.

8.3.2 Necessity of Placing Individual LUTs

Treating individual LUTs as movable objects during placement can result in superior placement quality [Che04] and some modern placement algorithms demonstrate that this is practicable at scale [Li19]. However, a typical FPGA CAD flow includes a packing stage before the actual placement [Bet99, Mur15], which groups LUTs together so that each group can be implemented by a logic cluster of the FPGA. Then, these clusters, instead of LUTs, become movable objects in the placement process, greatly reducing the time needed to complete it [Chi11].

To actually use the direct connections of the FPGA, the endpoint LUTs of a circuit's connection must be aligned with the endpoint LUTs of a direct connection. This is often impossible to achieve by moving entire clusters of LUTs, as Fig. 8.5

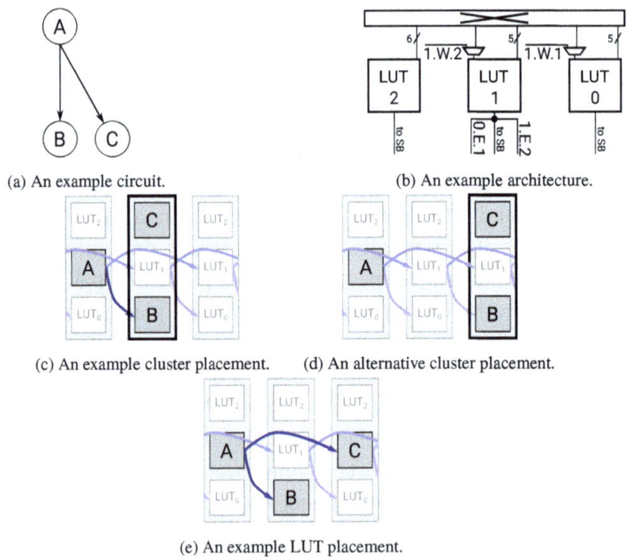

Fig. 8.5 Necessity of placing individual LUTs. (**a**) and (**b**), respectively, show a portion of a circuit and a simple FPGA architecture on which it is to be implemented. Two alternative cluster placements are shown in (**c**) and (**d**), both assuming $\{A\}$, $\{B, C\}$ for the initial packing of LUTs into clusters. Each cluster is represented by a vertical column of three LUTs, designated by the label of the circuit's LUT that it implements or $LUT_{0\text{–}2}$ when left unoccupied. The architecture's direct connections are depicted in blue: dark when used and light when unused. With this initial packing of LUTs, at most one connection of the circuit may be implemented using the direct connections of the FPGA, when only entire clusters are placed. If individual LUTs are able to move independently during placement, however, the outcome in Fig. 8.5e can be obtained, with both connections of the circuit implemented as direct and its critical path optimized

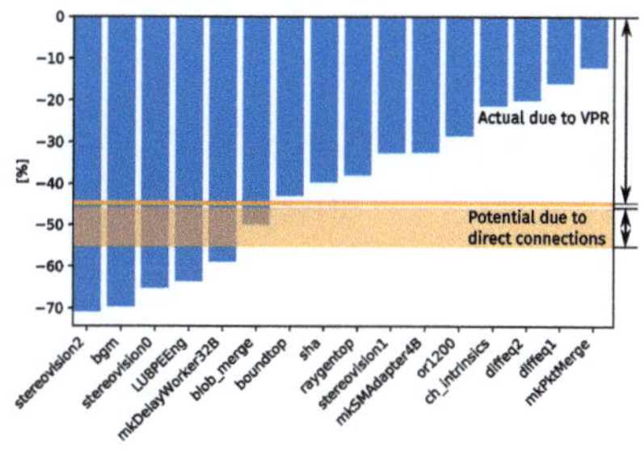

Fig. 8.6 Critical path delay improvement due to placement. Relative critical path delay improvement achieved by VPR at the end of the placement process, compared to the initial random placement of a subset of VTR benchmarks, is shown. A highly optimistic estimate of the average potential additional delay reduction due to proper use of direct connections of the target FPGA from Fig. 8.2 (Sect. 8.1) is superimposed in orange

illustrates. Our goal is to optimize the critical path delay of a circuit, which often requires implementing a small but precisely selected subset of the circuit's connections using the direct connections of the FPGA, and this can be met only by appropriately placing individual LUTs and not clusters.

Hence, for the currently popular cluster sizes of about ten LUTs, the problem we are facing could involve up to an order of magnitude more movable objects with an order of magnitude more candidate positions than what is usually tackled by a placer that follows a packing stage.

8.3.3 Global, Detailed, or Combined Placer?

General FPGA placers are very effective in optimizing the critical path delay of a circuit. To illustrate this, we measure the critical path delays of the subset of VTR circuits for which we previously computed the potential average critical path delay improvement due to direct connections (Sect. 8.1) at two instants of the VPR's placement process: (1) the very beginning—that is, when all clusters are placed randomly, and (2) at the end, when simulated annealing converges. The relative delays are plotted in Fig. 8.6. On average, they improve by almost 45%.

The average additional improvement over that achieved by VPR, obtainable through the implementation of connections between LUTs as direct, lies in the interval between ∼3%—the value confirmed in the last chapter—and ∼19%—the upper bound presented in Sect. 8.1. This means that the final combined improvement over the initial random placement will fall somewhere in the orange strip of Fig. 8.6. Whatever the actually obtained value of improvement due to direct connections may be, it is clear that it will be dwarfed by the improvement initially achieved by the general placer. Hence, it is meaningful to neglect the impact of direct connections on critical path delay until the critical path delay itself is reduced sufficiently for this additional optimization to become important. This enables runtime savings, since the initial part of the placement process can be performed at the cluster level.

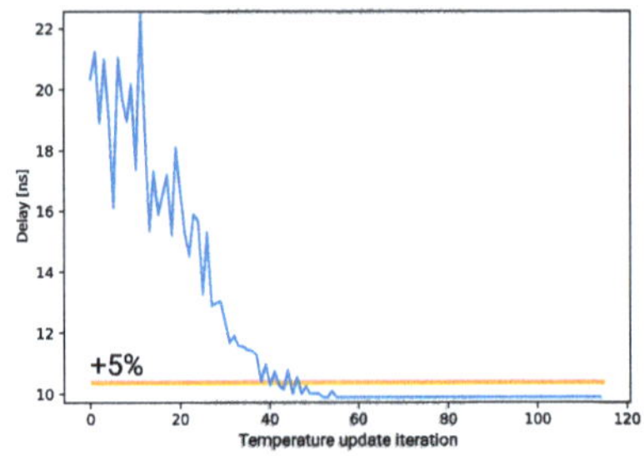

Fig. 8.7 Postplacement critical path delay evolution as a function of temperature update iteration while placing the *blob_merge* circuit. The orange line depicts the critical path delay which is 5% worse than the one obtained at the end of VPR's placement process. To achieve this final 5% improvement—roughly comparable to what one could realistically expect from appropriately using direct connections—a significant portion of iterations is used

8.3.4 Direct Connections at Low Temperature

Let us for the moment stay in the framework of simulated annealing used by VPR. An obvious solution to the problem of individual LUT placement that would partially mitigate the runtime increase would be to perform cluster placement until a certain temperature level, continuing with placement of individual LUTs afterward, until convergence.

8.3.4.1 Runtime Surge Persists

This approach has two issues, however. The first one is that it would not really resolve the problem of runtime surge. To illustrate this, we plot in Fig. 8.7 the evolution of the postplacement critical path delay over temperature update iterations during a placement of the *blob_merge* circuit [Luu14]. The orange line indicates the point where the critical path delay is 5% larger than the final postplacement critical path delay that VPR was able to achieve. This value was chosen as a reasonable estimate of what impact direct connections could have.

As Fig. 8.7 shows, a large portion of the temperature update iterations is spent on this final 5% delay reduction. Let us optimistically assume for the moment that, with some tuning of the exit criteria, the process would be able to end in 60 iterations. This would mean that about a third of the time would be spent on the final 5% of the delay reduction, and this would be the time when placement of individual LUTs would have to be performed if the direct connections are to be used appropriately. Given that the number of moves per iteration typically depends on the number of movable objects with $\Theta(n^{4/3})$ [Bet99], a tenfold increase in the number of movable objects when switching from placing clusters to placing LUTs would increase this third of the runtime almost 22×, increasing the overall time about 8×.

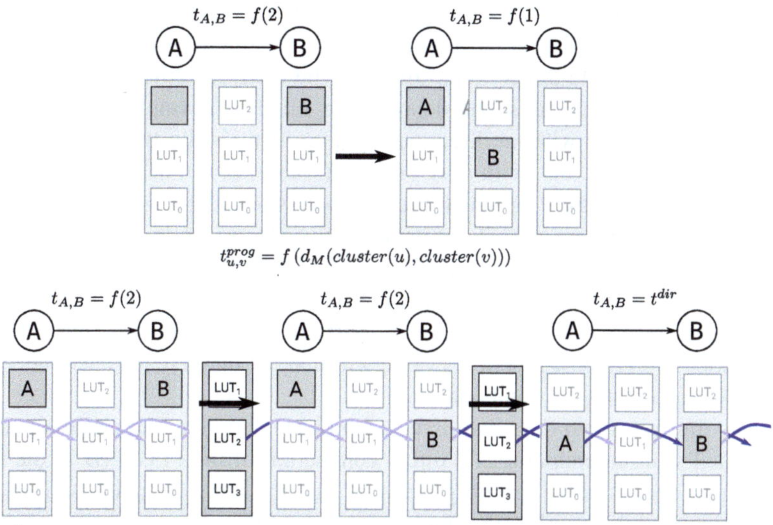

$$t^{prog}_{u,v} = f\left(d_M(cluster(u), cluster(v))\right)$$

$$t_{u,v} = \begin{cases} t^{dir}, & \text{if } u \mapsto LUT_1 \wedge v \mapsto LUT_1 \wedge cluster(v).x = cluster(u).x + 2 \wedge cluster(v).y = cluster(u).y \\ t^{prog}_{u,v}, & \text{otherwise} \end{cases}$$

Fig. 8.8 Difficulty of appropriately utilizing direct connections with simulated annealing-based placement. To implement a circuit connection as direct, both of its endpoints must be aligned with the endpoints of the direct connection. This is not easy to achieve using the standard moves of swapping pairs of randomly selected objects [Bet99]. Once a connection is implemented as direct, changing it back to programmable would be expensive, which could lead to suboptimal usage of scarce direct connections

Since runtimes of simulated annealing-based placers are already not competitive by today's standards [Van15], this would be prohibitive in a production setting.

8.3.4.2 Difficulties in Utilizing Direct Connections

The much more significant issue with this approach is its inaptness for the problem of aligning endpoints of circuit connections with the direct connections of the FPGA. As an illustration, let us take a look at Fig. 8.8. In the top part of the figure, a portion of an FPGA without any direct connections is shown, with a pair of LUTs that eventually need to be connected. For the sake of simplicity, let us assume that the delay of the implemented connection is some function of the Manhattan distance between the clusters in which the two endpoint LUTs reside at the given instant of the placement process (d_M in Fig. 8.8). Let us also assume that each move performed by the placer is a swap of two randomly selected LUTs [Bet99]. Each move will be reflected on the cost function, allowing the optimization to favor moves that improve it, as the temperature of the anneal decreases.

The bottom of Fig. 8.8 depicts the same architecture augmented with one type of direct connections. To appropriately model the impact that this has on the delay of the implemented connection of the circuit, we must introduce a discontinuity in the cost function. Namely, if A and B are positioned at LUT_1, two clusters apart horizontally, the direct connection may be used, resulting in a dramatic drop in delay. In all other cases, the delay is the same as it would have been in the original architecture without direct connections. This is illustrated by the formula at the bottom of Fig. 8.8, with $cluster(u).x$ ($cluster(u).y$) designating the x- (y)-coordinate of the cluster in which the node u resides and $u \mapsto LUT_1$ describing the fact that u is placed at LUT_1 of its respective cluster. Let us assume that a move of B was generated, resulting in the placement in the middle of the figure. In order for the LUTs to be properly aligned with the endpoints of the direct connection, a move bringing A to LUT_1 of its current cluster must be generated. If this happens, the sudden drop in cost function's value will make it unlikely for the connection to be broken again at a low temperature.

This illustrates two important issues: (1) If one endpoint of a circuit connection is aligned with an endpoint of a direct connection of the FPGA (B in the middle placement above), there are no guarantees that the other endpoint will be appropriately moved to complete the implementation of the circuit connection as direct, before the first endpoint moves again, and (2) once a connection is implemented as direct, unless the temperature is high, it is unlikely that it will move back to being programmable, which may prevent another, more critical connection of the circuit from using the particular direct connection of the FPGA.

While both of these issues could perhaps be partially mitigated by clever engineering of the cost function and adoption of directed moves [Vor07], it is evident that an approach better suited to the landscape created by the direct connections would be highly beneficial. Needless to say, appropriately capturing the discontinuities of the direct connections in other popular frameworks for large-scale placement, such as analytical [Mar19], would be difficult as well.

Adopting a detailed placement approach can resolve most of the above issues. This amounts to starting from a general placement produced by a placer which is unaware of the existence of direct connections and then strategically repositioning some of the nodes to improve the critical path delay through appropriate use of the fast direct connections. Another benefit of this approach is that it can be largely oblivious to which general placement algorithm is used to produce the starting placement. This would have been much more difficult if the direct connections were not strictly increasing the flexibility of the interconnect, as discussed in Sect. 8.2. Of course, the starting general placement does impact the ability of the detailed placer to improve the critical path delay. We will return to that issue in Sect. 8.9.2.3.

8.4 Prior Work on Detailed Placers

There is an abundance of published work introducing detailed placers, both for FPGAs and for ASICs [Li07, Cau11, Li12, Mih13, Mih13a, Dha16, Dha17]. Most of them operate on a sliding window principle, where a fixed region of the chip

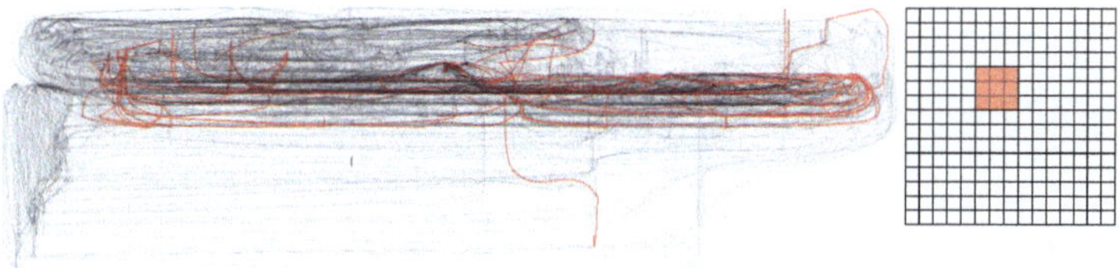

(a) Sliding window centered at cluster (9, 12).

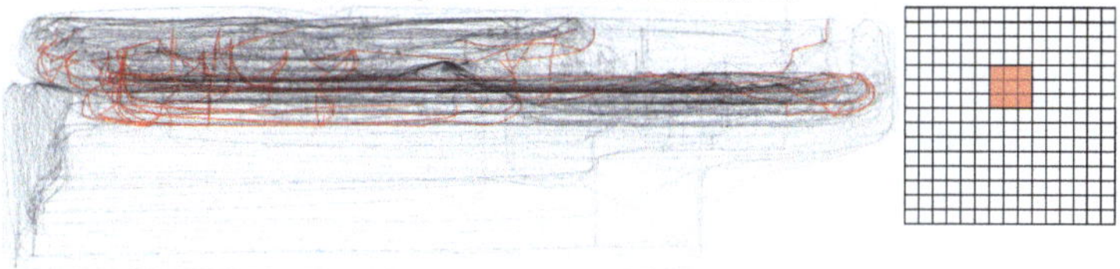

(b) Sliding window centered at cluster (10, 12).

Fig. 8.9 Sliding window-based movable node selection [Mar12]. The figure shows two different positions of a 3 × 3 cluster sliding window, on a portion of an FPGA containing a placement of the *blob_merge* circuit. All nodes inside the window are considered movable. Red edges in the timing graph of the circuit, shown on the left, connect different movable nodes. Clearly, the method gives little control over which edges may have their delays improved as a result of the moves. (**a**) Sliding window centered at cluster (9, 12). (**b**) Sliding window centered at cluster (10, 12)

is selected for optimization and then iteratively changed by sliding the window that determines it [Mar12]. One basic distinction between the various algorithms is how they optimize inside the window. Some of them rely on heuristics [Li07, Dha16], while others use exact optimization methods, such as ILP [Cau11, Li12], SAT [Mih13], or SMT [Mih13a]. The virtue of heuristics lies in their scalability which allows them to target larger windows at once, possibly increasing the improvement margin. Exact methods are usually not as scalable, so they are confined to smaller windows, with possibly smaller improvement margin but are guaranteed to actually meet it. We take a different approach to selecting which portion of the circuit will be optimized and describe it in more detail in Sect. 8.5. In this section, we analyze the existing approaches to provide motivation for introducing the proposed one. We also express the reasoning that led us to decide on using ILP as the optimization technique.

8.4.1 Movable Node Selection

The sliding window approach to guiding local optimization is illustrated in Fig. 8.9. One obvious downside of this approach is that it gives little control over which edges of the circuit's timing graph may suffer a change in delay as a result of moving the nodes inside the window. These edges are highlighted in red in Fig. 8.9.

Detailed placers that iteratively optimize edges of the critical path itself, thus avoiding this problem, have also been proposed [Dha17]. Such an approach naturally alleviates the potentially artificial spatial constraints imposed by the sliding

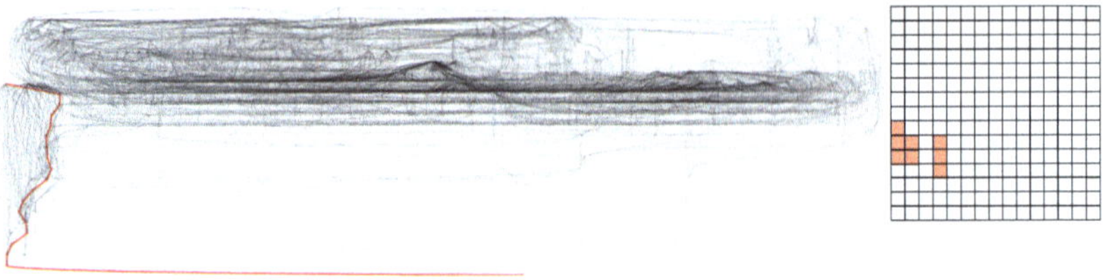

(a) Moving the nodes of the critical path.

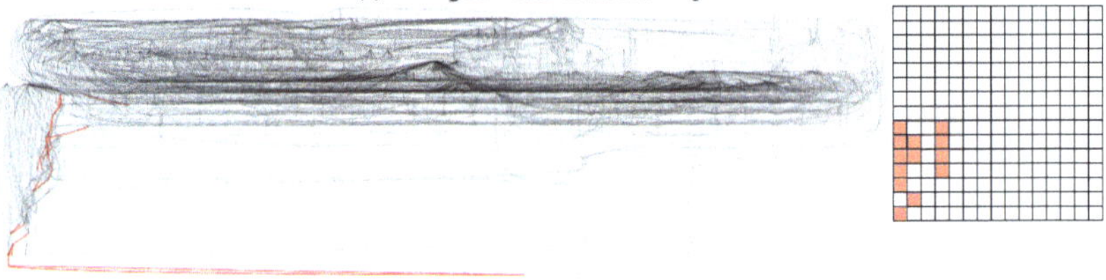

(b) Selecting a subset of nodes that maximizes delay reduction (Section 8.5).

Fig. 8.10 Movable node selection techniques without spatial constraints. (**a**) shows the results of deeming the nodes on a critical path of the circuit movable [Dha17], for the same example previously shown in Fig. 8.9. (**b**) shows results of applying the method introduced in Sect. 8.10b. This method also imposes no constraints on the spatial distribution of movable nodes, but it enables optimization of subgraphs of the timing graph of arbitrary complexity, maximizing the chance that the critical path delay is actually reduced once the selected nodes are moved. We note again that only a portion of the FPGA is shown; the movable nodes are in fact not in the corner

windows, as illustrated in Fig. 8.10a. However, optimizing only one simple path at a time may not be sufficient to actually decrease the critical path delay. Hence, we adopt a related but much more powerful approach, which selects a number of edges whose timing should be improved, regardless of their location in the timing graph, such that optimizing them maximizes the final critical path delay reduction. It then considers the endpoint nodes of the selected edges movable, regardless of the location of these nodes on the FPGA grid. An example of applying this method, described in Sect. 8.5, on the same circuit used to demonstrate the previous two approaches is shown in Fig. 8.10b. Clearly, the selected edges form a more complex topology in the timing graph than a simple path, while the movable nodes are distributed over a larger set of clusters than in either of the two previously described methods. Another example of spatial distribution of movable nodes obtained by the method proposed in Sect. 8.5 was shown in Fig. 8.1. Given that each cell represents a cluster of ten LUTs, that is, ten potentially movable objects, a sliding window approach would likely be limited to not more than a few cells in width and height [Mar12].

8.4.2 Movement Freedom

A sliding window-based selection method typically assumes that each node in the window can move anywhere within the window, as illustrated in Fig. 8.11a. We

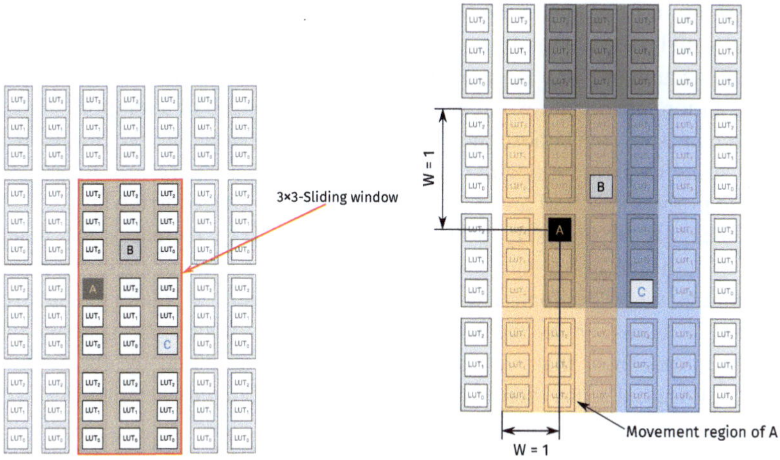

(a) Sliding window. (b) Adopted method.

Fig. 8.11 Movement freedom. (**a**) shows a portion of the FPGA with a sliding window of 3 × 3 clusters used to select movable nodes. Inside this window, all nodes are movable, and each of them can be placed anywhere in the window. (**b**) shows the approach that we take. Only a subset of nodes in each square region of the FPGA is selected for movement. Each of them can be placed at any position inside a square of half-width W, centered at its original cluster

take a similar approach by allowing each movable node to move anywhere within a square of half-width W, centered at its original cluster. This is illustrated in Fig. 8.11b.

The fact that the sliding window-based movable node selection assumes all nodes within the window to be movable has one important benefit: The subsequent optimization can guarantee that there are no overlaps between nodes. While the method that we use also guarantees that at the end of the optimization there will be no overlaps between the movable nodes, it leaves a possibility for a movable node to overlap with a stationary one, within its movement region. Such overlaps are only removed in a final postprocessing step, discussed in Sect. 8.7.2. The rationale is that if the set of movable nodes is appropriately selected, the benefit of optimally positioning them would most of the time far outweigh the penalty suffered from suboptimally moving some other, less critical nodes standing in their way. If all nodes within the movement regions of the originally selected movable nodes (Fig. 8.11b) were to be considered movable at the same time, the problem would quickly become impractically large, unless the number of originally selected movable nodes is itself severely restricted.

8.4.3 Choice of the Optimization Method

Direct connections are very sparse in a typical architecture considered here, thus requiring a high level of precision in placing the LUTs if the right connections of the circuit are to be aligned with them. As we have seen in Sect. 8.3.4, this leaves heuristics with less space for doing a good enough job at various points in the circuit that would accumulate to a large net improvement than they would have had when targeting a more abundant resource. For this reason, instead of attempting to design elaborate heuristics, we choose the exact approach. In particular, we opt for ILP, as it allows straightforward modeling of the timing information. Yet, we formulate the placement problem itself in a way that can be easily converted to SAT.

The necessity to precisely position individual LUTs increases the potential number of movable nodes and candidate locations for each of them by an order of magnitude when compared to the classical problem of placing entire clusters [Dha17]. However, as we will see shortly, it is exactly the sparsity of the dedicated interconnect that will help us resolve this problem, by enabling more efficient ILP formulations than in the case of general detailed placement [Mih13].

8.5 The LP-Based Node Selector

The first step in our placement flow is to determine the LUTs that will be moved from their initial positions. This problem is fundamentally linked to determining which connections of the circuit should have their delays reduced so that the reduction of the critical path delay is maximized.

8.5.1 Which Connections Should Be Improved?

Let T be the critical path delay that should be met after the detailed placement. Our goal is to select a minimal number of edges of the circuit's timing graph which should have their delays improved by their endpoint nodes being aligned with the endpoints of the direct connections of the FPGA, such that the postplacement critical path delay is reduced below T. The rationale is that the fewer edges there are to be improved, the fewer nodes will need to be moved, and more likely it is that the placement method of Sect. 8.6 will be able to find a solution in the allowed runtime budget.

Let $\tau_{u,v}$ be the initial postplacement delay of the connection $e = (u, v)$, as determined by the general placer. In the example of Fig. 8.12, this is illustrated using a simple model based on the Manhattan distance between the initial clusters of u and v, which we already saw in Sect. 8.3.4. In practice, any model used by the general placer can be used for obtaining the initial delays. Let us also assign to each edge $e = (u, v)$ a variable $imp_{u,v}$ describing how much its delay should be improved so that the critical path delay bound is met. Then, the final delay of the edge can be expressed as $t_{u,v} = \tau_{u,v} - imp_{u,v}$. In order to reduce the critical path delay below T, we need to find an assignment of *imp*-variables, which will appropriately reduce

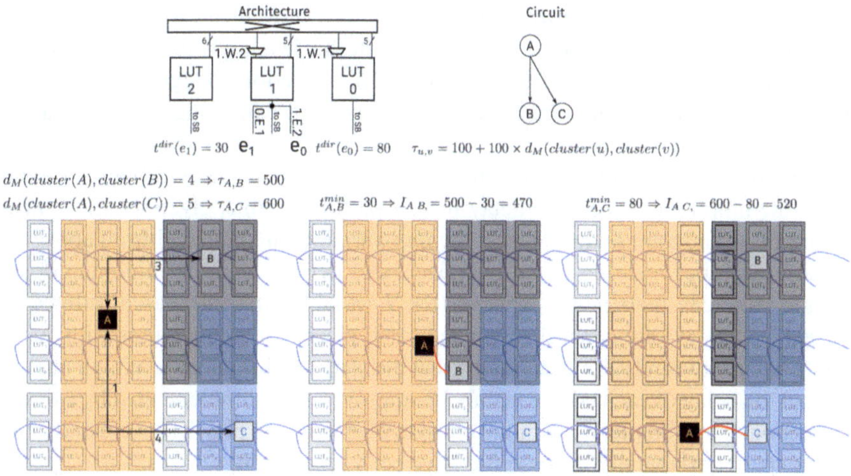

Fig. 8.12 Illustration of LP variables. Each edge in the timing graph of a circuit is assigned an *imp*-variable determining the amount by which its delay should be improved so that the target critical path delay is met. For the assignments of values to the *imp*-variables to reflect the restrictions on node movement, each variable is bounded from above by the difference between the initial postplacement delay and the minimum achievable delay, given the movement regions of its endpoints

the delay of each edge. We can achieve this by solving a *Linear Program* (LP) of the following form, introduced by Hambrusch and Tu [Ham97]:

$$\min \sum_{(u,v)\in E} imp_{u,v}, \tag{8.1}$$

$$\text{s.t.} \quad ta_u \leq T, \qquad \forall u \in V, \tag{8.2}$$

$$ta_v \geq ta_u + t_{u,v}, \qquad \forall (u,v) \in E, \tag{8.3}$$

$$0 \leq imp_{u,v} \leq I_{u,v}, \qquad \forall (u,v) \in E. \tag{8.4}$$

Here ta_u represents the arrival time of node u, and constraints (8.2)–(8.3) model the timing constraints in the usual manner [DeM94]. Note that to actually minimize the number of edges selected for improvement, the objective should be

$$\min \quad \left|\left\{(u,v) \in E : imp_{u,v} > 0\right\}\right|, \tag{8.5}$$

but representing it would require introduction of integral variables, which would render solving the program on the entire timing graph prohibitive.

The *imp*-variables must be nonnegative, as assigning a negative improvement to an edge with substantial slack could allow increasing the *imp*-variables of many other edges without changing the minimization objective. Representing the possibility of edges being slowed down due to node movement, to which negative *imp*-variables would correspond, is not needed at this level, where we only wish to determine which edges should be improved. This situation changes during actual movement of nodes, and hence in Sect. 8.6, we model the full range of delay values an edge can attain.

Similarly, the minimum values that can be assigned to the t-variables should reflect the minimum achievable delay for the particular edge, given the movement regions of its endpoint nodes. Hence, the *imp*-variables are bounded from above by the difference between the initial postplacement delay τ and the minimum delay that the edge can realistically achieve (the I-variables in (8.4)), as illustrated in Fig. 8.12. Since we are constructing a dedicated placer for architectures with direct connections between LUTs, we assume that the starting placement is of high quality and that the delay of each edge can only be improved if it is implemented by a direct connection. For example, if the initial clusters of A and C in Fig. 8.12 were one more cluster apart, horizontally, $I_{A,C}$ would have been zero, for $W = 1$.

8.5.2 Determining Movable Nodes

To extract the set of movable nodes, which we denote as V_m, from the solution of the above LP, we simply introduce a threshold θ on the minimum delay improvement. Then, the set of edges that should be improved and are thus candidates for implementation by the direct connections of the FPGA is $E_s = \{(u, v) \in E : imp_{u,v} \geq \theta\}$.

To actually implement these edges with direct connections, nodes incident to them must be moved and thus enter V_m. Controlling $|V_m|$ can be done only indirectly, by specifying the bound on the critical path delay, T. In general, the smaller the value of T, the more edges will have to be improved to meet it, and $|V_m|$ will rise accordingly. The fractional nature of the *imp*-variables, however, allows improvement to be spread among more edges than necessary, meaning that a more relaxed T does not necessarily result in smaller $|V_m|$. We comment on this further in Sect. 8.8.1, while the explanation of choosing the critical path delay bound is given in Sect. 8.7.

8.6 The ILP-Based Placer

In this section, we discuss various aspects of formulating the ILP that models the movement of the nodes selected by the process described in the previous section.

8.6.1 Naive ILP Formulation

Each LUT of the FPGA can be described by a triple (x, y, i), where x and y are the coordinates of its cluster and i the index within it. Let $P(u, W)$ be the set of positions within the square of half-width W, centered at the initial cluster of a movable node u (Fig. 8.11). Each LUT inside $P(u, W)$ is a candidate for placing u. To each node $u \in V_m$, we can assign the following set of variables: $x_{u,p} \in \{0, 1\}, \forall p \in P(u, W)$. The variable $x_{u,p}$ is 1 iff node u is placed at position p. The following set of constraints describes a valid placement, where we again note that overlaps with nodes outside V_m are removed in a postprocessing step:

$$\sum_{u \in V_m} x_{u,p} \leq 1, \qquad \forall p, \tag{8.6}$$

$$\sum_{p \in P(u,W)} x_{u,p} = 1, \qquad \forall u. \tag{8.7}$$

The first set of constraints prevents overlaps of movable nodes and the second makes sure that each movable node is assigned a unique position. Let $E_\psi = \{(u, v) \in E \setminus E_s : u \in V_m \vee v \in V_m\}$ be the set of edges which have at least one incident movable node and are thus affected by the placement but have not been selected for improvement. The delay of each edge in $E_s \cup E_\psi$ is determined by the location of its endpoints:

$$t_{u,v} = \sum_{p_u \in P(u,W), p_v \in P(v,W)} \tau_{p_u,p_v} e_{u,v,p_u,p_v}, \tag{8.8}$$

$$e_{u,v,p_u,p_v} \in \{0, 1\}, \qquad \forall p_u, p_v, \tag{8.9}$$

$$e_{u,v,p_u,p_v} \leq x_{u,p_u}, \tag{8.10}$$

$$e_{u,v,p_u,p_v} \leq x_{v,p_v}, \tag{8.11}$$

$$e_{u,v,p_u,p_v} + 1 \geq x_{u,p_u} + x_{v,p_v}. \tag{8.12}$$

Here, τ_{p_u,p_v} are constants corresponding to the least delay of the connection with its endpoints placed at p_u and p_v, respectively. Constraints (8.10)–(8.12) are merely a way to linearize a product of two binary variables [Wil13]. With the timing graph modeled as in the selection LP (constraints (8.2)–(8.3)), we have a complete formulation of the placement ILP. If (u, v) is from E_ψ, some variables may become constants, simplifying the corresponding constraints.

This formulation is generic and can be used to place circuits on architectures with or without direct connections. It is also rather intuitive and well known in the literature. In a very similar form, it has been used for SAT [Mih13] and SMT-based [Mih13a] timing-driven FPGA placement and for ILP-based wirelength-driven ASIC placement [Cau11].

The problem lies in the large number of position variables and quadratic length of delay assignment constraints (8.8) with respect to that number. Fixing W to 3—the length of the longest connection in the architecture of Fig. 8.2—leads to $7 \times 7 = 49$ clusters and 490 potential positions for each movable node. Any edge can have up to $490^2 > 240{,}000$ addends in the delay assignment constraint. This is clearly an issue, and we address it in the next section.

8.6.2 Exploiting the Sparsity of Dedicated Interconnect

Direct connections are sparse. If they were not, the width and count of the additional multiplexers and the increased loading of LUT outputs would greatly reduce their speed, slowing down the general routing as well. In the architecture of Fig. 8.2, there are only 14 connections originating in one cluster.

Let $E_d(u, v) \subseteq P(u, W) \times P(v, W)$ be the set of direct connections that can implement (cover) the edge $(u, v) \in E_s$. The exact position of a LUT matters only when an edge $e \in E_s$ incident to it is being covered. In all other cases, knowing its cluster is sufficient, since placement-time delay models of general placers rarely differentiate between different exact positions of LUTs [Mar00], which may be subject to change during routing, to reduce congestion [Lew03].

Instead of listing all exact positions for all movable LUTs and inspecting which edges are covered, we can list the edge covering possibilities and derive the LUT positions from them. Let $C(u, W)$ be the set of clusters within the square of half-width W, centered at the initial cluster of a movable node u. A binary variable $x_{u,c}$ indicates that u is placed in cluster $c \in C(u, W)$. We can now model the edge delays as follows:

$$t_{u,v} = \sum_{(p_u, p_v) \in E_d(u,v)} \tau_{p_u, p_v} e_{u,v,p_u,p_v} + \overline{c}_{u,v} \sum_{c_u \in C(u,W), c_v \in C(v,W)} \tau_{c_u, c_v} e_{u,v,c_u,c_v}, \tag{8.13}$$

$$\overline{c}_{u,v} = 1 - \sum_{(p_u, p_v) \in E_d(u,v)} e_{u,v,p_u,p_v}. \tag{8.14}$$

If there is a direct connection that covers the edge (u, v) in the current subproblem, the appropriate τ from the first sum will determine the delay because the *coverage indicator* $\overline{c}_{u,v}$ will be 0. In all other cases, the indicator will be 1, causing the second sum to determine the delay. The τ constants in that sum are the delays between the two appropriate clusters, as modeled by the general placer. The e_{u,v,c_u,c_v} variables are products of the cluster position variables, linearized using constraints similar to (8.10)–(8.12). Another level of linearization is applied to products with the coverage indicators. Note that constraints (8.13–8.14) are merely an ILP encoding of a generalized version of the delay model used in Fig. 8.8.

While the sparsity of direct connections created problems for convergence of common formulations of simulated annealing-based placement, it allows compact modeling of the problem as an ILP.

The maximum length of the delay assignment constraint for $W = 3$ and the architecture of Fig. 8.2 is now $49 \times 14 + 49^2 \ll 490^2$. The first addend corresponds to ≤ 49 ways to choose the starting cluster for the direct connection and 14 ways to choose the exact direct connection leaving it, while the second addend equals the number of cluster pairs that determine the edge delay if it is implemented as programmable. Similarly, the number of exact position variables for each node $u \in V_m$ is reduced from $|P(u, W)|$ to $| \cup_{\{u,v\} \in E_s} \{p_u : \{p_u, p_v\} \in E_d(u, v)\}|$, that is, to only those positions implied by covering of some edge incident to the given node.

Our model is still not complete. The linearizations (8.10)–(8.12) are of course kept, and they cause any e_{u,v,p_u,p_v} to imply exact positions of u and v. What is missing is that each exact position $x_{u,p=(x,y,i)}$ implies the corresponding cluster position $x_{u,c=(x,y)}$. The following set of constraints achieves that

$$x_{u,c=(x,y)} \geq \sum_i x_{u,p=(x,y,i)}, \quad \forall u \in V_m, c \in C(u, W). \tag{8.15}$$

Finally, we need to make sure that each node is assigned exactly one cluster, using constraints similar to (8.7).

8.6.3 Delay-Based Model Compaction

Further compaction of the model can be achieved by excluding irrelevant portions of the timing graph. Edges could be irrelevant either because their delay does not change during placement and they do not carry any information relevant for computing arrival and required times of the nodes affected by placement or because they can never become critical, under any feasible assignment of delays to the remaining edges.

8.6.3.1 Simplification of Fixed-Delay Subgraphs

An example of simplification of fixed-delay subgraphs is shown in Fig. 8.13, where the graphs are constructed merely for the purpose of illustration and have no further meaning. Edges selected for improvement (E_s) and the nodes incident to them (V_m) are shown in green, while other edges whose delay may change as nodes move (E_ψ) and the stationary nodes incident to them are shown in red. For sufficient timing information to be represented, all colored nodes from the graph in the figure are kept in its compacted version. Node 14 must also be kept, as it informs the solver that there is an alternative path between nodes 11 and 18, which may at some point become the determining factor for the arrival time of node 18. If that

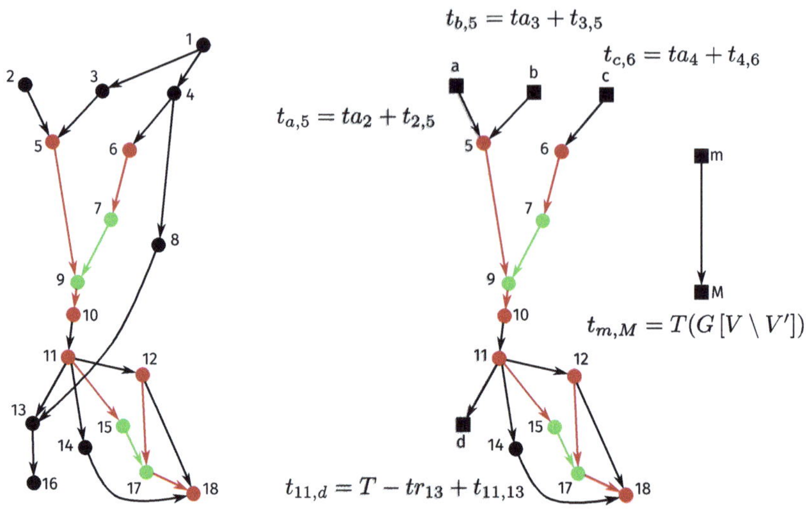

Fig. 8.13 Illustration of timing graph compaction through simplification of fixed-delay subgraphs. Green edges are selected for improvement ($\in E_s$), and their endpoint nodes are movable ($\in V_m$). Red edges are affected by the movement of nodes but were not selected for improvement ($\in E_\psi$). The remaining edges, shown in black, are unaffected by node movement. The graph on the right shows how the starting graph on the left can be simplified, by excluding some of its fixed-delay subgraphs and replacing them with additional constraining nodes, depicted as squares. Edge (m, M) is used to store the delay of the longest path in the part of the timing graph not represented in the compacted version $(G[V \setminus V'])$

happens, no more effort should be spent on trying to further decrease this arrival time, by decreasing the delay of edge (15, 17); optimizing other edges should be tried instead.

Let us now take a look at node 13. Since none of the movable nodes is reachable from it, it cannot affect the arrival time of any of them. It can, however, affect their required time, and it is important to represent this correctly in the compacted graph, as otherwise, critical subgraphs may be left out of optimization. To do this, it is not necessary to include any nodes from the transitive fanout of 13, because all the delays in it are unaffected by repositioning of the movable nodes. Instead, we can introduce an additional node, labeled d in Fig. 8.13, and connect it to 11 by an edge with a delay equal to the delay of the edge between 11 and 13 (fixed throughout the placement as neither 11 nor 13 is movable) increased by the difference between the original critical path delay and the required time of 13. This increase represents the total delay of the downstream portion of the graph unaffected by movement of nodes.

A similar approach can be applied to, e.g., node 3, but with its arrival time being relevant to represent the fixed delay of the upstream portion of the graph. This can be generalized as follows:

1. Find all nodes $V' \in V$ which can both reach and are reachable from nodes in V_m.

2. For each v added to V' in step 1 and each $u : (u, v) \in E$, if u was not added to V' in step 1, add a new node n to V', and connect v to it by an edge with delay $t_{n,v} = t_{u,v} + ta_u$, where ta_u is the arrival time of u.
3. For each u added to V' in step 1 and each $v : (u, v) \in E$, if v was not added to V' in step 1, add a new node n to V', and connect it to u by an edge with delay $t_{u,n} = t_{u,v} + T - tr_v$, where tr_v is the required time of v and T the original critical path delay.

Finally, we need to add another edge to the compacted graph which will represent the critical path delay of the portion of the original graph which was excluded from its compacted version. Otherwise, it may appear to the solver that the critical path delay can be reduced more than what is actually possible. This edge is labeled (m, M) in Fig. 8.13. Although the toy example of Fig. 8.13 fails to illustrate it, given that $|E_s|$ typically does not exceed a few tens or perhaps a few hundred, the above technique can provide great reduction in the size of the timing graph.

In principle, it would suffice to include additional nodes only for the most constraining parents/children, in steps 2 and 3, but savings from this would not be very high, so we do not do that in our implementation. Similarly, the path $11 \rightarrow 14 \rightarrow 18$ could be replaced by a single edge connecting nodes 11 and 18 and carrying the delay of the entire path. While this approach could result in significant further compaction, generalizing it to subgraphs with more complex connectivity is not as straightforward, so we chose to stay with the simple procedure described above.

8.6.3.2 Filtering Slow Edges

A lower bound on critical path delay achievable by solving the placement ILP can be easily computed from the solution of the selection LP (Sect. 8.5). The maximum delay of each edge can be easily computed by considering the allowed positions of its endpoint nodes. We annotate the timing graph with these maximum delays and compute the slacks of all edges, given the lower bound on the critical path delay. All edges that have a positive slack are guaranteed not to be critical for any valid solution of the ILP and can thus be safely removed from the timing graph.

8.6.3.3 Filtering Slow Position Pairs

Another very straightforward compaction method that we use is to compute the slacks of all edges on a timing graph annotated with the minimum achievable delay for each edge and then remove all e_{u,v,p_u,p_v} (e_{u,v,c_u,c_v}) position pairs for which τ_{p_u,p_v} (τ_{c_u,c_v}) exceeds the minimum delay of (u, v) increased by its slack.

8.7 The Complete Algorithm

In this section, we combine together the two stages of the placement flow, presented in Sects. 8.5 and 8.6, and introduce a postprocessing step that removes overlaps between the movable and the stationary nodes.

8.7.1 Composing the Detailed Placer

We use a simple binary search to minimize the target critical path delay specified when selecting the movable nodes (Sect. 8.5). The lower bound of the search range is determined by performing a timing analysis on the timing graph of the circuit with the delay of each edge replaced by the minimum it can attain, given the movement regions of its endpoint nodes. This is represented by line 3 of Algorithm 8.1. The upper bound is set to the critical path delay of the starting placement produced by the general placer, assuming that all connections are implemented as programmable. The search stops when the two bounds differ less than T_δ^{min}, which we set to 30 ps in the subsequent experiments.

The main loop consists of solving an improvement LP with the current target bound, to obtain the set of movable nodes (line 7; see Sect. 8.5) followed by solving the related placement ILP to actually position the movable nodes (line 9; see Sect. 8.6). The solver is instructed to minimize the critical path delay, T_{ILP}, as much as possible, given the allowed runtime budget. The solution is constrained to have a critical path delay at most as large as the smallest one encountered so far, $T_{incumbent}$. The lower bound on achievable critical path delay used for pruning the edges which can never become critical (see Sect. 8.6.3) is set to the critical

Algorithm 8.1 Detailed placer

1: $T_{incumbent} \leftarrow T_{start}$
2: $\text{gap}_{\text{incumbent}} \leftarrow \infty$
3: $T_{low} \leftarrow$ get_lower_bound()
4: $T_{high} \leftarrow T_{start}$
5: **while** $T_{high} - T_{low} > T_\delta^{min}$ **do**
6: $\quad T_{target} \leftarrow (T_{low} + T_{high})/2$
7: $\quad$ movable, $T_{LP} \leftarrow$ lp_select_movable_nodes(T_{target})
8: $\quad$ status, placement, T_{ILP}, gap $\leftarrow$
9: $\qquad\qquad$ ilp_place_nodes(movable, $T_{LP} \leq T \leq T_{incumbent}$, $T_{lb} = T_{LP} - 2T_\delta^{min}$)
10: $\quad$ **if** $(T_{ILP} < T_{incumbent}) \vee ((T_{ILP} = T_{incumbent}) \wedge (\text{gap} < \text{gap}_{\text{incumbent}}))$
11: $\qquad$ update_incumbent(T_{ILP}, gap, placement)
12: $\quad$ **if** $(\text{gap} < 1) \vee (\text{status} = \text{infeas})$
13: $\qquad T_{high} \leftarrow T_{target}$
14: $\quad$ **else**
15: $\qquad T_{low} \leftarrow T_{target}$
16: legalize(incumbent)

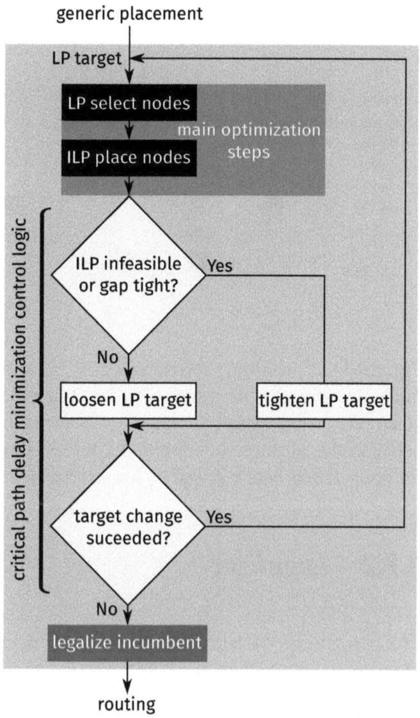

Fig. 8.14 Flowchart of the complete algorithm

path delay obtained after thresholding the LP solution, T_{LP} (see Sect. 8.5), reduced slightly to leave some margin for roundoff.

If the obtained critical path delay, T_{ILP}, is lower than the current best, $T_{incumbent}$, or they are equal, but T_{ILP} is proven by the solver to be closer to the optimum for the current set of movable nodes, the incumbent solution is updated (line 11). When the solution is proven to be within 1% of the optimum, the algorithm considers that the current placement problem was successfully solved and that an attempt to achieve more critical path delay reduction should be made. Hence, the binary search range is constricted from the right, on line 13. The same happens when the problem is proven infeasible. This means that the incumbent solution cannot be improved with the current selection of movable nodes. To resolve this, the set of movable nodes should likely be increased, which is achieved by reducing the target critical path delay so that more edges need to be improved to meet it. If the solver failed to provide a definitive answer in the allowed runtime budget (even if it did find a new incumbent solution but failed to prove it optimal), the problem is deemed too difficult to be solved in the allowed runtime budget, and the binary search range is constricted from the left (line 15) in a hope that a looser target critical path delay would result in an easier placement problem.

Once the binary search converges, any overlaps that may have occurred between the movable and the stationary nodes must be removed. This is done by the postprocessing step on line 16, discussed in more detail in the next section. The complete algorithm is visually illustrated in the flowchart of Fig. 8.14.

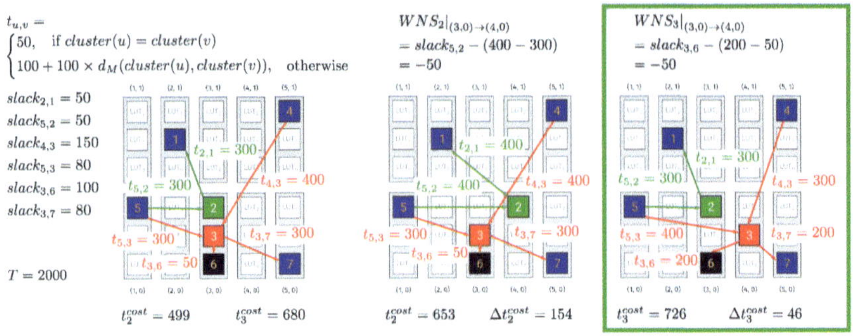

Fig. 8.15 Determining movement cost between a pair of clusters during legalization. The figure illustrates the cost of moving nodes 2 and 3 from the cluster (3, 0) to the cluster (4, 0). The postmove worst negative slack is the same for both nodes. Hence, the timing cost difference between the starting situation and after the move is used to break the tie. The difference is lower for node 3, and hence it will be moved by the legalizer

8.7.2 *Legalizer*

For removing overlaps between the movable and the stationary nodes, we adapt the algorithm of Darav et al. [Dar19]. Since our main goal is to optimize performance of the processed circuit, the legalizer must be timing-aware itself, not to undo the critical path reduction achieved by the detailed placer, unless that is necessary for achieving a legal placement.

8.7.2.1 Pricing LUT Movement

When faced with a decision about which LUT should be moved between clusters A and B, as a primary factor, we use the postmove worst negative slack of all connections incident to the LUT we are attempting to move and choose the LUT for which the magnitude of this slack is the smallest. This is illustrated in Fig. 8.15. In case of ties, we compute the difference in the timing cost of the LUT before and after the move and pick the LUT with the smallest increase in this cost. The timing cost is adopted from VPR's timing-driven placer [Mar00]:

$$crit_{u,v} = 1 - \frac{slack_{u,v}}{T}, \tag{8.16}$$

$$t_u^{cost} = \sum_{(u,v)\in E} t_{u,v} \times crit_{u,v}^{\alpha} + \sum_{(p,u)\in E} t_{p,u} \times crit_{p,u}^{\alpha}, \qquad \alpha \in \mathbb{R}^+. \tag{8.17}$$

Here, T designates the current critical path delay. For the selectivity parameter α, we use 8 in the subsequent experiments. We first run the legalizer without performing any slack updates, relying on the values obtained in the first static timing analysis after the detailed placement converged. If the legalized critical path delay exceeds the one computed by the ILP solver during detailed placement by more than 100 ps, we rerun the legalizer, committing each move to the timing graph as soon as it is decided and updating the slacks accordingly, to prevent suboptimal local move decisions from having a large cumulative effect.

8.7.2.2 Bounding Overlaps

The success of targeted application of the placement ILP to a limited set of movable nodes spread over wide regions of the starting placement relies on the observation that in many circuits, the gain from appropriately positioning a small number of critical nodes far exceeds the loss created by suboptimally moving other, less critical nodes that stand in their way. However, this only holds if not too many nodes need to be moved from their original positions during postprocessing. Otherwise, the timing information that was presented to the ILP solver, which assumes that all stationary nodes will retain their original positions, may be too significantly disturbed in the overlap removal process, leading to an inevitable loss in the achieved delay reduction. To prevent this from happening, we need to control the amount of overlap occurring in each cluster. This is easily achieved with the help of the following constraints:

$$\sum_{u \in V_m} x_{u,c} \leq \omega_c, \qquad \forall c. \tag{8.18}$$

The constant $\omega_c \in \mathbb{N}$ sets a limit on the number of movable nodes which can be placed in cluster $c = (x, y)$. We determine it as follows:

$$s = |\{V \setminus V_m\} \cap c| \tag{8.19}$$

$$\omega_c = N - s + \min(s, \delta). \tag{8.20}$$

Constant s holds the number of stationary nodes in cluster c, while N is the cluster size in the underlying FPGA architecture. The allowed overflow is determined by the parameter δ, which we set at 2 in the subsequent experiments, as we observed that this value does not limit achieved postplacement delay reduction and rarely leads to an increase in this delay after legalization. The $N - s$ part of Equation (8.20) specifies that all positions that were originally unoccupied, or were occupied by the movable nodes, can be filled by the movable nodes. The $\min(s, \delta)$ part guarantees that overlaps within the cluster can be resolved by removal of stationary nodes in the postprocessing step. As mentioned before, overlaps between movable nodes which are assigned exact positions are impossible due to constraints (8.6). However, some of the movable nodes may not be assigned an exact position but merely a cluster, in which case they could overlap with other movable nodes. Since the movable nodes are necessarily critical (albeit for the LP target critical path delay), as otherwise the minimum improvement solution of the improvement LP (Sect. 8.5) would not affect them, they should be positioned with care. Allowing them to overlap with other movable nodes, by letting the amount of overlap exceed the number of stationary nodes in the cluster, would leave their positioning to the fast but suboptimal legalizer.

8.8 Optimization

In the previous sections, we described the basic form of the proposed algorithm. It first solves an LP to determine which edges in the circuit's timing graph should have their delays reduced by moving their endpoint nodes to align them with the endpoint LUTs of the direct connections available in the FPGA architecture. Then it solves a related ILP to perform the actual placement. We focused on simplifying the ILP model to the extent that would allow for its solving in reasonable time. Until now, the solution of the improvement LP determined the formulation of the placement ILP, but we have done little to formulate the improvement LP itself in such a manner that its solution is more likely to produce a feasible ILP. In this section we focus on extending the formulation of the improvement LP to more tightly couple it to the placement ILP. We also extend the formulation of the placement ILP itself, so as to make it easier to solve.

8.8.1 Specialization of the Improvement LP to the Architecture

Keeping the set of edges selected for improvement (the number of movable nodes, $|V_m|$) reasonable is necessary for the related placement ILP to be solved in a reasonable amount of time. That is why the formulation of the improvement LP presented in Sect. 8.5 minimized the total delay improvement. Let us for the moment disregard the aforementioned fact that minimizing the total improvement does not necessarily translate to smaller $|V_m|$, nor does this necessarily translate to an easier-to-solve ILP. Let us assume instead that the obtained ILP can be solved in the allowed amount of time. The ILP can still be infeasible, for various different reasons, and we would like to predict and ideally prevent this already at the LP level. For example, simultaneous improvement of two different connections might imply two nodes being placed at the same position or one node being placed at two distinct positions at once.

Aside from the initial placement of the circuit and the allowed movement regions, the FPGA architecture strongly influences feasibility of the placement ILPs constructed from the solutions of the improvement LPs. Figure 8.16 shows a simple architecture and a piece of a circuit. With the current LP formulation, there is nothing that would prevent the solution from including both edges of the circuit, although it is clear that the architecture will not be able to improve both of

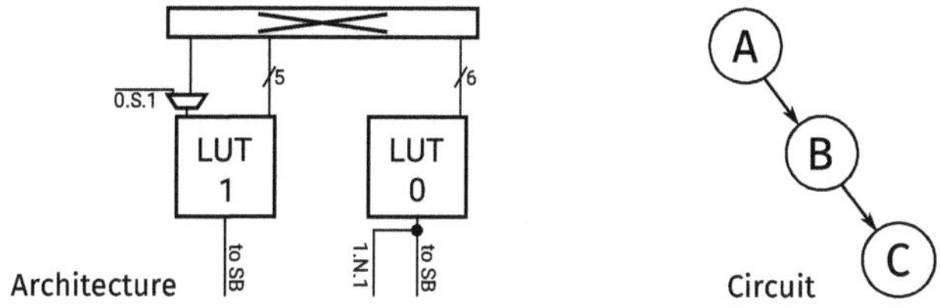

Fig. 8.16 Pitfalls of the basic selection LP. An architecture is shown on the left and a piece of a circuit on the right. A solution selecting both circuit connections for improvement is valid, but not supported by the architecture in which each LUT has only one incident direct connection

them. We cannot enforce exclusivity in choosing between these two edges without introducing integer variables, but we can use additional constraints to increase the chance of obtaining solutions that the architecture can support. To begin with, we can introduce a bound on the sum of the improvements of the two edges, equal to the maximum of the two individual bounds. This still does not prevent the solution from including both connections, but covering only one of them during placement will suffice for this short path to meet what is expected of it in terms of overall delay reduction. For that reason, we introduce pairwise improvement bounds, for each pair of edges sharing a common node. In general, this will not be the maximum of the two individual bounds, but the largest total improvement achievable within the movement regions of the three incident nodes. To further improve feasibility, we include bounds on the total improvement of the incoming, the outgoing, and all the edges incident to each individual node.

8.8.2 *Solving Successive ILPs*

During the binary search for the smallest achievable critical path delay, the placer may have to solve many ILPs. However, since all of them are describing a detailed placement problem of the same circuit, on the same FPGA architecture, and with the same starting general placement, they will inevitably be related. We can use this fact to make the solution of the ILPs simpler, as well as to improve the chances that they are feasible, by slightly adjusting the LP formulation.

8.8.2.1 Enforcing ILP Solution Overlaps

During experimentation, we observed that the number of covered edges rarely substantially decreases between two consecutive incumbent solutions. Moreover, in most cases that we have inspected, there was a substantial overlap between two consecutive incumbent solutions in terms of which edges were covered in them. We can use this fact to help the solver find feasible solutions more easily. Let E_s^i be the set of edges selected for improvement in the problem that led to the incumbent solution and $E_c^i \subseteq E_s^i$ the set of edges that are actually covered in the incumbent solution. We denote the set of edges selected for improvement in the current problem as E_s, like before. Similarly, E_c denotes the set of edges covered in a valid solution of the current problem. Then we add the following two constraints to the ILP:

$$|E_c| \geq \eta \left|E_c^i\right| \times \frac{|E_s|}{\max\left(|E_s|, \left|E_s^i\right|\right)}, \quad \eta \in (0, 1), \tag{8.21}$$

$$\left|E_c \cap E_c^i\right| \geq \quad \zeta \left|E_c^i\right| \times \frac{\left|E_s \cap E_s^i\right|}{\left|E_s^i\right|}, \quad \zeta \in (0, 1). \tag{8.22}$$

The first constraint specifies the minimum number of covered edges, with respect to the number of covered edges in the incumbent solution, appropriately scaled down if the number of edges selected for improvement is smaller than in the problem which produced the incumbent solution. The second constraint specifies the minimum amount of overlap between the set of covered edges in the incumbent solution and the solution of the current problem. Note that $\left|E_c^i\right|$ and $\left|E_s^i\right|$ are constants obtained by inspecting the incumbent solution, while $|E_s|$ and $\left|E_s \cap E_s^i\right|$ are also constants obtained from the solution of the improvement LP. Hence, the right-hand side of both constraints is constant. The left-hand side is a single sum, encoded using the coverage indicators $c_{u,v} = 1 - \overline{c}_{u,v}$, where $\overline{c}_{u,v}$ was introduced in Sect. 8.6.2. In the subsequent experiments, parameters η and ζ are set to empirically determined values of 0.7 and 0.3, respectively. In general, there is no requirement that $\eta + \zeta = 1$.

8.8.2.2 Using ILP Solutions to Improve LP Formulation

Anticipating which edges selected for improvement will not actually improve due to conflicts with improvement of other selected edges is difficult at the LP level. On the other hand, whenever the ILP returns a feasible solution, it is possible to inspect it for selected edges which did not actually get covered and discourage their repeated selection in the LP solution. To do that, we extend the objective of the LP to

$$\min \sum_{(u,v)\in E} \alpha_{u,v} imp_{u,v}, \quad \alpha_{u,v} \in \mathbb{R}^+ \cup \{0\}. \tag{8.23}$$

We temporarily set $\alpha_{u,v}$ to 0 for all edges covered by the incumbent solution (E_c^i), to prevent unnecessary restriction of possible overlap between it and the solution of the next problem (Sect. 8.8.2.1). For the remaining edges of the timing graph, the coefficients are initially set to 1. Each time a solution to the ILP is found, for every edge $(u, v) \in E_s$, we multiply $\alpha_{u,v}$ by 0.9 if it is covered by the solution and 1.1 if it is not. In this manner, we encourage the solution of the improvement LPs to include edges that were repeatedly shown to be successfully coverable and discourage it

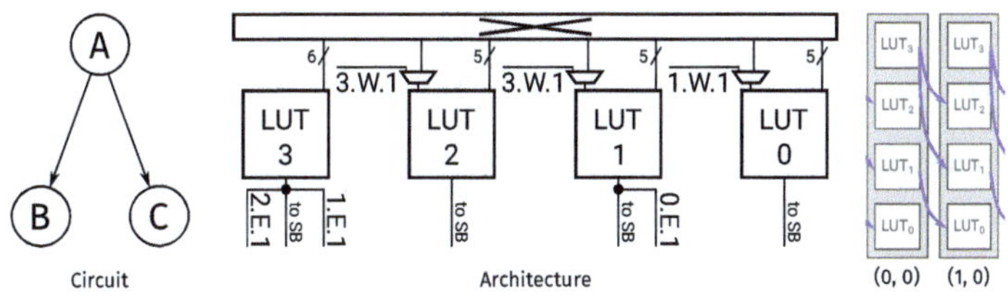

Fig. 8.17 Example used to illustrate formulation tightening through node degree matching

from selecting the ones which were repeatedly shown to be difficult to cover. Note that since we do not modify the bound on achievable improvement of any edge, but merely the way in which the invested improvement enters the objective, if the target critical path delay is small enough to require selection of edges which over time grew expensive, there is nothing that would prevent this from happening. Of course, exact values of the scaling parameters can be changed as required.

8.8.3 ILP Formulation Tightening

In Sect. 8.6.2, we have exploited a specific characteristic of FPGA architectures with direct connections to encode the placement ILP much more efficiently. We can exploit characteristics of a particular architecture further, but this time to produce additional constraints that will tighten the formulation of the ILP.

The approach that we use is to classify LUT positions based on the number of incoming and outgoing direct connections they have. Then, we introduce constraints which keep count of the incoming and outgoing direct connections that each movable node has under the current assignment of values to the variables. Finally, we use these counts in implications that help exclude the positions which cannot accommodate the required number of direct connections.

To illustrate this, let us take a look at the example in Fig. 8.17. Let us for the sake of simplicity neglect the timing information and assume that all nodes of the circuit are movable and that the objective is to maximize the number of covered edges. This can be described with the following ILP:

$$\max \quad c_{A,B} + c_{A,C},$$

$$c_{A,B} = e_{A,B,(0,0,3),(1,0,2)} + e_{A,B,(0,0,3),(1,0,1)} + e_{A,B,(0,0,1),(1,0,0)},$$

$$c_{A,C} = e_{A,C,(0,0,3),(1,0,2)} + e_{A,C,(0,0,3),(1,0,1)} + e_{A,C,(0,0,1),(1,0,0)},$$

$$e_{A,B,(0,0,3),(1,0,2)} \le x_{A,(0,0,3)}, x_{B,(1,0,2)}; \quad e_{A,B,(0,0,3),(1,0,2)} + 1 \ge x_{A,(0,0,3)} + x_{B,(1,0,2)},$$

$$e_{A,B,(0,0,3),(1,0,1)} \le x_{A,(0,0,3)}, x_{B,(1,0,1)}; \quad e_{A,B,(0,0,3),(1,0,1)} + 1 \ge x_{A,(0,0,3)} + x_{B,(1,0,1)},$$

$$e_{A,B,(0,0,1),(1,0,0)} \le x_{A,(0,0,1)}, x_{B,(1,0,0)}; \quad e_{A,B,(0,0,1),(1,0,0)} + 1 \ge x_{A,(0,0,1)} + x_{B,(1,0,0)},$$

$$e_{A,C,(0,0,3),(1,0,2)} \le x_{A,(0,0,3)}, x_{C,(1,0,2)}; \quad e_{A,C,(0,0,3),(1,0,2)} + 1 \ge x_{A,(0,0,3)} + x_{C,(1,0,2)},$$

$$e_{A,C,(0,0,3),(1,0,1)} \leq x_{A,(0,0,3)}, x_{C,(1,0,1)}; \quad e_{A,C,(0,0,3),(1,0,1)} + 1 \geq x_{A,(0,0,3)} + x_{C,(1,0,1)},$$
$$e_{A,C,(0,0,1),(1,0,0)} \leq x_{A,(0,0,1)}, x_{C,(1,0,0)}; \quad e_{A,C,(0,0,1),(1,0,0)} + 1 \geq x_{A,(0,0,1)} + x_{C,(1,0,0)},$$

$$x_{A,(0,0,3)} + x_{A,(0,0,2)} + x_{A,(0,0,1)} + x_{A,(0,0,0)} + x_{A,(1,0,3)} + x_{A,(1,0,2)} + x_{A,(1,0,1)} + x_{A,(1,0,0)} = 1,$$
$$x_{B,(0,0,3)} + x_{B,(0,0,2)} + x_{B,(0,0,1)} + x_{B,(0,0,0)} + x_{B,(1,0,3)} + x_{B,(1,0,2)} + x_{B,(1,0,1)} + x_{B,(1,0,0)} = 1,$$
$$x_{C,(0,0,3)} + x_{C,(0,0,2)} + x_{C,(0,0,1)} + x_{C,(0,0,0)} + x_{C,(1,0,3)} + x_{C,(1,0,2)} + x_{C,(1,0,1)} + x_{C,(1,0,0)} = 1,$$

$$x_{A,(0,0,3)} + x_{B,(0,0,3)} + x_{C,(0,0,3)} \leq 1,$$
$$x_{A,(0,0,2)} + x_{B,(0,0,2)} + x_{C,(0,0,2)} \leq 1,$$
$$x_{A,(0,0,1)} + x_{B,(0,0,1)} + x_{C,(0,0,1)} \leq 1,$$
$$x_{A,(0,0,0)} + x_{B,(0,0,0)} + x_{C,(0,0,0)} \leq 1,$$
$$x_{A,(1,0,3)} + x_{B,(1,0,3)} + x_{C,(1,0,3)} \leq 1,$$
$$x_{A,(1,0,2)} + x_{B,(1,0,2)} + x_{C,(1,0,2)} \leq 1,$$
$$x_{A,(1,0,1)} + x_{B,(1,0,1)} + x_{C,(1,0,1)} \leq 1,$$
$$x_{A,(1,0,0)} + x_{B,(1,0,0)} + x_{C,(1,0,0)} \leq 1.$$

While solving the continuous relaxation of the above program, the solver could yield the following fractional solution:

$$\begin{aligned} e_{A,B,(0,0,3),(1,0,2)} &= e_{A,C,(0,0,3),(1,0,2)} \\ = e_{A,B,(0,0,1),(1,0,0)} &= e_{A,C,(0,0,1),(1,0,0)} = 0.5 \end{aligned}$$

$$c_{A,B} = c_{A,C} = 1$$
$$x_{A,(0,0,3)} = x_{A,(0,0,1)} = 0.5$$
$$x_{B,(1,0,2)} = x_{C,(1,0,2)} = x_{B,(1,0,0)} = x_{C,(1,0,0)} = 0.5,$$

with all other variables at zero. Of course, this solution is not feasible for the ILP itself, since it implies that all nodes partially occupy two positions each. It would be good to make this solution infeasible for the relaxation too. To do so, let us start by introducing covered fanout counting variables, fo_u, and covered fanin counting variables, fi_u, for each movable node. In the running example, these would be

$$fo_A = c_{A,B} + c_{A,C}; \quad fi_A = 0, \tag{8.24}$$
$$fo_B = 0; \quad fi_B = c_{A,B}, \tag{8.25}$$
$$fo_C = 0; \quad fi_C = c_{A,C}. \tag{8.26}$$

Let us focus on fo_A, since no other variable in the current example is interesting, as will soon become apparent. Let the binary variable $x_{u,i} = \sum_{p_u \in \{p=(x,y,j) \in P(u,W) : j=i\}} x_{u,p_u}$ designate that the movable node u is placed at $\mathrm{LUT_i}$ in one of the clusters within its movement region. The following implication always holds: $(fo_A > 1) \implies x_{A,3} = 1$. To encode this, we can first introduce another binary variable $fob_{u,\theta}$, indicating that $fo_u \geq \theta$. We can assign a valid value to this variable with the help of the following two constraints [Wil13]:

$$\theta fob_{u,\theta} \leq fo_u, \tag{8.27}$$

$$(\mu - \theta + 1) fob_{u,\theta} + \theta - 1 \geq fo_u, \tag{8.28}$$

where μ is the largest number of connections originating at u which could potentially be covered (upper bound on fo_u). In the running example the constraints would be

$$2 fob_{A,2} \leq fo_A, \tag{8.29}$$

$$fob_{A,2} + 1 \geq fo_A. \tag{8.30}$$

In the previous fractional solution, $(c_{A,B} = c_{A,C} = 1) \implies fo_A = 2$. Hence, constraint (8.30) implies that $fob_{A,2} = 1$. To complete the implication that the fanout constraint has on valid placement positions, we merely need to add the following constraint:

$$fob_{A,2} \leq x_{A,3} = x_{A,(0,0,3)} + x_{A,(1,0,3)}, \tag{8.31}$$

which makes the previous fractional solution invalid in the continuous relaxation of the program as well. As can be seen in Fig. 8.2, in a typical architecture, the fanin and the fanout of LUTs are rather small, which means that not many values of the θ-threshold need to be considered. This also allows for combining the fanin and fanout constraints. For example, encoding $(fib_{u,2} \wedge fob_{u,1}) \implies x_{u,4}$ would constraint a movable node with at least two covered incoming edges and one covered outgoing edge to LUT_4, as it is the only one which can support that in the architecture of Fig. 8.2. It is important to note, however, that depending on the value of μ, degree matching may not be as effective as the above example illustrates. For instance, if μ and θ are 6 and 2, respectively, $fob_{u,2}$ can be as low as 1/5.

In principle this degree-matching approach could be recursively extended to counting the covered fanins/fanouts of predecessors and successors up to a certain distance [Nik19], to further constrain the set of valid positions in the continuous relaxation. We have not tried this in practice yet.

8.9 Results

In this section, we present the results of applying the proposed placement algorithm on the target architecture.

8.9.1 Experimental Setup

We inherit delay modeling from the previous chapter. We also retain the experimental methodology, along with its limitations. Notably, we do not support carry chains, fracturable LUTs, or sparse crossbars at the moment. One important restriction of the previous methodology is now lifted, however. We extended VPR to support cluster output equivalence specification after placement, independently for each cluster. As a result, there are no longer any constraints on route-time LUT permutation for the reference architecture, while for the one with the direct connections, only those LUTs that actually use a direct connection are kept fixed; the others may be freely permuted by the router. To further improve realism, we allow each cluster output in both architectures to reach all four adjacent routing channels. Thus we avoid the situation where different pins have access to different channels, which is not representative of industrial architectures, such as Agilex [Chr20]. At the moment, we do not have a sound method for legalizing the number of inputs to each cluster, so we increase the number of physical cluster inputs to 60 for both architectures (the maximum for a ten 6-LUT cluster). As this is not uncommon in industrial architectures [Fen12, Li19], we do not believe that it has any impact on the validity of the results.

All experiments were performed on a 20-core (40-thread) Xeon-based server with 256 GB of RAM, using CPLEX 12.10 with a timeout of 10 minutes for the solver. The reported results are medians of five different starting placements, and each circuit was routed by the *delay-targeted* routing algorithm of Rubin and DeHon [Rub11], implemented on top of VTR 7.0 [Luu14], with the channel width fixed at 300 tracks.

8.9.2 Delays

In this section, we present the impact of applying the proposed algorithm to the architecture of Fig. 8.2 on the critical path delay benchmark circuits.

8.9.2.1 Postplacement Delays

For the cases when the LUTs are allowed to move only within their initial clusters ($W = 0$) and in a region of 3×3 clusters centered at the initial clusters ($W = 1$), the delays obtained through solving the sequence of placement ILPs (Sect. 8.7) are shown in the columns labeled as *covered* in Table 8.1. The > 400 ps difference between the average delays of 10.09 ns and 9.68 ns is significant and translates to about $3\times$ greater relative average improvement over the reference, when LUTs are allowed to change clusters.

Table 8.1 Delays in nanoseconds. Each entry corresponds to a median of delays obtained for five different placement seeds. Entries have been computed independently of the corresponding entries in other columns. For instance, the routed critical path delay of *or1200* amounting to 12.20 ns does not necessarily correspond to the postrouting critical path delay of the placement for which the median postplacement critical path delay of 11.75 ns was obtained. Rather, the entries state that the median postplacement delay for this circuit was 11.75 ns, whereas the median routed delay was 12.20 ns. Similarly, the 230 ps of overhead due to direct-connection-selection multiplexers is the median penalty that was paid and does not necessarily correspond to the amount of overhead which contributed to the median routed delay being 12.20 ns

	Postplacement							Postrouting				
		Lower bound		Covered								
Circuit	Ref.	W = 0	W = 1	W = 0	W = 1	Legal.	− Δ[%]	Ref.	W/ dir.	Mux	− Δ[%]	W/o dir.
raygentop	4.70	4.70	4.70	4.70	4.70	4.70	0.00	4.87	4.88	0.02	−0.21	4.88
ch_intrinsics	3.15	3.15	2.84	3.15	3.15	3.15	0.00	3.28	3.27	0.03	0.30	3.27
mkDelayWorker32B	6.83	6.83	6.55	6.83	6.58	6.58	3.66	7.09	7.04	0.03	0.71	7.36
mkSMAdapter4B	5.16	5.11	4.97	5.11	5.02	5.02	2.71	5.38	5.26	0.05	2.23	5.63
bgm	23.56	23.56	22.21	23.56	22.66	22.66	3.82	23.66	23.04	0.20	2.62	26.33
boundtop	6.10	6.01	5.57	6.01	5.73	5.73	6.07	6.05	5.82	0.05	3.80	6.37
stereovision0	3.74	3.74	3.31	3.74	3.52	3.52	5.88	3.74	3.57	0.06	4.55	4.06
diffeq1	20.45	19.48	18.24	19.81	19.19	19.19	6.16	21.16	20.01	0.12	5.43	21.86
diffeq2	15.69	14.92	13.46	15.02	14.48	14.48	7.71	16.14	15.14	0.11	6.20	16.68
blob_merge	9.90	8.76	6.79	9.44	8.90	9.16	7.47	9.89	9.21	0.11	6.88	10.56
or1200	13.08	12.66	10.76	12.77	11.69	11.75	10.17	13.12	12.20	0.23	7.01	15.66
LU8PEEng	105.05	101.07	91.47	101.49	95.57	95.63	8.97	104.86	96.45	0.98	8.02	110.17
sha	11.89	11.02	9.15	11.25	10.65	10.83	8.92	11.88	10.86	0.15	8.59	12.59
Geomean	10.29	9.99	9.03	10.09	9.68	9.72	5.54	10.46	10.01	0.09	4.30	11.07

We may note that the 1.94% average improvement for the $W = 0$ case is noticeably lower than what was previously reported in Chap. 7. This could be an artifact of an inferior movable node selection method, although the lower bounds in Table 8.1 suggest a more fundamental cause. The cause is in fact of architectural nature: Because we use a 60- instead of a 40-input cluster architecture, a much denser packing is obtained, bringing some of the intercluster routing delay into the clusters. Since the architecture has no local direct connections, when $W = 0$, the placer cannot do anything to improve their delay, while when $W > 0$ it can. To verify the hypothesis, we reran the experiments for $W = 0$ on a subset of circuits for which the average relative improvement was 2.22%, using a 40-input architecture. The improvement rose to 3.55%, which is much closer to the previously reported results. This illustrates the importance of considering other architectural parameters when deciding which direct connections are the most beneficial.

Placements for $W = 0$ are legal by construction, but those for $W = 1$ are not. The postlegalization results are also reported in Table 8.1. The delay does sometimes deteriorate due to legalization, but in most cases by a modest amount.

8.9.2.2 Postrouting Delays

Postrouting delays are reported in the column designated as *w/ dir*. The postlegalization relative improvement is generally retained throughout the routing process. Many of the cases where a nonnegligible deterioration occurs can be explained by the delays of the additional multiplexers that are not modeled during placement. Those circuit connections that are implemented as direct are forced to suffer this additional delay, while the others are rarely impacted by it. This is due to the sparsity of the direct connections, which causes relatively few LUT inputs to be delayed. The difference that dedicated placement brings to the postrouting delay is shown in Fig. 8.18.

In the placement ILP formulation, we allow connection delays to decrease only when implemented as direct. However, it is possible that some of the delay improvements in Table 8.1 are due to shortening of programmable connections or packing improvement. To verify if this is the case, we also routed the circuits placed with the dedicated algorithm, but without actually using the direct connections. The results are reported in the *w/o dir.* column of Table 8.1. Clearly, it is not the overall improvement of placement that led to the positive results. In fact, the dedicated placer significantly distorts the general placement, in a way that makes sense only in the presence of direct connections.

8.9.2.3 Sensitivity to the Starting Placement

In Sect. 8.3.3, we argued that it is sufficient to apply a dedicated detailed placer to a general starting placement produced without knowledge of existence of the fast direct connections, because the additional delay decrease due to appropriate usage

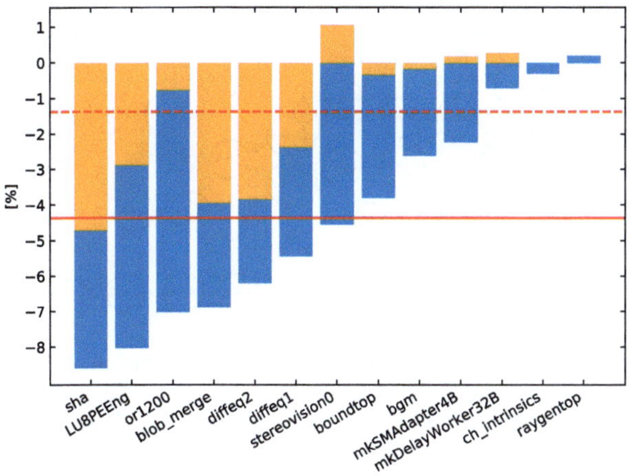

Fig. 8.18 Relative change in the postrouting critical path delay. The $W = 0$ and the $W = 1$ cases are shown in orange and blue, respectively. The dashed (solid) red line represents the relative change of the geomean critical path delay over all circuits for $W = 0$ ($W = 1$)

Fig. 8.19 Sensitivity to starting placement. Starting postplacement delays for all five starting placements of each circuit are shown in gray. Postlegalization delays for $W = 1$ are shown in blue. All values are normalized by the maximum starting postplacement delay for the particular circuit

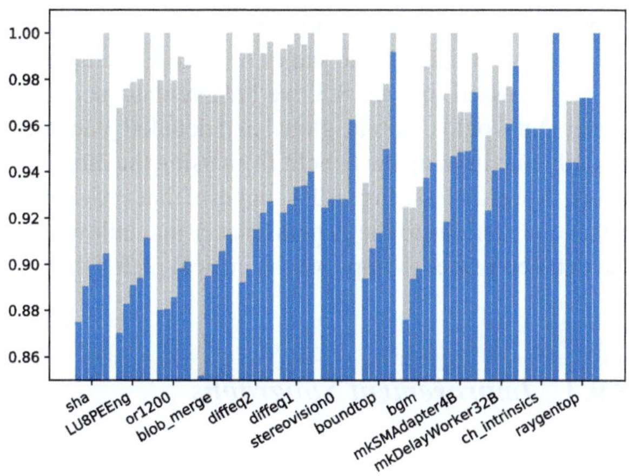

of these direct connections is small compared to the amount of optimization that the general placer can achieve with respect to a random, unoptimized placement. Given that in practice, during detailed placement, a small subset of movable nodes can be moved only to a limited distance, how much of this further delay reduction can actually be achieved could depend on the starting placement.

To assess this, we plot in Fig. 8.19 the starting postplacement delays (gray) and the final postplacement delays after legalization (blue) for all five starting placements of each circuit. All delays corresponding to the same circuit are normalized by the largest starting postplacement delay occurring for that circuit. We can see that there are significant differences in the achieved relative delay

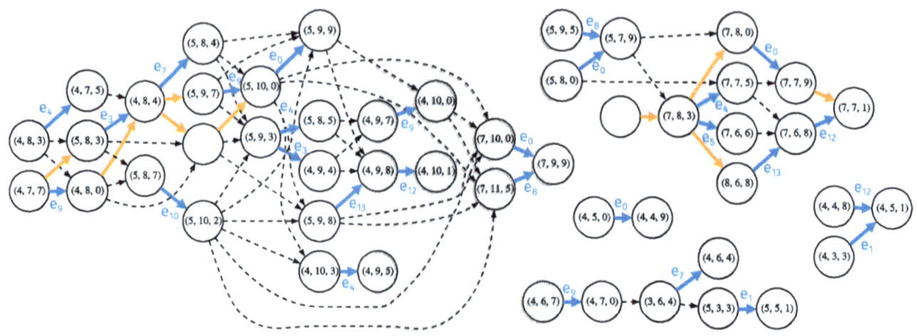

Fig. 8.20 Subgraph selected for improvement in the *blob_merge* circuit. All nodes are movable. The blue solid arrows correspond to the edges that were successfully implemented as direct after dedicated placement. The orange solid arrows represent the edges selected for improvement that were not successfully implemented as direct. Finally, the dashed arrows depict the edges that exist between the movable nodes but were not selected for improvement. Those nodes that have an incident direct connection have their final positions as labels. The label of each blue edge corresponds to the identifier used in Fig. 8.2 to mark the direct connection that implemented it

improvement between different placements of the same circuit, even if the starting postplacement delay is the same. A notable example is the *blob_merge* circuit.

As discussed in Sect. 8.3, we believe that successfully constructing a dedicated placer that would combine global and detailed placement in a scalable manner, while actually maximizing the benefit from using the direct connections, is not particularly likely. Nevertheless, Fig. 8.19 suggests that providing some information about the direct connections to the general placer may allow it to create more opportunity for the detailed placer to improve the critical path delay.

8.9.3 Improvement Subgraphs

The size and the structure of the circuit subgraphs induced by the connections selected for improvement (the solid edges in Fig. 8.20) influence both the time needed to solve the placement ILPs and the achievable critical path delay reduction. Some basic properties of the last successfully placed subgraph in the run resulting in the median postplacement delay are given in Table 8.2. The circuits that achieved a final delay improvement of $< 3\%$ are omitted, as either their subgraphs were very small or no successful placement was found for any of them.

Perhaps the most apparent feature of the subgraphs is their fragmentedness, visible in the *components* columns which show the sizes of the *weakly connected components* [Bon08] (maximal subgraphs where every node can be reached from all others when edge orientation is neglected). The *diameter* [Bon08] (the longest of the shortest paths between all pairs of nodes) often remains substantial, however. The node degrees are low, which is appropriate for the architecture of Fig. 8.2.

Table 8.2 Properties of the subgraphs induced by the connections selected for improvement. The bold highlighted rows show the corresponding properties of the subgraphs induced by the connections that were successfully improved by being implemented as direct. Columns W and H correspond, respectively, to the width and the height, in a number of clusters, of the region bounding the movable nodes. Angular brackets denote an average

	Size				Components			Degrees				
Circuit	$\|V\|$	$\|E\|$	W	H	#	$\langle\|V\|\rangle$	$\max\|V\|$	$\langle$total$\rangle$	max total	max in	max out	Diameter
boundtop	29	18	18	9	11	2.64	5	1.24	3	2	1	4
	25	**14**	**18**	**9**	**11**	**2.27**	**3**	**1.12**	**2**	**2**	**1**	**3**
stereovision0	51	39	39	13	12	4.25	16	1.53	5	4	2	6
	36	**20**	**39**	**12**	**16**	**2.25**	**3**	**1.11**	**2**	**2**	**2**	**3**
diffeq1	38	32	7	6	6	6.33	14	1.68	6	1	5	9
	31	**19**	**7**	**6**	**12**	**2.58**	**4**	**1.23**	**3**	**1**	**2**	**4**
diffeq2	38	33	6	4	5	7.60	17	1.74	4	3	2	11
	30	**17**	**6**	**4**	**13**	**2.31**	**3**	**1.13**	**2**	**2**	**1**	**3**
blob_merge	49	37	4	8	15	3.27	9	1.51	5	2	4	6
	47	**28**	**4**	**8**	**19**	**2.47**	**3**	**1.19**	**2**	**2**	**2**	**3**
or1200	97	71	14	16	26	3.73	11	1.46	3	3	2	11
	87	**50**	**14**	**15**	**37**	**2.35**	**4**	**1.15**	**3**	**3**	**1**	**3**
LU8PEEng	334	227	24	21	111	3.01	10	1.36	5	4	5	8
	294	**164**	**24**	**21**	**130**	**2.26**	**4**	**1.12**	**3**	**3**	**2**	**3**
sha	74	52	13	8	23	3.22	10	1.41	4	2	3	7
	59	**33**	**13**	**8**	**26**	**2.27**	**4**	**1.12**	**2**	**2**	**2**	**4**

The subgraphs induced by the connections that are actually implemented as direct (blue edges in Fig. 8.20) are noticeably smaller than the ones originally selected for improvement, but they still cover a large portion of their edges.

Without the information on how the individual connections selected for improvement are positioned within the entire circuit graph, it is not apparent how covering each of them influences the reduction of the critical path delay. We show one particular improvement subgraph in Fig. 8.20. Dashed arrows mark the edges between the movable nodes that were not selected for improvement. It should not be surprising that they often occur as intermediate edges of paths that were selected for improvement. The intention of the selection LPs of Sect. 8.5—although there is no guarantee that it will actually be realized—is to select a minimal subset of edges of a path as this directly influences the size of the placement ILPs. We can see that, in this case, the numerous small connected components are not merely pieces of unrelated paths but in fact constitute a carefully selected subgraph of a nontrivial graph. This showcases the generality of the movable node selection method anticipated in Sect. 8.4.

Another interesting observation that can be made from Fig. 8.20 is that the connections that were successfully covered use a wide variety of direct connections available in the architecture, with different span lengths and directions, both vertical and horizontal. This seems to confirm what we concluded in Chap. 7 that using only very simple patterns of direct connections, such as the vertical cascades, may not expose their full potential.

8.9.4 Runtimes

Runtime breakdowns for the placement run that resulted in the median postplacement delay of the given circuit are reported in Table 8.3. Circuits *ch_intrinsics* and *raygentop* are omitted because for them no improvement was possible in the median case, and this was detected immediately after computing the lower bound on achievable critical path delay (Sect. 8.7). In all cases, the number of ILPs solved until convergence is small (≤ 7). For majority of the solved ILPs, a solution at least as good as the previous best one was obtained, meaning that infeasible cases were often eliminated at the LP level, since the sought critical path delay was impossible to meet. LP solution time was generally very small, with a trend of increasing with increasing circuit size; that is expected since the LP models the entire circuit. This solution time can be further reduced by considering only the critical subgraphs of the timing graphs.

Solution times for the ILPs are displayed in the last two columns of the table. There seems to be no correlation between the size of the circuit and the solution time, which is expected, as the size of the movable node set has no a priori correlation with the circuit size either. Some of the ILPs are solved by merely solving the continuous relaxation of the problem in the root of the search tree ($\langle|\text{tree}|\rangle = 0$). Others, however, require substantial branching. In these cases, the

Table 8.3 Solution runtime. All runtimes are in seconds. Columns under *ILP preparation* hold the average time taken to set up each ILP and the average time taken to solve the preceding LP. Columns under *ILP status* hold the number of algorithm iterations which resulted in a feasible ILP, an infeasible ILP, and a timed-out ILP, respectively. Columns under *flexibility* hold the average number of positions per movable node (covering position pairs per edge selected for improvement). Finally, columns under ⟨ILP sol.t⟩ hold the average time taken by the ILP solver to find a feasible solution (prove infeasibility) and the average size of the branching tree. The remaining runtime entering the wall-clock time includes loading of data structures, setting up the initial LP, attempts to solve infeasible LPs, and final legalization

				ILP preparation		ILP status			Flexibility (last feas.)			⟨ILP sol. t⟩	
Circuit	#LUTs	Wall clock	CPU	⟨LP sol. t⟩	⟨setup t⟩	Feas.	Infeas.	Timeout	⟨pos./u⟩	⟨pair/e⟩	$\langle \lvert tree \rvert \rangle$	Feas.	Infeas.
diffeq2	322	92.05	370.63	0.05	3.12	4	0	0	42	81	73	12.88	–
diffeq1	485	156.86	810.27	0.07	3.69	5	0	0	43	78	215	22.93	–
mkSMAdapter4B	1982	52.82	34.92	0.33	2.38	2	1	0	32	37	0	0.06	0.04
sha	2280	2355.68	38358.80	0.21	6.44	5	0	2	38	88	4561	210.42	–
or1200	3054	279.96	1423.68	0.11	6.71	3	0	0	44	85	682	59.12	–
boundtop	3070	87.45	73.94	0.22	3.64	4	1	0	31	50	0	0.44	0.02
mkDelayWorker32B	5602	127.45	110.88	0.26	5.44	3	0	0	42	39	0	0.09	–
blob_merge	6019	2967.47	47072.96	0.46	4.76	5	0	2	42	93	9455	311.82	–
stereovision0	14779	274.80	271.00	0.41	10.13	3	1	0	36	79	0	2.13	0.02
LU8PEEng	26455	4637.07	55084.80	3.43	72.80	3	0	4	37	81	4666	264.04	–
bgm	36480	1624.23	3445.75	3.38	71.54	4	0	0	36	82	5412	43.03	–

capability of CPLEX to branch in parallel can be useful. For this size of the search trees, memory is, however, not a concern. The largest trees required on the order of a few hundreds of megabytes.

Each ILP also needs time to be constructed, which is reported under ⟨setup t⟩. It can be significant, but that is mostly due to a fairly inefficient Python implementation.

The average number of positions per movable node (⟨pos./u⟩) is substantial in most cases, though lower than the theoretical maximum of 90. The average number of covering pairs per edge (⟨pair/e⟩) is also much lower than the 8100 that would occur in the naive formulation.

Finally, the table also shows wall-clock times measuring the duration of the entire detailed placement run, from loading data structures with the output of VPR to storing the postlegalization results. These times also include a one-time LP setup, which is later updated by merely changing the target critical path delay bound and the objective of Sect. 8.8.2.2. As with ILP setup, the combined overhead of the aforementioned steps is tolerable, but not always negligible (maximum being 1140s for *bgm*); we believe this to be mostly due to a fairly inefficient Python implementation.

The reader may notice that the runtime spent on dedicated detailed placement is substantial, given the size of the circuits in Table 8.3. In some cases, it even overshadows the rest of the CAD flow; for example, standard VTR 7.0 flow takes only 22.34 s on the *blob_merge* circuit, with a fixed routing channel width. The fact that this single additional stage in the CAD flow requires 133× more runtime than the rest may seem daunting at first, but it is important to note that its most runtime-intensive phase—solving the placement ILPs—depends not on circuit size but on the size of the movable node sets. On the other hand, runtime expanded on the rest of the CAD flow is directly related to the circuit size, meaning that for larger circuits, the algorithm proposed here would likely take but a fraction of the total runtime. Furthermore, given that obtaining performance improvements is increasingly difficult, recent trends have called for CAD tools that can remove the last few tens of picoseconds of delay at the expense of runtimes that were previously considered intolerable [Kah21].

8.9.5 Independent Subpattern

In Sect. 7.7.4, we observed that the first four direct connections that were added to the pattern were responsible for 68% of the geomean routed critical path delay reduction achieved by the complete pattern of Fig. 8.2. The subpattern containing these four connections is shown in Fig. 8.21. A particularly interesting feature of this architecture is that each LUT has at most one incident direct connection, meaning that the set of edges selected for improvement that are covered by any valid solution of a placement ILP will constitute a *matching* in the circuit graph. Since matchings can be found efficiently [Gib85], we can intuitively expect that the

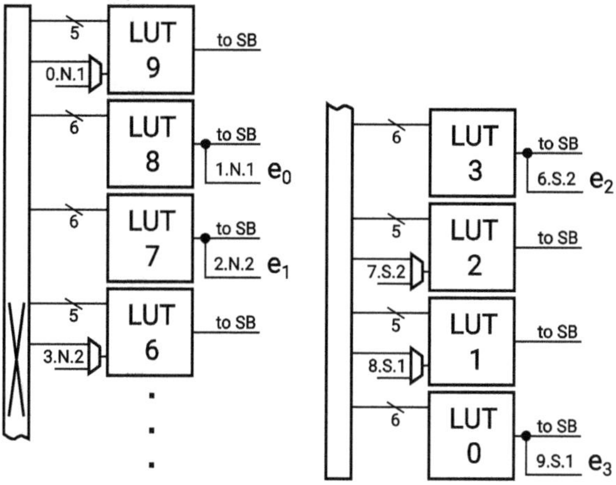

Fig. 8.21 The architecture composed of the first four direct connections added to the best pattern found in the architectural study of Chap. 7. Each LUT has at most one incident direct connection. LUTs without incident direct connections are omitted from the figure

placement problems formulated for this class of architectures are easier to solve in practice, although the timing dependence between the edges and the necessity to avoid overlaps between the movable nodes could mean that they are not necessarily easier in theory.

8.9.5.1 Optimization-Runtime Trade-Off

To assess whether the subpattern of Fig. 8.21 still causes bulk of the improvement achieved by the entire pattern of Fig. 8.2 when nodes are allowed to move across clusters, we repeat the experiments on it as well. The results shown in Table 8.4 indeed confirm this, with the architecture achieving a 2.87% geomean routed delay reduction, or 67% of the 4.30% achieved by the complete architecture. The total wall-clock runtime spent for median-postplacement-delay runs of all circuits was 2760 s, while the total CPLEX runtime amounted to only 336 s. For the complete architecture, the total wall-clock runtime was 12,720 s, while the solver alone used 8790 s. Hence, 67% of the delay improvement was achieved in 4.6× less total time and using 26× less solver runtime.

8.9.5.2 Increasing Movement Freedom

The much smaller solver runtime spent on the small independent subpattern of Fig. 8.21 allowed us to assess the potential benefits of increasing the movement freedom of the movable nodes on the obtained critical path delay reduction. Results for $W = 2$—allowing each node to move in a 5×5 cluster region with 250 candidate positions—are also shown in Table 8.4, as well as in Fig. 8.22. They show that a number of circuits experience a large additional improvement and that the average improvement is significantly increased as well. This suggests that future

Table 8.4 Critical path delays obtained on the architecture of Fig. 8.21

	Postplacement					Postrouting				
Circuit	Ref.	$W = 1$	$-\Delta$ [%]	$W = 2$	$-\Delta$ [%]	Ref.	$W = 1$	$-\Delta$ [%]	$W = 2$	$-\Delta$ [%]
raygentop	4.70	4.70	0.00	4.70	0.00	4.87	4.88	−0.21	4.88	−0.21
ch_intrinsics	3.15	3.15	0.00	3.15	0.00	3.28	3.30	−0.61	3.29	−0.30
mkDelayWorker32B	6.83	6.63	2.93	6.58	3.66	7.09	7.06	0.42	7.04	0.71
mkSMAdapter4B	5.16	5.02	2.71	5.02	2.71	5.38	5.31	1.30	5.25	2.42
bgm	23.56	22.85	3.01	22.66	3.82	23.66	23.20	1.94	22.84	3.47
boundtop	6.10	5.89	3.44	5.69	6.72	6.05	5.91	2.31	5.73	5.29
stereovision0	3.74	3.74	0.00	3.74	0.00	3.74	3.76	−0.53	3.76	−0.53
diffeq1	20.45	19.62	4.06	19.42	5.04	21.16	20.33	3.92	20.20	4.54
diffeq2	15.69	14.97	4.59	14.61	6.88	16.14	15.46	4.21	15.18	5.95
blob_merge	9.90	9.26	6.46	9.17	7.37	9.89	9.26	6.37	9.23	6.67
or1200	13.08	12.13	7.26	11.95	8.64	13.12	12.50	4.73	12.21	6.94
LU8PEEng	105.05	97.51	7.18	94.55	10.00	104.86	98.03	6.51	95.03	9.37
sha	11.89	10.97	7.74	10.87	8.58	11.88	11.01	7.32	10.92	8.08
Geomean	10.29	9.90	3.79	9.78	4.96	10.46	10.16	2.87	10.04	4.02

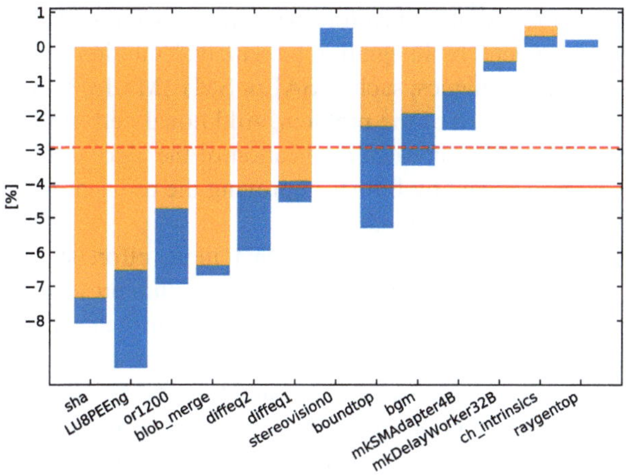

Fig. 8.22 Relative change in the postrouting critical path delay due to direct connections of the architecture of Fig. 8.21. The $W = 1$ and the $W = 2$ cases are shown in orange and blue, respectively. The dashed red line represents the relative change of the geometric mean critical path delay over all circuits, for $W = 1$, while the solid line represents the same for $W = 2$

effort invested in making the solution of the placement problems more scalable with respect to freedom of movement may pay off. As an illustration, with $W = 2$ the solver runtime rose to 7025s ($21\times$ increase).

It is also interesting to observe that by increasing W to 2, the simple architecture of Fig. 8.21 was able to achieve 93% of the geomean routed critical path delay reduction achieved by the architecture of Fig. 8.2 for $W = 1$ while still using less runtime. This illustrates that increased CAD effort may compensate for architectural inefficiencies. It could also make the direct connections more compelling, given that the architecture of Fig. 8.21 is significantly simpler to implement.

8.10 Conclusions and Future Work

In this chapter, we introduced an efficient ILP-based placement algorithm for FPGA architectures with direct connections between LUTs, which vastly improves their effectiveness compared to architecture-oblivious algorithms. We also removed some important limitations of the previously used experimental methodology and showed that the direct connections continue to bring benefits in this more realistic setting.

The fact that a simple change in the underlying architecture—increase in the number of cluster inputs—substantially altered the conclusions about the utility of a particular type of direct connections suggests that a comprehensive study of the mutual influence of direct connections and other architectural parameters is due.

Our experiments showed that increasing the freedom of movement beyond what was done in this chapter would lead to increased benefits. Another future step on the algorithmic front should therefore be to address the scalability issues that prevent this at the moment, by integrating incremental solution approaches [Li12], or even other solution techniques, such as SAT or SMT [Mih13, Mih13a], that could be better suited to the nature of the problem. Additional performance gains could perhaps also be achieved by repeated application of the algorithm with previously replaced nodes kept fixed and by further improving the problem formulation along the lines of the discussion of Sect. 8.8.3.

Finally, results of Sect. 8.9.2.3 suggest that more effort should be put into informing the general placer which precedes the algorithm presented in this section about the existence of dedicated connections, so that it can create more opportunities for their subsequent use. Ideally, one would consider the existence of fixed-connectivity patterns already at the synthesis and technology mapping stage, in the spirit of prior work by Ray et al. [Ray12], as well as the recent results of Tempia Calvino [Cal24]. Needless to say, whatever depth reduction can be achieved in logic synthesis and technology mapping, it is of little value if it cannot be retained during placement and routing [Van15]. Hence, the proposed algorithm would remain significant even if synthesis onto more complex patterns becomes tractable. One implementation of it is available at https://github.com/EPFL-LAP/fpl20-placement.

References

[Bet99] Betz, V., Rose, J., & Marquardt, A. (1999). *Architecture and CAD for Deep-Submicron FPGAs*. Kluwer Academic Publishers. ISBN: 0-7923-8460-1.

[Bon08] Bondy, J. A., & Murty, U. S. R. (2008). *Graph Theory*. 1st. Springer Publishing Company, Incorporated. ISBN: 1846289696.

[Cal24] Calvino, A. T. (2024). Technology Mapping and Optimization Algorithms for Logic Synthesis of Advanced Technologies. PhD thesis. École Polytechnique Fédérale de Lausanne.

[Cau11] Cauley, S., Balakrishnan, V., Charlie Hu, Y., & Koh, C.-K. (2011). A parallel branch-and-cut approach for detailed placement. *ACM Transactions on Design Automation of Electronic Systems, 16*(2), 18, 1–19.

[Che04] Chen, G., & Cong, J. (Aug. 2004). Simultaneous Timing Driven Clustering and Placement for FPGAs. In *Proceedings of the 14th International Conference on Field Programmable Logic and Applications*, Leuven, Belgium (pp. 158–167).

[Chi11] Chin, S. Y. L., & Wilton, S. J. E. (Feb. 2011). Towards scalable FPGA CAD through architecture. In *Proceedings of the ACM/SIGDA 19th International Symposium on Field Programmable Gate Arrays*, Monterey, CA, USA (pp. 143–152).

[Chr20] Chromczak, J., Wheeler, M., Chiasson, C., How, D., Langhammer, M., Vanderhoek, T., et al. (Feb. 2020). Architectural Enhancements in Intel®Agilex™FPGAs. In *Proceedings of the 2020 ACM/SIGDA International Symposium on Field-Programmable Gate Arrays*. Monterey, CA, USA, pp. 140–49.

[Dar19] Darav, N. K., Kennings, A. A., Vorwerk, K., & Kundu, A. (Feb. 2019). Multi-Commodity Flow-Based Spreading in a Commercial Analytic Placer. In *Proceedings of the 2019 ACM/SIGDA International Symposium on Field-Programmable Gate Arrays*. Seaside, CA, USA (pp. 122–31).

[DeM94] De Micheli, G. (1994). *Synthesis and Optimization of Digital Circuits*. McGraw-Hill. ISBN: 0070163332.

[Dha16] Dhar, S., Adya, S. N., Singhal, L., Iyer, M. A., & Pan, D. Z. (Nov. 2016). Detailed placement for modern FPGAs using 2D dynamic programming. In *Proceedings of the 35th International Conference on Computer-Aided Design*, Austin, TX, USA (pp. 1–8).

[Dha17] Dhar, S., Iyer, M. A., Adya, S. N., Singhal, L., Rubanov, N., & Pan, D. Z. (Mar. 2017). An Effective Timing-Driven Detailed Placement Algorithm for FPGAs. In *Proceedings of the 2017 ACM International Symposium on Physical Design*, Portland, OR, USA (pp. 151–570.

[Fen12] Feng, W. (Dec. 2012). K-way partitioning based packing for FPGA logic blocks without input bandwidth constraint. In *2012 International Conference on Field-Programmable Technology*, Seoul, Korea (South) (pp. 8–15).

[Gib85] Gibbons, A. M. (1985). *Algorithmic Graph Theory*. Cambridge University Press.

[Ham97] Hambrusch, S. E., & Tu, H.-Y. (1997). Edge weight reduction problems in directed acyclic graphs. *Journal of Algorithms, 24*(1), 66–93.

[Kah21] Kahng, A. B. (2021). Advancing Placement. In *Proceedings of the 2021 International Symposium on Physical Design*. Virtual Event (pp. 15–22).

[Lew03] Lewis, D., Betz, V., Jefferson, D., Lee, A., Lane, C., Leventis, P., et al. (Feb. 2003). The Stratix Routing and Logic Architecture. In *Proceedings of the 2003 ACM/SIGDA Eleventh International Symposium on Field Programmable Gate Arrays*, Monterey, CA, USA (pp. 12–20).

[Li07] Li, C., Xie, M., Koh, C.-K., Cong, J., & Madden, P. (2007). Routability-driven placement and white space allocation. *IEEE Transactions on CAD of Integrated Circuits and Systems, 26*(5), 858–871.

[Li12] Li, S., & Koh, C.-K. (Mar. 2012). Mixed integer programming models for detailed placement. In *International Symposium on Physical Design*, Napa, CA, USA (pp. 87–94).

[Li19] Li, W., Lin, Y., & Pan, D. Z. (Nov. 2019). elfPlace: Electrostatics-based Placement for Large-Scale Heterogeneous FPGAs. In *2019 IEEE/ACM International Conference on Computer-Aided Design (ICCAD)*, Westminster, CO, USA (pp. 1–8).

[Luu14] Luu, J., Goeders, J., Wainberg, M., Somerville, A., Yu, T., Nasartschuk, K., et al. (July 2014). VTR 7.0: Next generation architecture and CAD system for FPGAs. *ACM Transactions on Reconfigurable Technology and Systems, 7*(2). ISSN: 1936-7406.

[Luu14b] Luu, J., McCullough, C., Wang, S., Huda, S., Yan, B., Chiasson, C., et al. (May 2014). On Hard Adders and Carry Chains in FPGAs. In *Proceedings of the 2014 IEEE 22nd International Symposium on Field-Programmable Custom Computing Machines*, Boston, MA, USA (pp. 52–59).

[Mar12] Markov, I. L., Hu, J., & Kim, M.-C. (Nov. 2012). Progress and challenges in VLSI placement research. In *2012 IEEE/ACM International Conference on Computer-Aided Design*, San Jose, CA, USA (pp. 275–282).

[Mar00] Marquardt, A., Betz, V., & Rose, J. (Feb. 2000). Timing-Driven Placement for FPGAs. In *Proceedings of the 2000 ACM/SIGDA Eighth International Symposium on Field Programmable Gate Arrays*, Monterey, CA, USA (pp. 203–213).

[Mar19] Martin, T., Maarouf, D., Abuowaimer, Z., Alhyari, A., Grewal, G., & Areibi, S. (June 2019). A Flat Timing-Driven Placement Flow for Modern FPGAs. In *2019 56th ACM/IEEE Design Automation Conference (DAC)*, Las Vegas, NV, USA (pp. 1–6).

[Mih13] Mihal, A., & Teig, S. (July 2013). A Constraint Satisfaction Approach for Programmable Logic Detailed Placement. In *Proceedings of the 16th International Conference on Theory and Applications of Satisfiability Testing*, Helsinki, Finland (pp. 208–223).

[Mih13a] Mihal, A. (July 2013). A Difference Logic Formulation and SMT Solver for Timing-Driven Placement. In *Informal Proceedings of the 11th International Workshop on Satisfiability Modulo Theories*, Helsinki, Finland (pp. 16–25).

[Mur15] Murray, K. E., Whitty, S., Liu, S., Luu, J., & Betz, V. (Mar. 2015). Timing-driven titan: enabling large benchmarks and exploring the gap between academic and commercial CAD. *ACM Transactions on Reconfigurable Technology and Systems, 8*(2). ISSN: 1936-7406.

[Nik19] Nikolić, S., Zgheib, G., & Ienne, P. (Sept. 2019). Finding a Needle in the Haystack of Hardened Interconnect Patterns. In *2019 29th International Conference on Field Programmable Logic and Applications (FPL)*, Barcelona, Spain (pp. 31–37).

[Nik20a] Nikolić, S., Zgheib, G., & Ienne, P. (Aug. 2020). Timing-Driven Placement for FPGA Architectures with Dedicated Routing Paths. In *2020 30th International Conference on Field-Programmable Logic and Applications (FPL)*. Virtual Event (pp. 153–161).

[Nik22] Nikolić, S., Zgheib, G., & Ienne, P. (Dec. 2022). Detailed placement for dedicated LUT-Level FPGA interconnect. *ACM Transactions on Reconfigurable Technology and Systems, 15*(4). ISSN: 1936-7406.

[Ray12] Ray, S., Mishchenko, A., Eén, N., Brayton, R. K., Jang, S., & Chen, C. (Mar. 2012). Mapping into LUT structures. In *2012 Design, Automation & Test in Europe Conference & Exhibition*, Dresden, Germany (pp. 1579–1584).

[Rub11] Rubin, R., & DeHon, A. (Feb. 2011). Timing-Driven Pathfinder Pathology and Remediation: Quantifying and Reducing Delay Noise in VPR-Pathfinder. In *Proceedings of the ACM/SIGDA 19th International Symposium on Field Programmable Gate Arrays*, Monterey, CA, USA (pp. 173–176).

[Van15] Vansteenkiste, E., Kaviani, A., & Fraisse, H. (Dec. 2015). Analyzing the divide between FPGA academic and commercial results. In *2015 International Conference on Field Programmable Technology (FPT)*, Queenstown, New Zealand (pp. 96–103).

[Vor07] Vorwerk, K., Kennings, A., Greene, J., & Chen, D. T. (Aug. 2007). Improving Annealing via Directed Moves. In *2007 International Conference on Field Programmable Logic and Applications*, Amsterdam, The Netherlands (pp. 363–370).

[Wil13] Williams, H. P. (2013). *Model Building in Mathematical Programming*. Fifth. Wiley.

Part III
Additional Topics

The last chapter concluded the part of the book that is based on the results of original research. Nevertheless, the treatment of the topic of programmable interconnect design would not be complete without a mention of two additional important problems: (1) designing time-domain-multiplexed interconnect, including *Networks on Chip* (NoCs) and (2) 2.5D and 3D integration of multi-die devices. We discuss these topics in Chaps. 9 and 10, respectively.

Chapter 9
Time-Domain-Multiplexed Interconnect

Contents

In Sect. 3.5, we explained how reconfiguration of programmable interconnect essentially enables time-domain-multiplexing of a limited number of wires across different user designs. We also anticipated that wires could be time-domain-multiplexed between signals of one and the same design. In this chapter, significantly inspired by a lecture by Professor Vaughn Betz [Bet20a], we shall motivate the introduction of time-domain-multiplexed interconnect in reconfigurable architectures and attempt to explain how it interacts with the traditional programmable interconnect that has been the main focus of the book. Moreover, we will discuss various aspects of adoption of *Networks on Chip* (NoCs) [DeM10]—both hard and soft—in FPGAs.

© The Author(s), under exclusive license to Springer Nature Switzerland AG 2025

S. Nikolić, *Modern Programmable Interconnect Design*,
https://doi.org/10.1007/978-3-031-80629-2_9

9.1 The Tristated Bus

In the previous chapters, we have seen that wires, and especially the thick, fast ones, are a resource that is getting scarcer with technology scaling. Yet, even in an ASIC, a typical wire is actually used for conveying a transition in some signal <10% of the time [Dal01]. Would it then not be appealing to enable sharing of wires between multiple signals of the same user circuit? After all, this is something that hardware designers have been doing since long before integrated circuits were invented. In preceding chapters we have not made much use of the information about how FPGAs are actually used to implement other hardware. We shall change that now and learn from some classical concepts in hardware system design.

Let us start by observing the simplest possible bus that allows communication between four distinct modules, shown in Fig. 9.1. To simplify arbitration, we assume that the modules have a fixed priority, which decreases from left to right. This allows daisy-chain arbitration [Dal04]: For a module to be able to transmit over the common bus, it has to make a request to do so by raising the *req* signal. If at that time grant for transmission has not been given to any of the higher priority modules, the present one can transmit. Transmission occurs by reducing the impedance on the appropriate tristate buffer, at which point output T of the given module becomes broadcast to the R input of all other modules. Information about requests is propagated to all lower priority modules through the *carry* signal. Needless to say, this type of arbitration would not be appropriate for any but the simplest systems. Nevertheless, it is useful for illustration of some concepts that are of interest here.

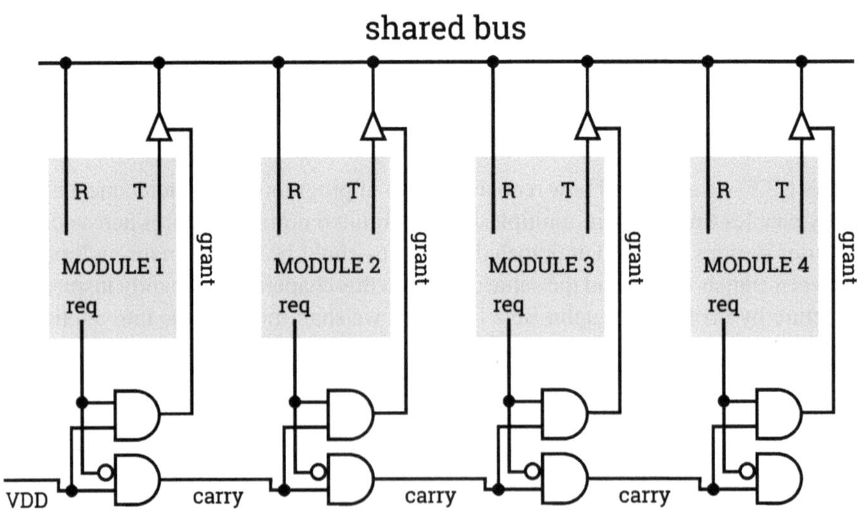

Fig. 9.1 A simple circuit composed of four modules communicating over a shared bus, using tristate buffers and simple daisy-chain arbitration (Adapted from Dally and Towles [Dal04])

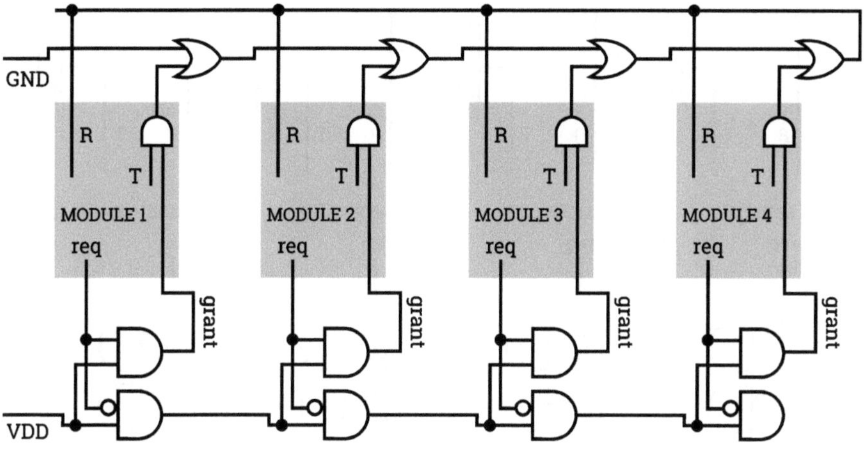

Fig. 9.2 A cascade of OR-2 gates recreates the bus of Fig. 9.1 using point-to-point connections. Modules without permission to transmit are outputting 0 to their OR-2 gate, thus letting the output *T* of the transmitting module to pass through the cascade (Adapted from Dally and Towles [Dal04])

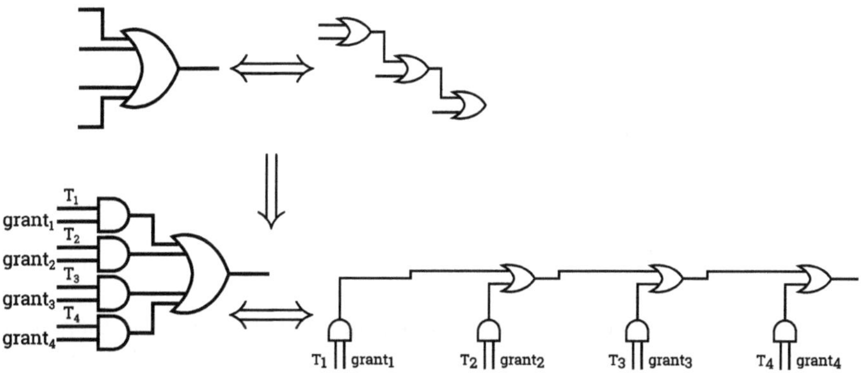

Fig. 9.3 The OR-2 cascade of Fig. 9.2 is nothing but a distributed 1-hot 4:1 multiplexer

To support this design style, early FPGAs incorporated user-controllable tristate buffers [Xil96, Gra00]. However, as technology progressed, tristate buffers became impractical and were no longer used to implement buses even in ASICs [Dal04]. A shared bus is easily emulated by a sequence of OR-2 gates depicted in Fig. 9.2 [Dal04]. When a module is given a grant to transmit, its output signal is passed through the sequence of OR-2 gates, whereas the output signals of other modules are blocked (reduced to the neutral zero value) by the corresponding AND-2 gate.

It is worth noting that the sequence of OR-2 gates is merely a distributed implementation of a 1-hot 4:1 multiplexer. This is depicted in Fig. 9.3. An alternative way to distribute a 4:1 multiplexer is to use a sequence of 2:1 multiplexers, as shown in Fig. 9.4.

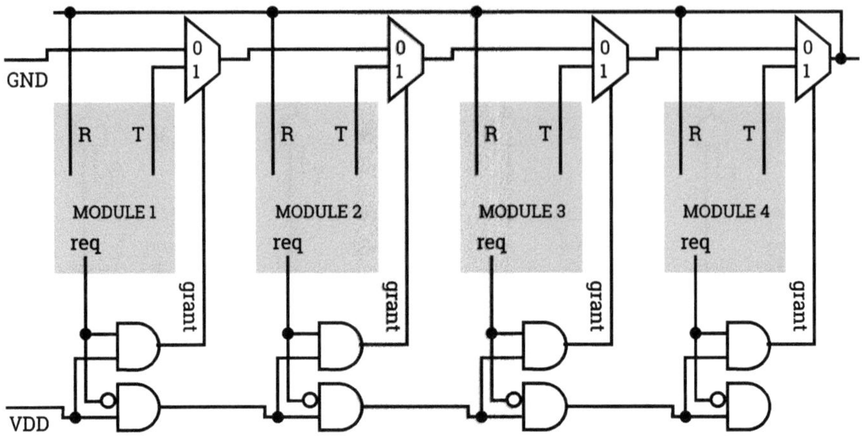

Fig. 9.4 Another way to break up a 1-hot 4:1 multiplexer. Note the similarity with segmented-channel routing

9.2 Does This Look Familiar?

A similar decomposition of a large multiplexer driving a shared wire led us to the fundamental concept of segmented-channel routing in Sect. 3.5.2. There is only one major difference: In the multiplexer-based bus of Fig. 9.4, select inputs are provided by the arbiter; that is, they are generated within the user circuit. On the other hand, as we have already mentioned on several occasions, since segmented-channel wires are meant to emulate fixed wires of an ASIC throughout the "life" of a user circuit (i.e., until the next reconfiguration of the FPGA), there the multiplexer inputs are driven by SRAM cells.

Of course, it would be possible to provide access to multiplexer select inputs to signals generated within the user circuit, for instance, by a set of 2:1 decoupling multiplexers driving the select inputs. In a way, that would be similar to how LUTRAM allows users to access LUT masks as read-write memories. Yet, it would be harder to justify the cost of the new hardware needed to allow access to routing multiplexer select inputs, since the individual routing multiplexers are much smaller than a LUT mask. More importantly, it would be difficult for the tools to guarantee that the user signals will not lead to a short circuit. For these reasons, that is not done in practice, although it could be appealing as a research question. Instead, multiplexers of Fig. 9.4 would be implemented in soft logic, often using special technology mapping techniques to minimize the number of occupied LUTs [Met05, Nac15]. Some FPGAs also contained special hardened logic inside their CLBs, which was intended for efficient implementation of wide multiplexers [Pet21].

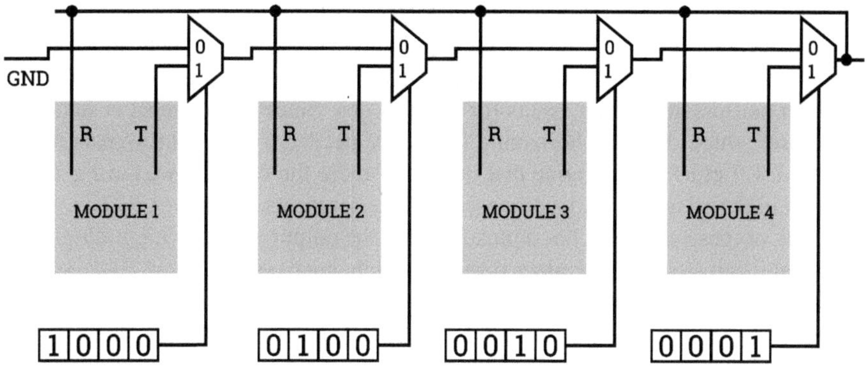

Fig. 9.5 If the modules communicate in a repetitive fashion, the appropriate sequence of multiplexer select input values could be stored in SRAM and replayed

9.3 Statically Scheduled Communication

Let us now imagine a particular communication pattern between the four modules of Fig. 9.4: Going from left to right, each module transmits its signal and then waits for the subsequent ones to finish, all in a round-robin fashion. In that case, we can store the sequence of select input values for each multiplexer in the appropriate number of SRAM bits, as shown in Fig. 9.5. Replaying the sequence in a round-robin fashion will lead to the desired behavior. This way, the select inputs of the multiplexers are no longer provided by signals generated within the user circuit; instead, as in classical stored-select multiplexers of programmable interconnect, they are provided by the appropriate SRAM cells. The only difference is that instead of one cell, each select input receives several of them.

9.4 Time-Domain-Multiplexed Interconnect

Extending the number of SRAM bits on multiplexer select inputs is key to enabling time-domain-multiplexed programmable interconnect. For the static scheduling of bus communication in Fig. 9.5 to be possible, it was necessary to know the transmission sequence a priori. In case of a user circuit implemented on an FPGA with classical programmable interconnect, it is assumed that all wires carrying signals can toggle in every clock cycle. Assuming that the user circuit has a single clock domain, to map it onto an architecture with time-domain-multiplexed interconnect, one can divide its clock periods into several time slots, during which a particular wire will carry different signals. This divided clock is usually called *micro clock* (μCLK) [Tri97a].

To clarify this further, let us turn to an example. Figure 9.6 shows a simple circuit implemented on an FPGA with statically configured programmable interconnect (top of the figure). It consists of three LUTs in three different horizontally adjacent clusters generating inputs for a fourth one. Since interconnect is statically configured, three different horizontal tracks are needed to carry the signals to the target LUT. Let us now imagine that instead of these three tracks, we have a single time-multiplexed track, again composed of three segments. Moreover, let us assume that each of these segments has a pulse latch at the output of its driving multiplexer, such that it can store signal values during multiple micro cycles.

One way to schedule the three connections onto the time-multiplexed wires is shown in the bottom three figures of Fig. 9.6. Dashed latches are operating in transparent mode, and only the horizontal wires are assumed to be time-multiplexed. In the first micro cycle, all three wires are driven by the vertical wires connected to the appropriate LUT outputs; thus, the output of LUT_3 is available at the input of LUT_4. Once the next active edge of the micro clock arrives, the output of LUT_2 is registered in the latch of the middle wire and becomes available at the input of LUT_4. Simultaneously, configuration for the second micro cycle is brought to the select inputs of the multiplexers driving the horizontal wires, causing the two rightmost ones to take inputs from their predecessors in the same track. Upon arrival of the third active edge of the micro clock, the output of LUT_1 is registered and becomes available at the input of LUT_4; this completes the serialization of the three connections.

9.4.1 Interconnect Pipelining

Note that registers shown in Fig. 9.6 were not really necessary for functionality. On the other hand, registers at the inputs of LUT_4, which are not shown in the figure, are necessary because input signals arrive in a serialized fashion and must be deserialized for the circuit to behave properly. Unlike the toy example of Fig. 9.5 which tried to simplify the circuit by profiting from a specific communication pattern, time-domain-multiplexed interconnect must transparently provide correct implementation of any user circuit that is implementable on an equivalent statically configured FPGA.

The interconnect registers are, however, useful for pipelining, without which the target period of the micro clock may not be met. For instance, if the latch of the middle wire in Fig. 9.6 was not used, its driver could have been changed only in the third micro cycle, causing the output of LUT_1 to traverse two switch-blocks before reaching the input of LUT_4, instead of just one.

Interconnect pipeline registers exist in some modern commercial FPGAs with statically configured programmable interconnect as well, most notably, the Altera/Intel Stratix 10 [Lew16] and Agilex [Chr20] families. While they are very useful for register retiming [Lei91], because they allow existing registers to move

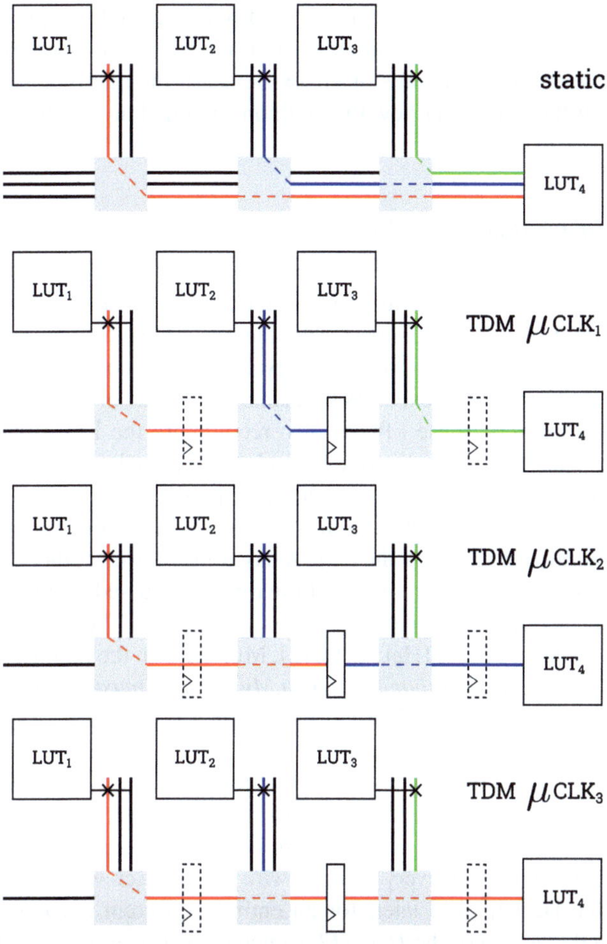

Fig. 9.6 Scheduling connections on time-domain-multiplexed (TDM) wires. The bottom three figures show the multiplexer configurations in the three micro cycles that fit into one cycle of the user clock, whereas the top figure represents the equivalent circuit on an FPGA with statically configured interconnect. For the circuit to function correctly, deserialization registers at the inputs of LUT_4 must be present, even though they are not shown in the figure. Wire latches that are drawn with dashed lines are operating in transparent mode. Ideally, dedicated algorithms for simultaneous (placement and) routing and scheduling would be used to map a circuit onto an architecture with TDM interconnect [Tei12], although alternative algorithms that perform only the scheduling, starting from a placement and routing solution produced on a comparable statically configured architecture, have also been proposed [Francis13]

into the middle of long routing paths, for user designs that are not latency insensitive [Mur14], it is less easy to introduce additional registers into the circuit in order to further increase the effectiveness of interconnect pipelining: Essentially, this must be done manually by the user because of the required alterations to the rest of the design, and CAD tools can at most provide valuable suggestions [Int24]. On the other hand, in case of time-domain-multiplexed interconnect discussed here, user designs are entirely oblivious of the existence of the micro clocks, and hence the tools have full control over how the interconnect pipeline registers are used.

9.4.2 Multi-context FPGAs

Although some researchers have tried to tackle the problem of incorporating time-domain-multiplexed interconnect into otherwise statically configured architectures [Francis13, Lin96], that problem usually arose in the context of designing and utilizing time-domain-multiplexed FPGAs. Namely, instead of merely reconfiguring the routing multiplexers, these FPGAs also reconfigure the LUTs in every micro cycle [Tri97a]. The primary motivation behind this approach has been to extend the capacity of a device, by letting it swap different partitions of a circuit in a round-robin fashion. While in the 1990s this was appealing due to the impossibility to fit the required number of LUTs on the chip, today it may have another benefit, which is particularly of interest in the context of the topics discussed in this book: Instead of connections having to traverse long distances in space, they may travel over a much shorter distance *in time* [Hal10, Tei12]. Much like in the famous saying: *If the mountain will not come to Muhammad, then Muhammad must go to the mountain,*[1] a signal does not need to travel to its destination LUT, but the destination LUT can be dynamically reconfigured closer to the signal's source, as soon as some physical LUT has computed its previous function. For that to be possible, outputs of disappearing LUTs must be stored in additional registers [Tri97a]. According to Professor Betz, the idea of placing these additional registers in the form of a pulse latch at the output of every routing wire multiplexer was first proposed by Tabula [Bet20a, Hal10][2] and later, for an entirely different purpose, adopted by Altera/Intel in the design of the *HyperFlex* interconnect pipeline registers [Lew16]. Typically, each preloaded configuration that is cycled through during the operation of a time-domain-multiplexed FPGA is called a *context*, and hence time-domain-multiplexed FPGAs are also called *multi-context* FPGAs [Hau07]. Both the early Xilinx experiments [Tri97a] and the Tabula devices which came closest to achieving commercial success [Hal10] supported up to eight contexts, although, of course, other numbers could be envisioned as well.

---

[1] The saying was often used by Professor László F. Nagy of the University of Novi Sad, during his exceptional explanations of fundamental concepts in electronics. It has no religious connotation.

[2] Most likely, Francis independently came to the same idea [Fra08], but for a different reason; namely, she did not attempt to propose another full time-multiplexed FPGA (also called multi-context FPGA [Hau07]) but rather investigate the potential benefits of time-domain-multiplexing some of the channel wires inside an otherwise statically configured FPGA.

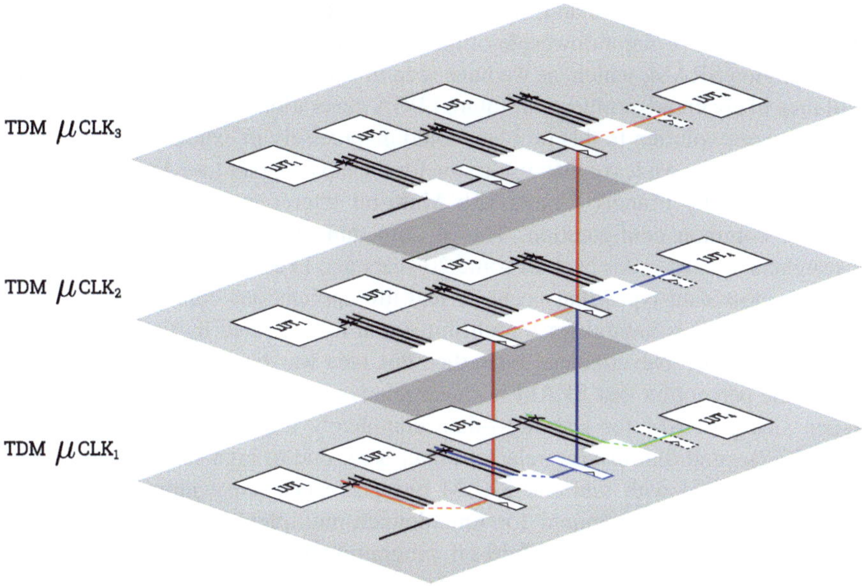

Fig. 9.7 Time as the third dimension. Different contexts can be perceived as different layers in a 3D IC [Sch99, Hal10]. Registering values at routing wires makes them available in the next context, which is equivalent to creating a via between two adjacent layers in a 3D IC; the only major difference is that signals cannot travel down (back in time), but they can wrap around from the top to the bottom layer (persist through completion of the sequence of round-robin context switches) [Hal10]. This analogy formed the basis of Tabula's CAD tools for multi-context FPGAs [Tei12]

9.4.3 Time as the Third Dimension

Placing a register in the driving multiplexer of each wire has another interesting benefit. Namely, if different contexts are perceived as layers in the third spatial dimension, then the interconnect registers provide a way of conveying values of signals from one layer to the next. In such a model, depicted in Fig. 9.7, registers hence act like vias and enable the CAD tools to solve the problem of 2D placement and routing simultaneously with reconfiguration scheduling, by placing and routing a circuit in three spatial dimensions [Hal10, Tei12]. To the best of our knowledge, this space-time concept was first proposed by Schmit [Sch99, Sch05]. A similar idea later appeared in the domain of *Coarse-Grained Reconfigurable Array* (CGRA) CAD, in the form of *Modulo Routing Resource Graphs* (MRRGs) [Mei03].

9.4.4 A Note on SRAM

We will dedicate some attention to spatially 3D devices in the next chapter. As of now, it is important to note that time-domain-multiplexed interconnect could be used to reduce wiring demand, by serializing some non-timing-critical signals of the user circuits. A major downside of the approach is the necessity to replicate the configuration SRAM, which, as we have seen in Sect. 4.8.9, is difficult to scale and will thus likely create problems in future FPGA generations, even if used sparingly. However, one solution to this problem could lie precisely in extending FPGAs into three dimensions; we leave that discussion for the next chapter too. Another solution was observed by Francis [Francis13]: Multi-bit interconnects can be combined to share common configuration. This is something that has drawn attention of researchers at least as far back as some 20 years ago [Ye06]. The idea is appealing because most data types that occur in real-life applications contain more than a single bit, and it is natural for a multi-bit signal to be routed by the same bundle of wires [Gay20]. Nevertheless, back then this idea was largely dismissed, mostly due to the perception that SRAM is cheap relative to the multiplexers themselves. When sizing SRAM for stability—then completely neglected in academia (see Sect. 4.2.2)—and diminishing relative driver area, due to high series resistance of wires, is coupled with multiple SRAM bits per every select input of a routing multiplexer, which is required for time-domain-multiplexing, conclusions could become drastically different. Multi-bit programmable interconnect has recently become a topic of interest for other reasons as well [Cou25].

9.4.5 A Way to Achieve Design Portability

Time-domain-multiplexing of interconnect would also allow for another possibility. Namely, when designing a CPU microarchitecture for a certain ISA, a vendor has many trade-offs at their disposal. Some instructions can be made faster and others slower; what matters is that all of them can be executed eventually. FPGAs as of now have nothing resembling an ISA, which would make user designs portable between different devices without any further investment of manual labor. This is certainly a major deterrent for companies to step in and create more *Intellectual Property* (IP). One way to ensure that a design requiring certain connectivity would route, while still allowing vendors the flexibility to invest resources in different portions of the device as they see fit, would be to serialize surplus signals using TDM interconnect.

9.5 A Need for Bus Standardization

In the preceding sections, buses served us as an example motivating the introduction of time-multiplexed programmable interconnect to enable sharing of physical wires between different signals of the user circuits. In their essence, buses implemented on top of an FPGA—i.e., in soft logic—serve the same purpose, although without the possibility to multiplex signals at a clock rate faster than the fastest clock in the user domain. To be efficient, they typically require more elaborate arbitration mechanisms, pipelining, split-transactions, etc. [Dal04], all of which goes beyond simple time-domain-multiplexing of physical wires.

As these additional aspects grow more sophisticated, designing and verifying buses becomes more and more time consuming. In Chap. 1, we have seen that main uses of reconfigurable platforms range from cheap alternatives for entities that cannot finance design and production of an ASIC to rapid deployment of accelerators of algorithms that evolve at a very high pace. This second case is especially appealing to large companies that have large budgets, but that can also benefit more from cost savings available by rapid deployment of new hardware at scale [Put23]. Nevertheless, development time is highly important for smaller entities as well. Not only does securing market share depend on timely entry, but engineer salaries also contribute a large fraction of new hardware development, and if they have to be paid for too long a period, FPGAs cease to represent an affordable alternative.

To solve these issues, engineers adopted a *System on Chip* (SoC) design approach, where instead of developing each new hardware block from scratch, they can procure verified implementations in the form of third-party IP and simply instantiate them in their system as needed [Kea02]. However, for that to be possible, communication between different modules has to be standardized; otherwise, significant effort and deep knowledge of internal functioning of third-party IP blocks would be required to integrate them into a working system. As in many other cases, this concept first appeared in the realm of discrete components [Dal04]. It was then brought into ASIC SoCs through the trend of increasing integration with miniaturization and then finally into soft SoCs intended for operation on FPGAs.

9.6 Hardening Buses

Supporting advanced features of high-performance busses in soft logic requires significant resources. This is made even worse by the common need to make the soft buses much wider than their ASIC counterparts, to maintain the required bandwidth despite the lower maximum achievable clock frequency [Abd14]. The disparity between the cost of soft and hard implementations of advanced buses is vividly captured by the following quote from a 2000 paper by Jan Gray [Gra00]:

> *The title of this paper is "Designing a Simple FPGA-Optimized RISC CPU and System-on-a-Chip." While there are many strengths to the two emerging industry-standard on-chip buses, AMBA and CoreConnect, they are neither simple nor FPGA-optimized, and you will probably never live to see their annotated Verilog source code in a paper such as this. Instead we are going to use a bus that is simple, extensible, and very efficient in its use of programmable logic.*

Yet, merely 2 years later, Xilinx tools started supporting the *OPB* and *PLB* buses that were part of IBM's *CoreConnect* [Xil02]. Once IBM PowerPC hardened CPUs were replaced by ARM ones, OPB was also replaced by AXI (part of AMBA).

This illustrates how important user adoption and ease of use is when designing a reconfigurable architecture and a tool and IP ecosystem around it.

Given that AXI is today the de facto standard bus for soft SoCs and that the underlying FPGAs have not changed that much since the days of the above Gray's comment, high resource consumption in soft logic remains an important issue. We have already seen in Chap. 1 that a highly effective way of increasing FPGA density is to harden functionality that is used very often; hence, it is logical to ask whether one should not attempt hardening the standard buses as well [Bet15, Bet20a]

9.6.1 *Issues with Bus Hardening: Topology*

Buses in ASICs and soft buses on FPGAs are traced to fit the floorplan of the specific system which they serve. The first question when attempting to harden a bus is therefore how to determine the right topology? At first, this could seem like an unsolvable problem because each user circuit will have at least a slightly different floorplan. However, the question is essentially the same as the one where should the wires of the programmable interconnect be traced, given that every user circuit has different connectivity requirements; the difference is mostly in granularity. In Chap. 1, we have seen that, much in the spirit of the saying about Muhammad and the mountain, early FPGAs solved this by creating a mesh of prefabricated wires and relying on placement to bring the circuit elements close to the wires, rather than the other way around. The key was that the regularity of a mesh allowed very flexible placements, with little need for the elements of the technology-mapped user circuits to move far away from their optimal placement in an empty plane. At least in principle, one could attempt a similar approach of letting the floorplanning algorithms appropriately position blocks of a user circuit with respect to the hardened bus.

9.6.2 *Issues with Bus Hardening: Width*

Nevertheless, that would not resolve another fundamental problem: Different circuits require different bus widths. Moreover, a typical SoC contains multiple different buses of varying widths [Kea02]. Of course, one could think of hardening the widest possible bus, but that would be very wasteful for the circuits that do not need it; for instance, some Xilinx IPs can support AXI4 buses that are up to 1024 bits wide [AMDXil22]. On the other hand, if a narrower bus is hardened, some modules would have to transmit data over multiple cycles. During those cycles, a traditional bus would not be able to support transmission by any other modules. To resolve this limitation, already in early 2000s engineers started to introduce message packetization, which quickly led to adoption of packet-switched *Networks on Chip* (NoCs) [Ben02].

9.7 Networks on Chip

For a long time, ASIC designers were able to design systems around a single clock source. Yet, as clock frequencies kept increasing along with increasing chip size, it became natural to expect that assumptions of instantaneous synchronization of all elements on the chip would no longer hold. While asynchronous circuit design never became mainstream, as may have been expected in late 1980s and early 1990s [Var90], effects of noninstantaneous clock propagation started to be felt in practice by the early 2000s, which triggered wider adoption of *Globally Asynchronous, Locally Synchronous* (GALS) design [Ben02]. That brought forward another important question: How does one implement reliable communication between a large number of components and across multiple clock domains, in the presence of transmission delay and even transmission success uncertainties? Luckily, ASIC designers did not have to solve this problem from scratch. Network engineers already faced it much earlier, and by the time transmission issues started to occur on a microscale, they had already developed a large body of knowledge and design techniques for solving it. Most notably, instead of trying to make the transmission medium more certain on the physical level, networks enable tolerating the inevitable uncertainties in the medium by resolving errors using higher level communication protocols [Dal01]. Although not necessarily the first ones to propose the idea of transferring macroscopic networks onto a chip [DeM10], works of Dally and Towles [Dal01] and Benini and De Micheli [Ben02] were probably the most instrumental in establishing *Networks on Chip* (NoCs) as the predominant way of organizing system-level communication.

Much like integration of cache memories in the first RISC processors [Pat80a], that highly successful approach is another instance of how adoption of established solutions from macroscopic systems can propel single-chip systems forward. As is often the case, it did not take long before researchers started to implement NoCs on top of FPGAs [Set05, Kap06], and not long after that, first proposals to harden them came about as well [Francis13, Abd16]. In subsequent sections, we will explore some benefits, downsides, and design trade-offs of integrating hardened networks on chip with programmable interconnect of FPGAs.

9.7.1 What Exactly Is a NoC?

Before delving into FPGA-specific topics related to networks on chip, let us first introduce some basic concepts and definitions. They were mostly adapted from Dally and Towles [Dal04], and interested readers should refer to their text for an in-depth discussion that goes beyond the scope of this book.

Much like a macroscopic network, a network on chip is a collection of *routers* connected together by direct *links* or *channels*. The channels abstract the physical wires that implement them, while it is the routers' task to implement the chosen

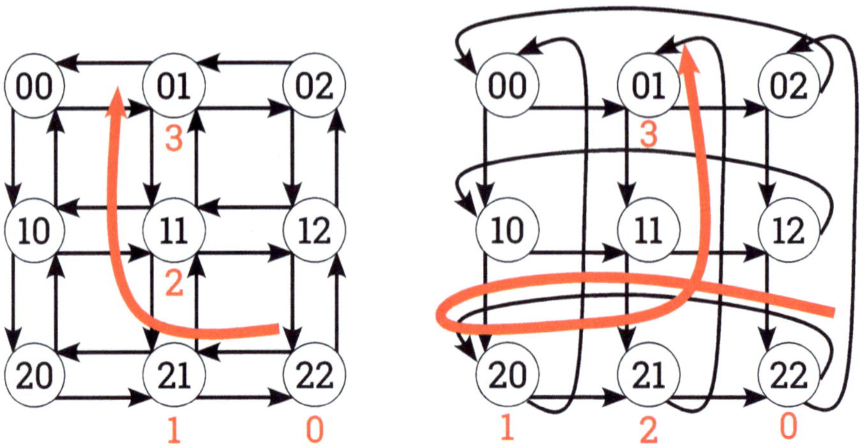

Fig. 9.8 Basic NoC topologies and *dimension-ordered* routing [Dal04]. Figure on the left shows a bidirectional *mesh* network topology, where each router is connected to its four immediate neighbors. A directional torus, shown on the right, allows not only connections to routers with the increasing coordinate in the given dimension but also wrap around from the last router to the first. In dimension-ordered routing, each path is uniquely determined from the source and the target coordinates, by composing the shortest horizontal-only path that reaches the column of the target router with the shortest vertical-only path that reaches the row of the target router. Routers forming those paths for the example of connecting coordinates 22 and 01 are marked by their hop indices in red, in both figures. Note that although the number of hops is identical in this case, the length of links in the directional torus is variable and significantly longer than in the mesh

transmission protocol for sending packets over these channels. Each message that a particular module may send through the network must first be packetized, while each packet might in turn be divided into several fixed-size *flow control digits (flits)*.

9.7.1.1 Topology

Each NoC is characterized by three main ingredients: (1) *topology*, (2) *routing*, and (3) *flow control*. Dally and Towles describe these concepts very vividly by an analogy with, respectively, a road network, a vehicle routing algorithm, and a set of traffic lights actually allowing the cars to take certain roads [Dal04]. In the case of NoCs, topology describes how different routers are connected together by the routing channels. Deciding the topology of a NoC is hence similar to deciding where to construct roads and junctions. Perhaps unsurprisingly, one of the most common topologies, shown on the left of Fig. 9.8, is very similar to organization of streets around urban building blocks: a rectangular mesh, where every router—that is, every junction in the analogy—is connected to the four neighboring routers by a pair of channels that form a bidirectional link, much like the two opposing lanes of a street connecting two junctions. In previous chapters, we have seen that the mesh topology is also the most common way of organizing programmable interconnect in FPGAs. As we have already mentioned, one of the main benefits is that it puts very few constraints on placement of user circuits. For the same reason, it is very appealing for embedded FPGA NoCs as well, although we will see in Sect. 9.7.4 that some other practical constraints may lead designers away from adopting it.

It is also not uncommon to see one-way streets. Sometimes when space between buildings is very narrow, that is simply the only possibility, but even if that is not the case, reserving all lanes of a street for a single direction has certain benefits; for instance, flow control is easier, because one can seamlessly take left turns without having to take care of the right-hand rule and pedestrians have to keep watch of only one incoming side of the junction. Following the analogy of Dally and Towles, the right side of Fig. 9.8 shows a *directional torus* topology that is akin to a road network composed exclusively of one-way streets.

9.7.1.2 Routing Algorithms

Once a packet arrives at one of its input ports, the router needs to dispatch it further. Which output port it will choose depends on the *routing algorithm*. Routing algorithms come in two basic flavors: *deterministic* and *adaptive*. Deterministic routing algorithms will always output one and the same path between two points in the network, regardless of traffic conditions in it. Assuming that all car drivers follow instructions from a GPS navigation system, which today is very often indeed the case, continuing the road network analogy of Dally and Towles, this would be equivalent to the offline standalone navigation devices of some 20 years ago, which came with preloaded static maps and determined the routes based solely on information stored in them. On the other hand, adaptive routing algorithms strive to minimize contention in the network by taking into account not only the length of the path that a packet has to travel but also the congestion, at least in the channel that the current router is to dispatch it to [Dal04]. This is more akin to the modern service offered by, e.g., Google Maps, which gathers traffic congestion information in different streets and adjusts the offered routes accordingly.

Due to simplicity of implementation and better control over *Quality of Service* [Swa19], NoCs used in FPGAs typically employ deterministic routing algorithms. A particularly popular one is the so-called *dimension-ordered routing* [Dal04] that is also illustrated in Fig. 9.8. It operates by first routing a packet along one axis, usually horizontal, and then upon reaching the coordinate of the target router, proceeds with routing along the other axis.

The most flexible way of implementing a routing algorithm in a router is to use *routing tables* which store (sets of) precomputed routes for each destination [Dal04]. Simple algorithms such as dimension-ordered routing can also be implemented directly in dedicated circuitry [Dal04].

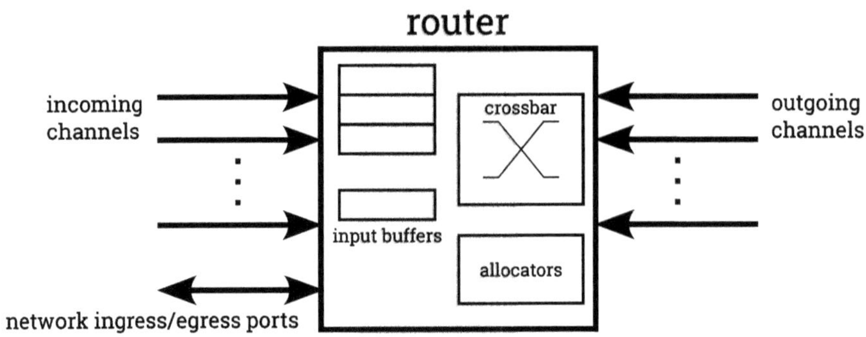

Fig. 9.9 A highly simplified schematic of a typical NoC router. Incoming packets are buffered at the inputs until the resources that they require become available—a decision made by the various allocators. There could be more buffers than the physical incoming channels, in case *virtual channels* [Dal04] are used to maximize their bandwidth utilization. Optionally, a router can have ingress/egress ports that allow it to connect to processing elements outside of the network. Very often, frequency conversion is required at these transition ports, as is conversion to a communication protocol different from the one used by the network itself [Swa19]

9.7.1.3 Flow Control

A highly simplified schematic of a NoC router is shown in Fig. 9.9. Once a packet arrives from one of the incoming routing channels, it is temporarily stored until the resources required for its further transmission become available (they could be in use for transmitting another packet). The number of incoming physical channels will typically be determined by the topology of the network: For instance, a router in a mesh network will have four incoming physical channels, whereas a router in a directional torus will have two. Very often, *virtual channels* are used to maximize the utilization of the bandwidth available in the physical channels [Dal04]. In that case, buffers and channel control circuitry are replicated for each virtual channel.

Allocation of resources to packets is done by various allocators in the router, according to a *flow control mechanism.* Various flow control mechanisms are explained in detail by Dally and Towles [Dal04], and for the interest of space, we shall not repeat their explanations here.

Each router can optionally have ingress and egress ports which allow it to communicate with modules outside of the network. Since these could operate at a different frequency (typically (much) smaller) than the network itself, frequency conversion has to be performed. Networks also often use a different communication protocol than the rest of the system, so that conversion must be done at the ingress/egress ports as well [Swa19]. When every router is associated with precisely one module, that is typically called a *direct interconnect network* [Ben02].

9.7.2 Sources of Interconnect Uncertainty in FPGAs

We have seen that a standardized high-performance bus already solves many of the problems related to ease of SoC design. We have also mentioned that a major remaining difficulty with buses in ASICs was timing closure, due to unpredictable parasitics [Ben02]. This was particularly complicated in older bus designs that used shared transmission medium, which assumed that all communicating modules are attached to one and the same electrical node. However, that problem was resolved in FPGAs as soon as tristate buffers were fully replaced by directional drivers [Lew03, Lem04]. Characterization of wires is greatly simplified by prefabrication [Mai19]; as a matter of fact, Benini and De Micheli even considered statically configured programmable interconnect of an FPGA a special embodiment of a network on chip [Ben02]. Yet, despite all this resulting in a better starting position, NoCs were introduced into FPGAs. It is important to understand what motivated that.

We have already seen in Sect. 2.9.1.1 that connection delays in island-style FPGAs are highly dependent on floorplan and placement of the circuit, which determines not only the minimum achievable length of particular connections but also detours necessitated by congestion induced in various channels. Whenever custom interconnect topology appears in a user circuit, that problem will arise. Of course, this is also a major source of uncertainty in ASIC design [Dal01].

Hardening a NoC entirely eliminates the problem. However, is that really necessary? A fixed-topology NoC would resolve the issue whether hardened or soft coupled with a fixed, precomputed floorplan and placement. Although with the current FPGA CAD tools it is not trivial to incorporate a preplaced component into a design [Gue24], it is doable and such approaches have been proposed. For example, the *SpiderWeb* soft NoC from Intel utilized special floorplanning and placement techniques to keep the NoC layout fixed [Lan20], while Kapre and Gray used a set of floorplanning constraints to stabilize the floorplan of the *Hoplite* soft NoCs [Kap17]. With some manipulation of precompiled bitstreams [Kas23], even the runtime that FPGA CAD tools spend on implementing a fixed NoC could be avoided.

9.7.3 School of Hard NoCs

So if the main motivation for hardening a NoC—improving predictability and easing timing closure—could already be satisfied by a fixed soft NoC, then why harden it? This is another instance of the classical problem of deciding when to harden components in a reconfigurable architecture: Under the premise that most modern high-performance designs will require a NoC, if substantial area savings could be obtained from hardening, then there is little reason not to do it.

9.7.3.1 Measuring the Hardening Benefits

By comparing various designs of NoC routers implemented in soft logic and in standard cells, Abdelfattah and Betz concluded that hardening a router would bring a 22× area reduction, on average [Abd12]. In another paper, they also concluded that a high-performance hardened NoC would increase the FPGA core area by about 1.4% [Abd14]. That means that an equivalent soft NoC would consume about 31% of the resources available on the same FPGA. As a comparison, Putnam et al. report that a fixed shell used in their *Catapult* system consumed about 23% of a Stratix V FPGA [Put14]. This is substantially less than what Abdelfattah and Betz measure for a soft NoC, and given that Xilinx decided to harden the elements of a similar shell in their latest Versal devices [Gai19], the case for hardening seems very clear. As we have already mentioned several times until now, the obtained area savings would not only be reflected on the device cost but also on its performance, by increasing the logic density and hence making the average wires shorter.

9.7.3.2 FPGA Designs Are Not ASIC Designs Implemented in Soft Logic

In their studies, Abdelfattah and Betz inspected the impact of many router microarchitecture parameters on the difference between its soft and hard implementations. In fact, to the best of our knowledge, their work has thus far been the most comprehensive treatment of the topic of hardened NoCs in FPGAs, and interested readers are highly recommended to refer to it [Abd16]. Nevertheless, all of the routers that they considered made use of extensive buffering and supported virtual channels and advanced control flow mechanisms. As a result, even in the most beneficial setting observed for the soft router, it was 13× larger than its hardened counterpart [Abd12]. On the other hand, the hardened router chosen for the subsequent studies was reported to have required an area equivalent to nine clusters of ten 6-LUTs [Abd14]. From this we can compute the area of the smallest soft router to have been 1170 6-LUTs.

Kapre and Gray, however, acknowledged that inheriting a circuit design meant for ASIC implementation and simply porting it to an FPGA is not always the wisest option a designer can make [Kap17]. Hence, instead of following the path of the previous authors, they tasked themselves with developing a NoC that is highly optimized for connecting a very large number of relatively small cores on an FPGA. Instead of using a bidirectional mesh, they switched to a directional torus, thus cutting the number of incoming physical channels in half; they also dropped the use of virtual channels and complicated flow control mechanisms and even cast away the input buffers [Kap17]. As a result of highly optimized RTL and custom technology mapping, they managed to pack a router into merely 60 6-LUTs [Kap17]. Not only is that almost 20× smaller than the smallest soft router used by Abdelfattah and Betz, but it is also 1/3 more area efficient than their hard router. Once this is taken into account, the case for hardening a NoC is not as compelling as it would seem otherwise.

Of course, hardening the router of Kapre and Gray would have brought further area savings, but they would likely not be as dramatic as those measured by Abdelfattah and Betz. This is not to object to the final conclusions about NoC hardening or to investing larger area in implementing a more advanced router. Rather, it is to point out that when deciding what blocks to harden, it is also important to keep in mind the sensitivity of the conclusions to the chosen references.

It is also worth noting that HLS-produced versions of the Hoplite routers required 576 6-LUTs, which is much closer to what Betz and Abdelfattah measured, even if the architecture is much simpler [Kap17]. The almost 10× area reduction obtained through a combination of RTL and custom technology mapping demonstrates the benefits that could be obtained through development of custom, domain-specific EDA tools [Bol22]. It also yet again reminds us of how sensitive conclusions of architectural studies can be to various unexpected sources of noise [Yan02].

9.7.3.3 Speedup Beyond Area Reduction

It has been a recurring statement in this book that optimizing area efficiency reduces the length of wires and thus improves performance of the FPGA as well. This clearly speaks in favor of hardening NoCs. However, unlike when hardening many other commonly occurring features, NoC optimization opportunities do not stop there. Fixed tracing of relatively few wires allows for better allocation of tracks in the metal stack, optimizing spacing to reduce sideways capacitance, and, perhaps most interestingly, introduction of highly optimized low-swing signaling circuits that can increase signal propagation speeds by as much as 3× [Dal01].

9.7.3.4 What About Adaptive Topology?

Link hardening also has downsides: Once done, topology of the network cannot be altered, and one important degree of freedom for optimizing the performance of implemented user circuits is lost [DeM10]. To obtain the best of both worlds, Abdelfattah and Betz have proposed a hybrid NoC in which the otherwise highly area inefficient routers are hardened, while links are still implemented by the regular channel wires of the programmable interconnect [Abd14]. They measured that the speed advantage of hardened versus the fully soft NoC drops from 6× in case of fully hardened implementation to 5× when using hardened routers and soft links. Such a small difference makes it appealing to consider the more flexible hybrid approach. However, it seems that Abdelfattah and Betz have not used the opportunity to introduce low-swing signaling into the hardened links, which, by a simple extrapolation of the plausible speedups reported by Dally and Towles [Dal01], would increase the speed advantage of the fully hardened NoC to 18×, while the hybrid one would stay at 5×. It is also worth noting that the small speed difference of the two approaches in the case when specialized signaling is not feasible is likely due to the fact that the long channel wires that are used for connecting the sparsely placed routers in the hybrid approach are traced at thicker metal and are not loaded by the numerous multiplexers of the connection blocks [Pet21]. This makes their performance much closer to dedicated wires of the same length than was the case for short-haul and medium-haul wires that we have been analyzing in most of the preceding chapters.

9.7.3.5 The Ultimate Use of NoCs

In principle, NoCs are not meant to guarantee latency of packet delivery. This means that they are mostly suited to providing high-bandwidth communication for transactions that can tolerate variable latency [Bet20a]. While latency-insensitive design is becoming increasingly popular [Mur14], hardened NoCs still predominantly serve communication with memory controllers and peripherals, which naturally exhibit variable latency themselves [Abd14]. Moreover, hardened memory and periphery controllers are usually located at the edges of large commercial FPGAs [Gai19], meaning that to access them, signals typically have to cross significant portions of the entire chip. As we have already mentioned, transmission over large distances at which reliable synchronization is not possible was one of the primary motivators for introduction of NoCs in ASICs as well [Ben02]. Hence, it should not be surprising that both Xilinx/AMD and Altera/Intel cite communication with memory controllers as the most significant use of their hardened NoCs [Gai23, Alt24].

9.7.3.6 Logical Wires

While scheduling traffic on NoCs is often done with guaranteed upper bounds on specific communication flow latency [Swa19], it is also possible to reserve NoC resources for a particular flow. For example, Abdelfattah et al. do that using the mechanism that they call *Permapaths* [Abd15]. In this situation, links of the NoC act like simple pipelined wires, with some additional router-incurred delay, while the ingress and egress ports (*FabricPorts* in the work of Abdelfattah et al.) still serve the purpose of packetization and frequency adaptation. Conceptually, this is similar to the *Logical Wires* proposed by Dally and Towles [Dal01].

We note however that this is not the optimal use of a NoC. Rather than substituting programmable interconnect of the FPGA which already offers fixed wiring, NoCs should be considered complementary, enabling high-bandwidth long-haul connectivity at which programmable interconnect is inferior.

9.7.4 Constraints of an Actual Device

Let us now briefly analyze a very interesting example of a commercial hardened NoC. Figure 9.10 shows a sketch of a floorplan of a plausible device from

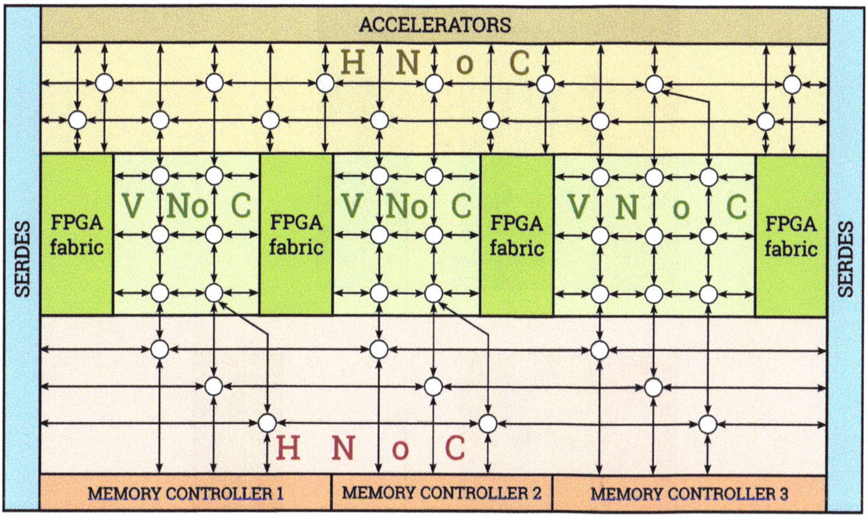

Fig. 9.10 An example floorplan of a plausible FPGA from the Versal family [Gai19], showcasing the hardened NoC. Hardened NoCs in Versal family of devices are composed of a collection of *vertical NoCs* (VNoCs) inserted in between columns of programmable fabric and *horizontal NoCs* which run along the top and the bottom edges of the programmable fabric and connect it to various large hardened blocks such as the memory controllers and AI engines [Gai19]. As the width of the device increases, additional VNoCs are inserted, while the bandwidth of both the HNoCs and the VNoCs can be increased or reduced according to needs of the particular device (typically such that they can saturate the bandwidth of the memory controller bellow the VNoC) [Swa19]. Note that various elements are not drawn to scale (The figure was adapted from [Swa19])

the Xilinx/AMD *Versal* family of FPGAs [Gai19]. Unlike most hardened NoCs explored in academia, Versal's NoC is not based on a regular topology such as a bidirectional mesh. Instead, it consists of a number of so-called *vertical NoCs* (VNoCs) and two *horizontal NoCs* (HNoCs) [Swa19]. VNoCs are inserted between columns of the programmable fabric, while HNoCs run along its top and bottom edges, providing connectivity with the memory controllers and AI engines, among other blocks [Gai19]. Each edge in the figure represents two 128-bit-wide channels that provide a bidirectional connection between the adjacent routers. In case a particular memory controller can sustain a higher bandwidth, it is possible to widen the mesh in the particular VNoC by adding an additional column of routers. Egress and ingress ports implement a conversion between AXI on the side of the fabric and a custom protocol used within the NoC, as well as any required frequency adaptation [Swa19].

9.7.4.1 Impact of Device Floorplan on NoC Topology

One may wonder why Versal designers chose not to use a regular topology such as a mesh or a torus, if the adopted solution complicates the routing algorithms [Swa19]?

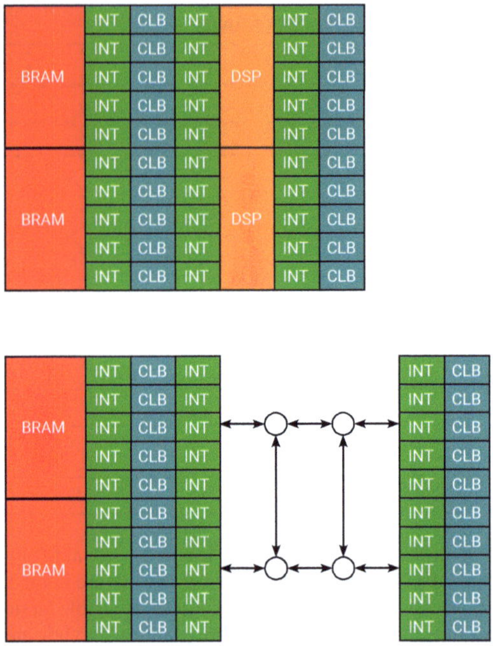

Fig. 9.11 Integration of VNoCs in ASMBL columns. Making VNoCs vertical-only allows seamless integration into the ASMBL floorplan while retaining the interconnect (INT) tiles as they are. Links of the VNoC are traced on the metal layers that would normally be reserved for vertical routing wires. Horizontal wires of the programmable interconnect can freely cross the VNoC columns, without any channel bandwidth reduction or a need to redesign the INT tiles, which would have been required if horizontal links between different VNoCs were introduced as in a standard mesh or torus [Swa19]. HNoCs, on the other hand, are positioned outside of the programmable fabric where they do not interfere with programmable routing and where the ASMBL floorplan is anyway not respected by the large hardened blocks

To answer this, let us remember the columnar organization of the FPGA floorplan, which is shown in Fig. 2.15 and is repeated with more relevant detail in Fig. 9.11.

All blocks out of which the columns of the programmable fabric are composed (e.g., memory blocks and DPSs) are constructed such that their height is an integral multiple of the height of the CLB [Shr23]. For example, in Xilinx/AMD 7-Series FPGAs, both DSPs and BRAMs have a height that is 5× larger than that of a CLB [Xil18, McM23]. The width of the different columns can and does vary, however [McM23], but that has no impact on the connectivity of the programmable interconnect: Horizontal wires that cross different columns still have the same logical (hop) length, although the differences in their physical length induced by the different column widths are reflected on the delay [Mai19].

To ease composability of the fabric, Xilinx/AMD FPGAs actually extract the interconnect multiplexers into separate CLB-high interconnect (*INT*) tiles [Zho21], as shown at the top of Fig. 9.11. In a sense, such separation almost makes the programmable interconnect a standalone IP that is to be integrated into a complex SoC that we still call an FPGA.

Integrating any number of routers of a VNoC that can fit into the height of the programmable fabric and any number of links in such a floorplan is simple, provided that no links leave the VNoC's bounding rectangle, apart from connecting to the immediately adjacent INT tiles (or perhaps logic tiles). Since all links within a VNoC are nearest-neighbor, they do not intersect and can thus be traced in a single layer above the VNoC's bounding rectangle, normally reserved for the vertical channel wires. Horizontal wires of the programmable interconnect are not impacted by the presence of the VNoC anymore than by the presence of, e.g., a DSP: At most, their delay will be increased, but connectivity will not change. On the other hand, if there were horizontal links between different VNoCs, horizontal programmable routing channel bandwidth would be impacted, or additional metal layers would have to be added to the stack. Additionally, due to these changes, the INT tile would likely have to be redesigned as well.

HNoCs, on the other hand, do not have to respect any such constraints. Since they are positioned at the outer edge of the programmable fabric, their metal requirements do not conflict with those of the programmable interconnect. Moreover, the large hard blocks at the top and the bottom of the chip already violate the ASMBL floorplan.

All this once again illustrates the importance of taking into account the various layout constraints when designing interconnect for a commercial device, no matter if it is a NoC like in this section, or classical programmable interconnect as in the previous chapters. In the particular case, this is not without repercussions: Since 2D coordinates are not uniformly defined, simple routing algorithms such as dimension-ordered routing are no longer applicable. Instead, the authors had to develop a new set of exact tools to schedule the different flows onto the NoC [Swa19]. Poor CAD can clearly fail to make proper use of architecture with a good layout, thus making its performance appear bad. However, even the best CAD cannot make up for a bad layout of the architecture. Hence, if codesign of CAD and architecture is not possible, it is usually better to opt for architecture that allows a very efficient layout and then let more complicated tools compensate for its other deficiencies. One notable exemption to this is when reducing tool runtime is the primary objective [Shr23].

9.8 Some Open Question

For a more in-depth discussion of various influences on the Versal NoC design decisions, besides consulting the paper of Swarbrick et al. [Swa19], one may also wish to consult the lecture of Dr. Gaitonde [Gai23]. Among other things, the interested reader will learn there that the problem of designing an embedded NoC that provides the best balance between area efficiency and flexibility required for supporting a large variety of user circuits is still very much open. That should not be a surprise, for this problem is not so much unlike the problem of designing programmable interconnect architectures; as we have seen in Chap. 1, about 15 years were required for people to learn to do that well. Moreover, this is bound to be a process without a real end, since both the fabrication technology and the designs implemented on FPGAs will continue to evolve. It was difficult to design the first hardened NoCs, when there was little real data about how the end users will be using them (one cannot measure what does not exist), but over the years, significant knowledge will certainly be gained and used to improve the future generations of NoC architectures.

What is perhaps more relevant to the topic of this book is how to codesign a NoC and a programmable interconnect architecture? As we have already explained, these two types of interconnect are largely complementary and meant to cater to different kinds of connections of the user circuits. Hence, one could simply design them separately, while assuming orthogonality, or fixing one and designing the other. In the analogy with logic synthesis that we used in Chap. 3, this would be equivalent to performing classical combinational synthesis, with registers simply removed from the circuit. However, it is known that vastly superior results can be obtained through sequential synthesis [Mis08]. It is likewise probable that the best results could be obtained by designing statically configured programmable interconnect, classical time-domain-multiplexed interconnect, and packetized interconnect of future reconfigurable architectures together. An important problem on the FPGA CAD front that is required for that endeavor is how to automatically partition connections of the circuit among the three types of physical interconnect, without any interference of the user. Most likely, this would call for raising the level of abstraction at which user circuits are specified, so that latency insensitivity can be automatically ensured.

References

[Abd12] Abdelfattah, M. S., & Betz, V. (2012). Design tradeoffs for hard and soft FPGA-based networks-on-chip. In *Proceedings of the International Conference on Field-Programmable Technology*. Seoul (pp. 95–103).

[Abd14] Abdelfattah, M. S., & Betz, V. (2014). The case for embedded networks on chip on field-programmable gate arrays. *IEEE Micro, 34*(1), 80–89. ISSN: 1937-4143.

[Abd15] Abdelfattah, M. S., Bitar, A., & Betz, V. (2015). Take the highway: Design for embedded NoCs on FPGAs. In *Proceedings of the 2015 ACM/SIGDA International Symposium on Field-Programmable Gate Arrays*. Monterey, CA (pp. 98–107).

[Abd16] Abdelfattah, M. S. (2016). Embedded Networks on Chip for Field-programmable Gate Arrays. Ph.D. Thesis. University of Toronto.

[Alt24] Altera (2024). *Altera®Agilex FPGAs Network-on-Chip (NoC): Overview & Introduction*. https://learning.intel.com/developer/learn/courses/15919/alterar-agilex-fpgas-network-on-chip-noc-overview-introduction. Accessed: 15 Oct 2024.

[AMDXil22] AMD Xilinx (2022). *AXI DataMover v5.1 LogiCORE IP Product Guide*.

[Ben02] Benini, L., & De Micheli, G. (2002). Networks on chips: A new SoC paradigm. *Computer, 35*(1), 70–78.

[Bet15] Betz, V. (2015). *The Case for Embedding Networks-on-Chip in FPGA Architectures.* FPT'15 Keynote. Queenstown.

[Bet20a] Betz, V. (2020) *Lecture 13: Embedded NoCs, Time-Multiplexed FPGAs, and Configuration Robustness Alternatives.* https://www.youtube.com/watch?v$=$67KILx2Y3XA. Accessed: 11 Oct, 2024.

[Bol22] Bolsens, I. (2022). *CHIPS Alliance FPGA Interchange Format.* Accessed on 15 Oct 2024. https://www.linkedin.com/pulse/chips-alliance-fpga-interchange-format-ivo-bolsens

[Chr20] Chromczak, J., Wheeler, M., Chiasson, C., How, D., Langhammer, M., Vanderhoek, T., et al. (2020). Architectural enhancements in Intel®Agilex™FPGAs. In *Proceedings of the 2020 ACM/SIGDA International Symposium on Field-Programmable Gate Arrays.* Monterey, CA (pp. 140–149).

[Cou25] Coulon, L., Ramirez, L., Anderson, J., Stojilović, M., & Ienne, P. (2025). FRIDA: Reconfigurable arrays for dynamically scheduled high-level synthesis. In *Proceedings of the 2025 ACM/SIGDA International Symposium on Field Programmable Gate Arrays,* Monterey, CA, USA (p. 12).

[Dal01] Dally, W. J., & Towles, B. (2001). Route packets, not wires: On-chip interconnection networks. In *Proceedings of the 38th Design Automation Conference.* Las Vegas, NV (pp. 684–689).

[Dal04] Dally, W. J., & Towles, B. P. (2004). *Principles and practices of interconnection networks.* Morgan Kaufmann. ISBN: 978-00-80497-80-8.

[DeM10] De Micheli, G., Seiculescu, C., Murali, S., Benini, L., Angiolini, F., & Pullini, A. (2010). Networks on chips: From research to products. In *Proceedings of the 47th Design Automation Conference.* Anaheim, CA (pp. 300–305).

[Fra08] Francis, R., Moore, S., & Mullins, R. (2008). A network of time-division multiplexed wiring for FPGAs. In *Proceedings of the Second ACM/IEEE International Symposium on Networks-on-Chip.* Newcastle Upon Tyne (pp. 35–44).

[Francis13] Francis, R. M. (2013). *Exploring networks-on-chip for FPGAs.* Technical Report. UCAM-CL-TR-828. University of Cambridge, Computer Laboratory. https://www.cl.cam.ac.uk/techreports/UCAM-CL-TR-828.pdf

[Gai19] Gaide, B., Gaitonde, D., Ravishankar, C., & Bauer, T. (2019). Xilinx adaptive compute acceleration platform: Versal architecture. In *Proceedings of the 2019 ACM/SIGDA International Symposium on Field-Programmable Gate Arrays.* Seaside, CA (pp. 84–93).

[Gai23] Gaitonde, D. (2023). *AMD Versal Adaptive SoC Architecture (Cornell ECE 6950 Guest Lecture).* https://www.youtube.com/watch?v$=$ogtmQ9i2Qb4.

[Gay20] Gaydadjiev, G. (2020). *FPL'20 Panel Discussion 2: How would one define Reconfigurable Computing?* Virtual Event.

[Gra00] Gray, J. (2000). *Designing a Simple FPGA-Optimized RISC CPU and System-on-a-Chip.*

[Gue24] Guerrieri, A., Guha, S., Lavin, C., Hung, E., Josipović, L., & Ienne, P. (2024). DynaRapid: Fast-tracking from C to routed circuits. In *Proceedings of the 34th International Conference on Field-Programmable Logic and Applications (FPL).* Torino (pp. 24–32).

[Hal10] Halfhill, T. R. (2010). Tabula's time machine. In *Microprocessor report* (pp. 1–14).

[Hau07] Hauck, S., & DeHon, A. (Eds.) (2007). *Reconfigurable computing: The theory and practice of FPGA-based computation.* Morgan Kaufmann. ISBN: 978-01-23705-22-8.

[Int24] Intel (2024). *Intel®Hyperflex™Architecture High-Performance Design Hand-book.* https://www.intel.com/content/www/us/en/docs/programmable/683353/21-3/hyper-pipelining-add-pipeline-registers.html. Accessed on 11 Oct 2024.

[Kap06] Kapre, N., Mehta, N., deLorimier, M., Rubin, R., Barnor, H., Wilson, M. J., et al. (2006). Packet switched vs. time multiplexed FPGA overlay networks. In *Proceedings of the 14th Annual IEEE Symposium on Field-Programmable Custom Computing Machines.* Napa, CA (pp. 205–216).

[Kap17] Kapre, N., & Gray, J. (2017). Hoplite: A deflection-routed directional torus NoC for FPGAs. *ACM Transactions on Reconfigurable Technology and Systems (TRETS), 10*(2), 1–24. ISSN: 1936-7406.

[Kas23] Kashani, S. (2023). *Building Chips Faster: Hardware-Compiler Co-Design for Accelerated RTL Simulation*. PhD Thesis. École Polytechnique Fédérale de Lausanne.

[Kea02] Keating, M., & Bricaud, P. (2002). *Reuse methodology manual for system-on-a-chip designs* (3rd ed.) Kluwer Academic Publishers. ISBN: 978-0-387740-98-0.

[Lan20] Langhammer, M., Baeckler, G., & Gribok, S. (2020). Spider-web-high performance FPGA NoC. In *Proceedings of the 2020 IEEE International Parallel and Distributed Processing Symposium Workshops (IPDPSW)*. New Orleans, LA (pp. 115–118).

[Lei91] Leiserson, C. E., & Saxe, J. B. (1991). Retiming synchronous circuitry. *Algorithmica, 6*(1), 5–35. ISSN: 1432-0541.

[Lem04] Lemieux, G., Lee, E., Tom, M., & Yu, A. (2004). Directional and single-driver wires in FPGA interconnect. In *Proceedings of the 2004 IEEE International Conference on Field-Programmable Technology*. Brisbane (pp. 41–48).

[Lew03] Lewis, D., Betz, V., Jefferson, D., Lee, A., Lane, C., Leventis, P., et al. (2003). The stratix routing and logic architecture. In *Proceedings of the 2003 ACM/SIGDA Eleventh International Symposium on Field Programmable Gate Arrays*. Monterey (pp. 12–20).

[Lew16] Lewis, D., Chiu, G., Chromczak, J., Galloway, D., Gamsa, B., Manohararajah, V., et al. (2016). The Stratix™10 highly pipelined FPGA architecture. In *Proceedings of the 2016 ACM/SIGDA International Symposium on Field-Programmable Gate Arrays*. Monterey, CA (pp. 159–168).

[Mai19] Maidee, P., Neely, C., Kaviani, A., & Lavin, C. (2019). An open-source lightweight timing model for rapidwright. In *Proceedings of the 2019 International Conference on Field-Programmable Technology (ICFPT)*. Tianjin (pp. 171–178).

[McM23] McMaster, J. (2023). *Xilinx Artix 7 XC7A50T Die Shots*. https://siliconpr0n.org/archive/doku.php?id$=$mcmaster:xilinx:xc7a50t. Accessed on 03 Oct 2024.

[Mei03] Mei, B., Vernalde, S., Verkest, D., De Man, H., & Lauwereins, R. (2003). Exploiting loop-level parallelism on coarse-grained reconfigurable architectures using modulo scheduling. In *Proceedings of the 2003 Design, Automation and Test in Europe Conference and Exhibition*. Munich (pp. 296–301).

[Met05] Metzgen, P., & Nancekievill, D. (2005). Multiplexer restructuring for FPGA implementation cost reduction. In *Proceedings of the 42nd Design Automation Conference*. San Diego, CA (pp. 421–426).

[Mis08] Mishchenko, A., Case, M., Brayton, R., & Jang, S. (2008). Scalable and scalably-verifiable sequential synthesis. In *Proceedings of the 2008 IEEE/ACM International Conference on Computer-Aided Design*. San Jose, CA (pp. 234–241).

[Mur14] Murray, K. E., & Betz, V. (2014). Quantifying the cost and benefit of latency insensitive communication on FPGAs. In *Proceedings of the 2014 ACM/SIGDA International Symposium on Field-Programmable Gate Arrays*. Monterey, CA (pp. 223–232).

[Nac15] Kapre, N., & Gray, J. (2015). Hoplite: Building austere overlay NoCs for FPGAs. In *Proceedings of the 25th International Conference on Field Programmable Logic and Applications*. London (pp. 1–8).

[Lin96] Lin, C.-C., Chang, D., Wu, Y.-L., & Marek-Sadowska, M. (1996). Time-multiplexed routing resources for FPGA design. In *Proceedings of the IEEE 1996 Custom Integrated Circuits Conference*. San Diego, CA (pp. 152–155).

[Pat80a] Patterson, D. A., & Sequin, C. H. (1980). Design considerations for single-chip computers of the future. *IEEE Journal of Solid-State Circuits, 15*(1), 44–52.

[Pet21] Petersen, M. B., Nikolić, S., & Stojilović, M. (2021). NetCracker: A peek into the routing architecture of Xilinx 7-series FPGAs. In *Proceedings of the 2021 ACM/SIGDA International Symposium on Field Programmable Gate Arrays*. Virtual Event (pp. 11–22).

[Put14] Putnam, A., Caulfield, A. M., Chung, E. S., Chiou, D., Constantinides, K., Demme, J., et al. (2014). A reconfigurable fabric for accelerating large-scale datacenter services. In *Proceedings of the 2014 ACM/IEEE 41st International Symposium on Computer Architecture (ISCA)*. Minneapolis, MN (pp. 13–24).

[Put23] Putnam, A. (2023). *If you build it, will they come? (Cornell ECE 6950 Guest Lecture)*. https://www.youtube.com/watch?v$=$TuS6bAZ5hps.

[Sch05] Schmit, H. (2005). Extra-dimensional Island-Style FPGAs. In P. Lysaght & W. Rosenstiel (Eds.), *New algorithms, architectures and applications for reconfigurable computing*. Springer (pp. 3–13). ISBN: 978-1-402031-28-1.

[Sch99] Schmit, H. (1999). Extra-dimensional island-style FPGAs (abstract only). In *Proceedings of the 1999 ACM/SIGDA Seventh International Symposium on Field Programmable Gate Arrays*. Monterey, California (p. 247).

[Set05] Sethuraman, B., Bhattacharya, P., Khan, J., & Vemuri, R. (2005). LiPaR: A light-weight parallel router for FPGA-based networks-on-chip. In *Proceedings of the 15th ACM Great Lakes Symposium on VLSI*. Chicago, IL (pp. 452–457).

[Shr23] Shrivastava, S., Nikolić, S., Ravishankar, C., Gaitonde, D., & Stojilović, M. (2023). IIBLAST: Speeding up commercial FPGA routing by decoupling and mitigating the intra-CLB bottleneck. In *Proceedings of the 2023 IEEE/ACM International Conference on Computer Aided Design (ICCAD)*. San Francisco, CA.

[Swa19] Swarbrick, I., Gaitonde, D., Ahmad, S., Gaide, B., & Arbel, Y. (2019). Network-on-chip programmable platform in versal ACAP architecture. In *Proceedings of the 2019 ACM/SIGDA International Symposium on Field-Programmable Gate Arrays*. Seaside, CA (pp. 212–221).

[Tei12] Teig, S. (2012). Going beyond the FPGA with spacetime. In *22nd International Conference on Field Programmable Logic and Applications (FPL)*. Oslo.

[Tri97a] Trimberger, S., Carberry, D., Johnson, A., & Wong, J. (1997). A time-multiplexed FPGA. In: *Proceedings of the 5th Annual IEEE Symposium on Field-Programmable Custom Computing Machines*. Napa, CA (pp. 22–28).

[Var90] Varshavsky, V. I. (Ed.) (1990). *Self-timed control of concurrent processes*. Mathematics and its Applications. Kluwer Academic Publishers. ISBN: 978-0-792305-25-5.

[Xil96] Xilinx Inc. (1996). *The programmable logic data book*.

[Xil02] Xilinx Inc. (2002). *Designing custom OPB slave peripherals for microBlaze*.

[Xil18] Xilinx Inc. (2018). *7 Series DSP48E1 Slice User Guide*. UG479, ver. 1.10.

[Yan02] Yan, A., Cheng, R., & Wilton, S. J. E. (2002). On the sensitivity of FPGA architectural conclusions to experimental assumptions, tools, and techniques. In *Proceedings of the 2002 ACM/SIGDA Tenth International Symposium on Field-Programmable Gate Arrays*. Monterey, CA (pp. 147–156).

[Ye06] Ye, A., & Rose, J. (2006). Using bus-based connections to improve field-programmable gate-array density for implementing datapath circuits. *IEEE Transactions on Very Large Scale Integration (VLSI) Systems, 14*(5), 462–473. ISSN: 1557-9999.

[Zho21] Zhou, Y., Maidee, P., Lavin, C., Kaviani, A., & Stroobandt, D. (2021). RWRoute: An open-source timing-driven router for commercial FPGAs. *ACM Transactions on Reconfigurable Technology and Systems (TRETS), 15*(1), 1–27. ISSN: 1936-7406.

Chapter 10
Interconnect Design for Multi-die Devices

Contents

2.5D FPGAs are already almost 15 years old, and with intensive research and fabrication breakthroughs, 3D FPGAs are coming ever closer to commercial adoption as well. Hence, a book on designing programmable interconnect would not be complete without at least a brief discussion of multi-die devices. Since these topics heavily depend on advanced fabrication, integration, and packaging technologies that are far beyond the scope of this book, our discussion will indeed be brief and focused on how the main developments are changing the perspective on programmable interconnect design and FPGA CAD as we have described it until now.

10.1 LUT Count Goes Up, Yield Goes Down

As capacity of FPGAs kept getting larger and larger, a point would have been reached when it would no longer be possible to produce them at a high-enough yield that would prevent the part cost from becoming prohibitively high [Com17]. However, that would only occur if they were produced on a single die of ever-increasing size. To resolve this issue, Xilinx developed the so-called *Stacked Silicon Integration* (SSI) technology when moving to the 28 nm node, around

© The Author(s), under exclusive license to Springer Nature Switzerland AG 2025

S. Nikolić, *Modern Programmable Interconnect Design*,
https://doi.org/10.1007/978-3-031-80629-2_10

2010 [Des10, Bol11]. Instead of basing the FPGA on a single large monolithic die, SSI combines multiple smaller dice that can be produced at a much higher yield [Cha12] that allows continued scaling of device capacity without surging part costs.

10.2 Interposers

A cross-cut view of a 7-Series FPGA—the first 2.5D FPGA family—is shown in Fig. 10.1. Multiple dice are positioned side by side on a common *passive* silicon *interposer*, typically implemented in an older, more mature technology [Rav18a]. For example, 7-Series FPGAs had dice (in Xilinx/AMD terminology, typically called *Super Logic Regions* (SLRs)) implemented in a 28 nm TSMC process, whereas the interposer used a 65 nm node [Bol11]. By looking at Fig. 10.1, one may wonder what is the role of the interposer, when the package substrate can already provide the mechanical support for the dice, as well as connectivity between them? The answer is that the pitch of connections achievable inside the package substrate is far too large to enable the bandwidth between the dice that would be anywhere close to what would exist in a monolithic FPGA [Cha12]. Moreover, introduction of the interposer reduces the depth that the interdie connections have to traverse, thus improving their speed [Cha12]. As we have seen in Chap. 4, older nodes also have more relaxed pitches in the metal stack, which makes the connections through them faster over a given distance. Nevertheless, reaching interposer wires still requires a signal to cross a deep via stack and a microbump, and wires themselves have to cross the horizontal space between the two adjacent dice [Rav18a]; all this leads to significant delays. Moreover, microbump pitch is still very large compared to

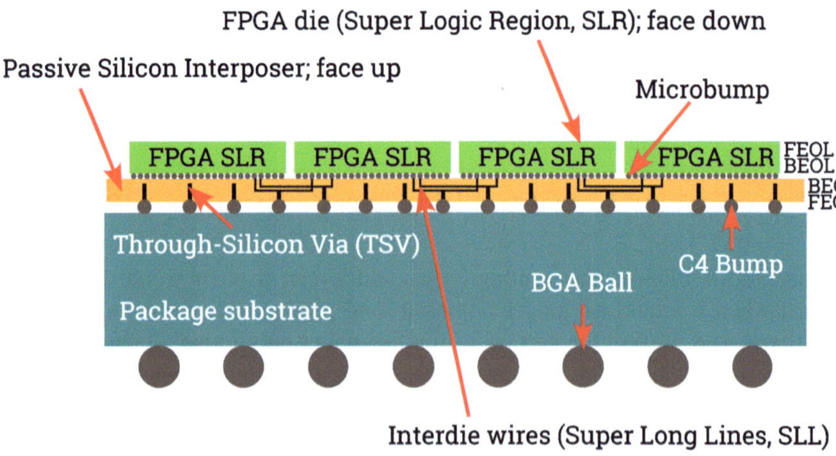

Fig. 10.1 Cross-cut view of a 7-Series 2.5D FPGA (Adapted from Chaware et al. [Cha12])

the pitches typically occurring in metal stacks, meaning that the number of signals that can cross the die boundary is relatively small. For instance, Ravishankar et al. compute the interdie wire (in Xilinx/AMD terminology typically called *Super Long Line* (SLL)) count to be about 25% of the number of vertical wires in an equivalent monolithic chip [Rav18]. For this reason, new placement algorithms have been proposed to minimize the demand for die boundary crossing [Rav18, Rai24].

Because the BEOL side of the interposer is facing up, to connect its wires to the package substrate, *Through Silicon Vias* (TSVs) are required. Although this already constitutes 3D integration, dice are still stacked side by side and the interposer itself is passive, providing only connectivity and mechanical support for the dice. Apart from a drastically increased yield, these devices have few other advantages over monolithic 2D FPGAs; at least they do not bring any that vertical stacking could potentially achieve (e.g., shorter wirelength [Aba04]). Hence, the name 2.5D FPGAs was adopted to designate them [Bol11].

10.2.1 Interdie Interconnect Pipelining

We discussed interconnect pipelining in Sect. 9.4.1. Wires that cross the die boundary can perhaps be treated as a special case. On one hand, they incur a much larger delay than other wires, and hence it would be beneficial to pipeline them. On the other hand, since relatively few signals cross the die boundary, it is more likely to expect that the user will be willing to invest further effort into accommodating the additional registers into their design. However, who will provide the physical registers that are necessary for this?

In UltraScale+ family of Xilinx/AMD FPGAs, special *Laguna* tiles were introduced in place of CLBs in certain columns, spreading 60 INT tiles away from the die boundaries [Rav18a]. This is illustrated in Fig. 10.2. Each Laguna tile provides optionally registered access to 24 bidirectional interdie wires, ending in Laguna tiles 60 INT tiles away, in the adjacent die. This enables significant delay reduction of interdie interconnect, allowing 500 MHz frequencies on signals that cross the die boundary [Gai19]. However, concentrating access to interdie connections in a few spread out locations puts significant pressure on programmable routing needed to reach these locations. Hence, the later Versal family of FPGAs opted for a more evenly distributed solution enabled by a redesign of the CLB [Gai19]. This once more illustrates the difficulty of designing interconnect for completely new technologies.

10.2.2 Heterogeneous Integration

First 2.5D FPGAs integrated several FPGA dice to improve device yield, but it soon became possible to also integrate heterogeneous dice in different technologies,

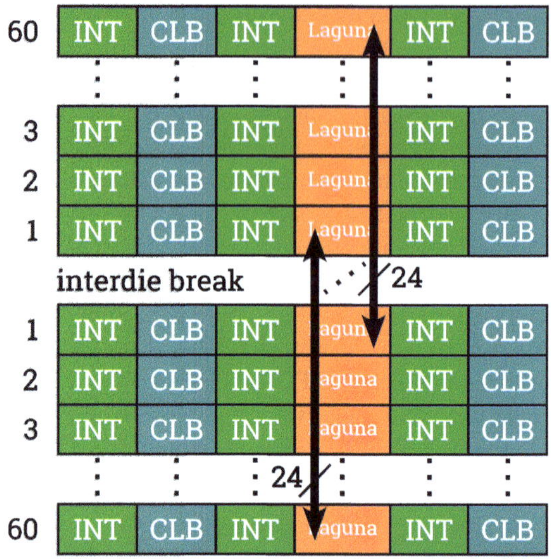

Fig. 10.2 Laguna tiles in UltraScale+ FPGAs. To provide optional pipelining for interdie wires, UltraScale+ FPGAs introduced *Laguna* tiles, replacing certain columns of CLBs up to 60 tiles away from the die boundaries. Each Laguna tile drives 24 bidirectional wires spanning 60 tiles (Adapted from Ravishankar et al. [Rav18a])

thus benefiting from their individual strengths. For instance, transceiver [Chr20] and *High-Bandwidth Memory* (HBM) [Gau17] dice became common in high-end FPGAs. It is worth noting that rather than relying on the passive silicon interposer, Intel introduced what they call *Embedded Multi-die Interconnect Bridge* (EMIB) [Chr20, Int21] that allows cheaper integration of heterogeneous dice, without the need of using any through-silicon vias. The idea is simple: Instead of having the interposer run underneath the entire surface area of all dice, a much smaller piece of passive silicon is embedded into the package substrate, underneath the edges of the dice that are to be connected together [Int21]. For instance, if UltraScale+ FPGAs were to use this integration technology, positioning the bridge under adjacent dice, about 60 INT tiles deep into the fabric would have been a reasonable choice, as this equals the span of the SLL wires. The embedded bridge then provides connectivity only between the dice, whereas the circuit I/Os are provided directly by the package substrate. Hence, TSVs are not needed in the bridge, since it suffices to reach it from the top. Although at first TSVs through a passive interposer could seem inferior, they could also be considered an important step toward developing technology for vertical stacking [Bol11].

10.3 From 2.5 to 3D

A true 3D FPGA comes from stacking active dice on top of each other. This requires use of TSVs passing through active silicon, which in early technologies presented a significant challenge [Com17], but with technology advancement, it may soon become standard in high-end FPGAs [Bou23]. Rather than diving into the details of the various possible integration technologies that go beyond the scope of this book, we shall list several ways in which future FPGAs could benefit from vertical stacking. We note that theoretical proposals and first prototypes of 3D ICs are in fact older than the FPGAs themselves [Kaw84]. Hence, it should not be surprising that many authors have already explored options of 3D FPGA design many years before they came to the verge of being economically viable [Lee98, Aba04]. It is also worth noting that HBM has been using active dice stacking for about as long as 2.5D FPGAs have been using passive interposers [Kan09]. In fact, it would not be surprising to see HBM placed above FPGA dice in the first generations of 3D FPGAs. That would help meet the ever-increasing memory bandwidth requirements and also likely reduce the technological risk compared to other integration possibilities discussed below.

10.3.1 A Stack of FPGA Dice

The most obvious way of profiting from the possibility to stack active dice on top of each other is to simply overlap the dice of a homogeneous 2.5D FPGA. In principle, a thus created 3D FPGA would have much shorter average connections than an equivalent-capacity 2D/2.5D FPGA [Aba04, Hal10]. Yet, due to large pitch of TSVs and their comparably high delay, it is very likely that for a long time to come, crossing dice is going to be avoided whenever possible, much like in today's 2.5D FPGAs. However, with maturing of stacking technology on one hand and increasing wire resistance in a single die on the other, it is possible that this will change.

In a recent detailed study of benefits of homogeneous FPGA die stacking in a plausible technology, Boutros et al. explore the possibility to connect two dice on top of each other—certainly the first step toward creating higher stacks [Bou23]. They measure that the average critical path delay reduction that can be obtained in this way ranges between 2 and 3%, depending on the exact architecture. That is comparable to, and even lower than the gains that we obtained through the various different 2D interconnect optimizations presented in the previous chapters. At the same time, vertical stacking makes the fabrication process considerably more expensive and reduces yield. Hence, integration technology will probably have to become more mature before vertical stacking of programmable fabric dice becomes practical.

10.3.2 NoC Separation

One particularly interesting integration possibility, especially in the context of the discussions of the previous chapter, is placing a die containing a hardened NoC on top of a programmable fabric die. A similar idea has been explored before in ASICs [Pav07], but Boutros et al. perceive it from a different angle. Namely, since the NoC does not consume the entire space available on its die, they use it to provide access to high-capacity SRAM banks [Bou23]. Upon compacting the programmable fabric after removal of the large NoC routers, they observe a 23% average improvement of the critical path delay. This is almost an order of magnitude more than what can be expected from the individual detailed optimizations of programmable interconnect introduced in the previous chapters (although combined, they could lead to comparable results), which demonstrates in practice the importance of improving logic density to reduce the length of wires in the programmable interconnect. It also points to the role that heterogeneous integration could play in achieving this objective.

10.3.3 Configuration SRAM Separation

A highly appealing possibility of separating configuration SRAM on top of the programmable fabric was explored by Lin et al. [Lin07]. As we have described in Chap. 4, configuration SRAM consumes a very significant portion of the programmable fabric and in most recent technologies its area scaling dramatically slowed down. If it could be placed on top of the rest of the fabric, that would drastically increase logic density, thus bringing considerable performance increases. However, the density of configuration SRAM means that it would be extremely difficult to separate it through die stacking and using TSVs for the numerous connections. Rather, advanced monolithic 3D IC fabrication technologies would likely be required to make this approach practical [Lin07]. Nevertheless, attempts to separate configuration SRAM on top of the rest of fabric have already been made in a commercial setting as well [Hal10]; at some point in future, they may actually succeed.

10.4 Conclusions

This chapter merely scratched the surface of the challenges and opportunities provided by advanced integration technologies that are still rapidly evolving. It will be very interesting to follow the architectural developments that will enable FPGAs to profit from that evolution. Interested readers may refer to Boutros et al. for a recent treatment of the problem of modeling and designing vertically stacked reconfigurable architectures [Bou23]. Their modeling and CAD tools could be of great use for those interested in joining the research effort in this very exciting domain.

References

[Aba04] Ababei, C., Maidee, P., & Bazargan, K. (2004). Exploring potential benefits of 3D FPGA integration. In *Proceedings of the 14th International Conference on Field Programmable Logic and Application*. Leuven (pp. 874–880).

[Bol11] Bolsens, I. (2011). *2.5D ICs: Just a Stepping Stone or a Long Term Alternative to 3D?* Accessed on 15 Oct, 2024. https://www.xilinx.com/publications/about/3-D_Architectures.pdf

[Bou23] Boutros, A., Mahmoudi, F., Mohaghegh, A., More, S., & Betz, V. (2023). Into the third dimension: Architecture exploration tools for 3D reconfigurable acceleration devices. In *Proceedings of the 2023 International Conference on Field Programmable Technology (ICFPT)*. Yokohama (pp. 198–208).

[Cha12] Chaware, R., Nagarajan, K., & Ramalingam, S. (2012). Assembly and reliability challenges in 3D integration of 28 nm FPGA die on a large high density 65nm passive interposer. In *Proceedings of the 2012 IEEE 62nd Electronic Components and Technology Conference*. Diego, CA (pp. 279–283).

[Chr20] Chromczak, J., Wheeler, M., Chiasson, C., How, D., Langhammer, M., Vanderhoek, T., et al. (2020). Architectural enhancements in Intel®Agilex™FPGAs. In *Proceedings of the 2020 ACM/SIGDA International Symposium on Field-Programmable Gate Arrays*. Monterey, CA (pp. 140–149).

[Com17] Computer History Museum (2017). *Oral History of Steve Trimberger*. https://archive.computerhistory.org/resources/access/text/2018/07/102740229-05-01-acc.pdf

[Des10] Design & Reuse (2010). *Xilinx 7 Series FPGAs Slash Power Consumption by 50% and Reach 2 Million Logic Cells on Industry's First Scalable Architecture*. https://www.design-reuse.com/news/23745/xilinx-virtex-7-kintex-7-artix-7.html. Accessed on 20 March, 2023.

[Gai19] Gaide, B., Gaitonde, D., Ravishankar, C., & Bauer, T. (2019). Xilinx adaptive compute acceleration platform: Versal architecture. In *Proceedings of the 2019 ACM/SIGDA International Symposium on Field-Programmable Gate Arrays*. Seaside, CA (pp. 84–93).

[Gau17] Singh, G., Ahmad, S., Wittig, R., Mittal, M., Arbel, Y., Arun, V. R., et al. (2017). Xilinx 16 nm datacenter device family with in-package HBM and CCIX interconnect. In *HotChips*.

[Hal10] Halfhill, T. R. (2010). Tabula's time machine. In *Microprocessor Report* (pp. 1–14).

[Int21] Intel (2021). *Video: Intel EMIB Technology Explained*. Accessed on 15 Oct, 2024. https://www.intel.com/content/www/us/en/corporate/usa-chipmaking/news-and-resources/video-intel-emib-technology-explained.html

[Kan09] Kang, U., Chung, H.-J., Heo, S., Ahn, S.-H., Lee, H., Cha, S.-H., et al. (2009). 8Gb 3D DDR3 DRAM using through-silicon-via technology. In *Proceedings of the 2009 IEEE International Solid-State Circuits Conference*. San Francisco, CA (pp. 130–131a).

[Kaw84] Kawamura, S., Sasaki, N., Iwai, T., Mukai, R., Nakano, M., & Takagi, M. (1984). 3-dimensional gate array with vertically stacked dual SOI/CMOS structure fabricated by beam recrystallization. In *Proceedings of the 1984 Symposium on VLSI Technology*. San Diego, CA (pp. 44–45).

[Lee98] Leeser, M., Meleis, W. M., Vai, M. M., Chiricescu, S., Xu, W., & Zavracky, P. M. (1998). Rothko: A three-dimensional FPGA. *IEEE Design & Test of Computers, 15*(1), 16–23.

[Lin07] Lin, M., Gamal, A. E., Lu, Y.-C., & Wong, S. (2007). Performance benefits of monolithically stacked 3-D FPGA. *IEEE Transactions on Computer-Aided Design of Integrated Circuits and Systems, 26*(2), 216–29.

[Pav07] Pavlidis, V. F., & Friedman, E. G. (2007). 3-D topologies for networks-on-chip. *IEEE Transactions on Very Large Scale Integration (VLSI) Systems, 15*(10), 1081–90.

[Rai24] Raikar, R., & Stroobandt, D. (2024). LiquidMD: Optimizing inter-die and intra-die placement for 2.5D FPGA architectures. In *Proceedings of the 14th International Symposium on Highly Efficient Accelerators and Reconfigurable Technologies*. Porto (pp. 90–98).

[Rav18a] Ravishankar, C., Fraisse, H., & Gaitonde, D. (2018). SAT based place-and-route for high-speed designs on 2.5D FPGAs. In *Proceedings of the 2018 International Conference on Field-Programmable Technology (FPT)*. Naha (pp. 118–125).

[Rav18] Ravishankar, C., Gaitonde, D., & Bauer, T. (2018). Placement strategies for 2.5D FPGA fabric architectures. In *Proceedings of the 28th International Conference on Field Programmable Logic and Applications (FPL)*. Dublin (pp. 16–20).

Chapter 11
Conclusions and Future Work

Contents

We have started this book by describing the challenges that modern reconfigurable architectures are facing due to recent changes in technological evolution. In this chapter, we summarize the contributions of this book aimed at helping to overcome these challenges. We also list some potential avenues for future work, inspired by the conclusions of the preceding chapters, that could be beneficial for ensuring that FPGAs and other reconfigurable architectures [Sch22a] deliver the performance that is expected from them.

11.1 What Have We Done?

In Chap. 2, we explained in detail the evolution of FPGA architecture design and, in particular, why design automation algorithms comparable with the ones that exist for ASIC development have thus far not been devised for programmable interconnect architectures. By describing the different integration and application contexts in which reconfigurable fabrics can be expected to be required, we have also

© The Author(s), under exclusive license to Springer Nature Switzerland AG 2025
S. Nikolić, *Modern Programmable Interconnect Design*,
https://doi.org/10.1007/978-3-031-80629-2_11

anticipated the need for rapid design of optimized programmable interconnect. Yet, our main premise, drawn from the recent developments of AMD and Intel FPGAs, was that the primary driver of the need to automate programmable architecture exploration at a much larger scale than has been done before are the changes in technology scaling. In order to empirically verify that this is indeed the case, as well as to enable such wide architectural exploration in advanced technologies for which it is relevant, in Chap. 4, we have introduced a new fast physical modeling framework.

11.1.1 Modeling Programmable Interconnect in Advanced Technologies

Initial experiments using this new framework demonstrated that technological changes trigger a need to revisit even such common rules of thumb of FPGA architecture design as the optimal cluster sizes. This also gave empirical support for an explanation of recent changes in the Intel Agilex FPGA family. Besides explaining the recent commercial developments, we were able to predict what may come in near future, and the picture was not pretty: With only local optimization around a previously known architectural solution, as has been customary in the past two decades, FPGAs are unlikely to be able to profit from transitioning to technology nodes beyond 5 nm. This is even if modifications applied to Agilex, such as removal of intermediate taps from channel wires or cluster size reduction, are retained. In fact, our predictions (see Fig. 4.19) show that with only minor optimizations, moving to a new technology node would not only bring no benefit, but would result in performance deterioration.

These conclusions of Chap. 4 clearly demonstrated the need for a significantly more extensive architectural exploration. Due to rising resistance, some of the problems that have to be tackled to achieve this goal become easier. In particular, the problem of optimizing channel segmentation can even be solved by exhaustive exploration, once the constraints of modern technologies and design principles are taken into account. On the other hand, rising resistance increases the importance of solving some other difficult combinatorial optimization problems, due to a need to reduce the distance that signals travel at low metal layers, as well as the capacitive load on shared wires, exerted by the numerous stored-select multiplexers of the programmable interconnect.

11.1.2 Letting the Router Automatically Design Switch-Blocks

One such problem is that of designing switch-patterns in switch-blocks, which we tackle in Chap. 5. The classical approach to solving this problem is to propose a set of explicit solutions and test each one of them by routing appropriate circuits. Whether the solutions are listed by hand, or produced by some automated search method, their number that can be evaluated in practice is inevitably very small, due to the high computational cost of running a routing algorithm. By observing that in the process of using the router as a mere black box to output one single performance measurement for each proposed architecture a large amount of valuable information is lost, we were able to altogether avoid explicitly listing solutions. Namely, instead of running the router on one particular proposed architecture, we extended the routing-resource graph on which it operates such that it contains an implicit representation of the entire search space. By slightly modifying the router's cost function, using the same *congestion negotiation* principles that it normally uses [McM95, Kap12], only applied in reverse, we were able to essentially leave it to the router itself to choose the switches that should be fabricated so that the performance of the circuits routed during exploration is maximized. We call this new algorithm *avalanche search*, due to the avalanche effect that occurs on the costs of different switches during negotiation.

Although avalanche search overcomes the fundamental scalability issue of having to list an exponential number of different potential solutions, in its present implementation, it suffers from some scalability issues of its own. Namely, the avalanche costs that enable switch presence negotiation are currently rendering A* routing—a crucial ingredient that enables the classical PathFinder algorithm to scale to the size of modern FPGAs—almost entirely ineffective. When coupled with a large increase of the routing-resource graph size, caused by embedding the entire search space in it, this limits the applicability of the algorithm in practice to designing FPGAs comprising only a few thousands of LUTs. Nevertheless, we strongly believe that these limitations are surmountable, unlike the previously existing fundamental barrier of having to explicitly list individual solutions; we suggested some possible remedies toward the end of Chap. 5. In recent years, the focus of FPGA architecture and CAD research has mostly been geared toward supporting large circuits, of the order of hundreds of thousands and even millions of LUTs, which are representative of flagship products of the main FPGA vendors [Mur20]. Yet, in Chap. 2, we have listed several reasons why design of highly optimized, custom reconfigurable architectures of relatively small size may again be of interest. In particular, relatively small eFPGAs are entering more and more SoCs, whereas standalone FPGAs with unprecedentedly low cost are being introduced for use in edge applications. For all these cases, avalanche search is relevant even as is. Furthermore, as hardened NoCs are taking over long-haul communication in large FPGAs, it may happen that regions left to programmable interconnect to cater to will reduce to about the size that avalanche search is already capable of exploring [Kap17].

11.1.3 A General Method to Project Layout and CAD Constraints on Architecture

Besides scalability, there is another reason that could limit the practical relevance of avalanche search as presented in Chap. 5 when designing switch-patterns for an actual product. Namely, avalanche search as such produces highly irregular switch-patterns and has no mechanism for ensuring that the output solutions respect any of the numerous layout constraints that may have to be imposed on a switch-pattern intended for actual fabrication. To rectify this, in Chap. 6, we proposed a general method for designing switch-patterns that respect arbitrary constraints encodable as ILP, with which we extended avalanche search of Chap. 5. We then measured the impact of enforcing many of the regularity constraints that can be observed in some commercial architectures, as well as others for which we believed that they could improve the efficiency of CAD tools that map user circuits on the produced architectures. We concluded that in almost all cases, with the help of the new constraint-enforcing method, avalanche search can find regular solutions with essentially equivalent performance.

It is, however, the generality of this extension that makes it the most relevant: A given FPGA vendor could encode any constraints that may be related to the specificities of their layout process or the limitations and capabilities of their CAD tools, of which we were naturally not aware while performing the study of Chap. 6.

11.1.4 Making the Fastest Connections Nonprogrammable

When one observes the difference between a direct wire connecting two terminals in an ASIC and a long programmable connection with several stored-select multiplexers along its way that achieves the same in a reconfigurable architecture, the question of whether this is truly always necessary naturally arises. Of course, it is self-evident that the numerous multiplexers are often necessary, as otherwise the architecture would not be reconfigurable in the sense defined in Chap. 3, but the question of whether they are *always* necessary remains. To answer that question, in Chap. 7 we introduced an efficient algorithm for automatically designing fixed-connectivity patterns of direct connections between individual LUTs, which enable timing-critical signals to avoid passing through multiple levels of stored-select multiplexers.

11.1.5 Making the Timing-Critical Signals Use the Direct Connections

Fixed-connectivity patterns are only as useful as the CAD tools are able to leverage them to reduce the critical path delay of a given circuit. This depends on successfully aligning timing-critical signals of the circuit with the fast direct connections of the interconnect architecture. In Chap. 7, we tried to avoid the complex task of developing dedicated CAD tools that would support this before that was justified by demonstrating that direct connections have a measurable utility. We did so by, for the most part, only aligning the direct connections of the constructed fixed-connectivity pattern with the timing-critical signals of the placed user circuits. In Chap. 8, we added support for alignment in the other direction—that is, aligning timing-critical signals of a user circuit with the direct connections of any single fixed-connectivity pattern—by introducing a dedicated, ILP-based detailed placement algorithm.

11.2 Where Has This Brought Us?

Impact of different architectural exploration as well as CAD algorithms presented until now has been measured through extensive experimentation and reported in the chapter that introduced each respective algorithm. Due to different underlying technologies and baseline architectures that were used in these experiments, as a result of research taken to develop the algorithms spanning several years, it is difficult to unify the impact of each individual contribution, without reimplementing several of the algorithms such that they could be integrated with newer and more capable versions of third-party modeling and CAD tools that we rely on (i.e., the VTR project [Mur20]). This, however, would take a significant amount of additional time, which, unfortunately, is no longer at the author's disposal. Nevertheless, these difficulties do not exempt us from at least trying to use some plausible projections and make predictions of where all of the presented contributions could have potentially led us on our quest to overcome the technological challenges faced by future reconfigurable architectures, had they been successfully combined in a single design automation flow.

Figure 11.1 captures our best attempt at this. It repeats a part of the plot of Fig. 4.19, showing that in the original measurements, moving an architecture with a cluster of eight 6-LUTs from a 5 nm to a 4 nm technology would result in 19.5% deterioration of the average critical path delay. A lower-bandwidth channel segmentation inspired by Agilex [Chr20] that we adopted in Chap. 5 improved the performance of the $N = 8$ architecture in 4 nm by 4.6%, demonstrating the opportunities created by a more extensive segmentation exploration, which, as explained in Sect. 5.12.2, can be performed exhaustively in new technologies.

Applying the avalanche search algorithm of Chap. 5 to produce a switch-pattern more appropriate for the given technological and target application setting than the parametric one of Chap. 4 led to a further 5.5% reduction of the average critical path delay. Such a significant improvement—on par with performance increase that addition of time-borrowing support produced on UltraScale+ FPGAs [Gan16]—demonstrates the effectiveness of design automation in overcoming the limitations of prior intuition; we once more note that the parametric switch-pattern of Chap. 4 represented our best effort at capturing the connectivity of recent FPGAs, before

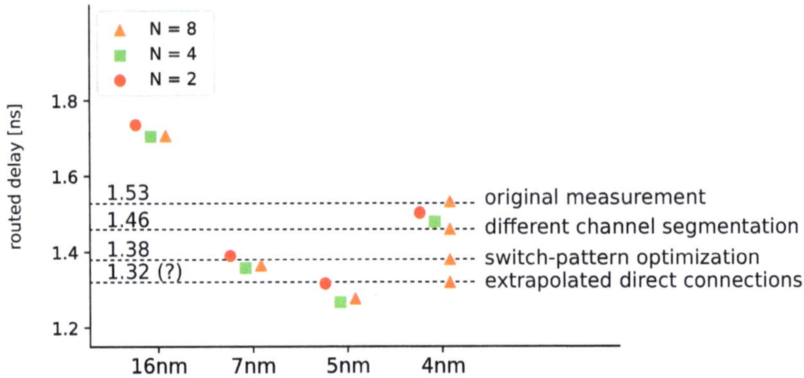

Fig. 11.1 Estimate of combined improvement of all proposed algorithms. Original measurements correspond to Fig. 4.19. Direct connection impact is a projection of the results of Chap. 8

we even isolated switch-pattern design as an important candidate for automation and started developing a solution for it, which ultimately resulted in the algorithm presented in Chap. 5. Of course, there is little doubt that commercial switch-patterns are capable of achieving higher performance than the parametric design that we were able to envision, which may make it more difficult for avalanche search to make as significant a difference in a production setting as we can observe here. Furthermore, it is likely that commercial user circuits, which are of much larger size than the ones that we used in Chap. 5, require richer connectivity and do not allow as significant switch-pattern sparsification as avalanche search was able to achieve in Chap. 5. Nevertheless, we believe that the experimental results clearly demonstrate the effectiveness of automated switch-pattern design in finding more optimal solutions than what relying on intuition can produce. This is not only important for designing reconfigurable architectures for different fabrication technologies, which may change the validity of usual assumptions like we have seen in Chap. 4, but also for different SoC-integration contexts and different target applications, where carrying over prior results [Koc21] clearly leads to suboptimal performance.

As we have mentioned in Chap. 7, experiments related to exploration of fixed-connectivity patterns have unfortunately been performed in a reasonably old planar technology and their results cannot be immediately compared to the ones of Fig. 4.19. Yet, in order to make a rough estimate of what their contribution would have been, we for the moment assume that the 4.3% average critical path delay improvement of Chap. 8 is maintained. Although it is difficult to assess how the utility of direct connections would really change in new technologies without performing actual experiments, we believe this assumption to be reasonable. Namely, we can observe once more that there are two main effects contributing to the perceived effectiveness of direct connections, which are impacted by technology scaling in an opposing manner: (1) the speed of multiplexers that the direct connections replace increases with scaling, which makes direct connections less appealing and (2) sensitivity to capacitive loading of shared programmable wires also increases with scaling, making the nonshared direct connections that do not suffer from this load more appealing.

When all contributions are combined in this manner, the overall performance of the $N = 8$ architecture at 4 nm is improved by 13.7%. This still did not make it faster than the 5 nm implementation, but the gap was reduced from 19.5 to 3.1%, which is well within the reach of other modifications, such as optimizing the local interconnect, as evidenced by the 3.3% improvement of the average critical path delay when cluster size is reduced to four.

Notwithstanding the remaining uncertainties of these projections, in this book we have demonstrated that design of programmable interconnect can be successfully automated despite inherent difficulty of various combinatorial optimization problems that have to be solved to this end. It is only necessary to carefully analyze each problem and leverage its specificities to avoid explicit evaluation of a large number of possible design points. We have also demonstrated that this can lead to a great improvement of programmable interconnect performance in advanced technologies, making such techniques highly appealing for helping FPGAs successfully transition to new nodes. As mentioned in Chap. 4 and as exemplified by our hypothetical F3b node, leveraging *Design Technology Co-optimization* (DTCO) can lead to even greater improvements. Rapid architectural exploration using algorithms presented in this book could be of great use there as well. Finally, although we have not attempted it in this work, design of customized programmable interconnect architectures for different target applications, which we have mentioned in Chap. 2 as potentially key to continued performance improvement of SoCs equipped with reconfigurable fabrics, is hard to envision without use of design automation algorithms such as the ones presented in this book.

11.3 Future Work

Many reasonably obvious extensions to the proposed algorithms that would make for natural research directions in near future have been proposed in the preceding chapters. For instance, resolving A* issues of avalanche search is imperative for making it scale to larger FPGA fabrics, as well as for leveraging the method to simultaneously design a larger fraction of programmable interconnect of an island-style FPGA (for example by embedding both the switch-block and the connection-block search space in the routing-resource graph). Rather than repeating other such directions that can be easily found at the end of each chapter, in this Section, we list some problems whose solution is at the moment further from reach, but for which the experiments of the preceding chapters led us to believe that solving them could be of great importance for future developments of reconfigurable architectures.

11.3.1 Separating High-Performance and High-Bandwidth Interconnect

As we have already stated many times over, reduction of capacitive loads, distance traversed at low metal layers, and tile area in general, so that the lengths of channel wires can also be reduced, is imperative for enabling further performance increase of reconfigurable architectures. Yet, it is highly unlikely that such reductions will be achievable uniformly in all areas of the programmable interconnect architecture. Even if that were possible, additional increase in performance could be obtained by making resources commonly catering to timing-critical signals particularly fast. We have explored this using one possible approach in Chap. 7, by adding fast direct connections between LUTs on top of an existing programmable architecture that could provide flexible interconnect for the majority of the remaining, noncritical signals. Another approach was mentioned in Sect. 4.8.4, in relation to a plausible explanation of the construction of a 32 6-LUT cluster in Versal FPGAs [Gai19]. Namely, we conjectured that the large cluster is in fact a hierarchical supercluster of several smaller clusters, which allows low-performance, high-bandwidth local interconnect between them, leaving the few timing-critical signals to be routed through a faster combination of cluster-level local interconnect and lower-resistance channel wires. Further development of such architectures, where largely disjoint networks are optimized respectively for minimum area/maximum bandwidth, and minimum delay/maximum usability by timing-critical signals of typical circuits, rather than attempting to strike a balance between the two objectives within the same network, seems to be a very promising direction for future research. We note that using avalanche search in routability-only mode (Sect. 5.13.5), driven by a large number of circuits with high Rent's exponents, could be used for developing architectures that meet the minimum area/maximum bandwidth criterion. Then, a second run in timing-driven mode could be used to augment the architecture with an appropriate set of routing resources optimized for performance.

11.3.2 Routing Comes After Synthesis but Cannot Be an Afterthought

As we have mentioned in Sect. 3.9, FPGA CAD algorithms rarely consider routing in early stages of the flow, other than attempting to minimize some metric derived from the total number of connections that remain to be routed. While minimizing the total number of connections is important, it is not sufficient to fully profit from the existence of a clear distinction between the high-bandwidth resources geared toward satisfying the needs of the vast majority of signals, and the high-performance ones, which should be used for preventing delays on the timing-critical signals. Instead, CAD tools in the previous stages of the CAD flow (especially synthesis and technology mapping) should plan ahead the split of signals between the two networks, so that the scarce fast resources can be properly utilized. As discussed in Sect. 7.3.1, when fast resources do not offer flexible access, they cannot simply be accounted for in terms of density in a region; rather, they have to be perceived as combinatorial structures, for which advanced algorithms in early CAD flow stages are particularly important.

11.3.3 LUT and Multiplexer SRAM Sharing

One of the primary ways of reducing area that we identified in Sect. 4.8.9 is reduction of the number of SRAM bits required by the stored-select multiplexers. This can be achieved by using more encoding-efficient multiplexer designs [DeH95], or sharing SRAM bits between multiplexers that form a multi-bit bus [Ye06]. However, solving the above problem of accounting for inflexible connectivity patterns in all CAD flow stages could open up a rather interesting new possibility: SRAM bits could be shared between LUTs and routing multiplexers. Oftentimes, the LUT mask is only partially used, and the remaining bits could be utilized for configuring routing multiplexers. The difficulties arise, however, when the bits used by the multiplexers are also used by the LUT. Can we simultaneously use both the LUT and the multiplexer? An important observation is that configuring the LUT fixes a connectivity pattern formed by the multiplexers that share SRAM bits with it. If we assume existence of algorithms that can take into account fixed-connectivity patterns at scale during synthesis, then they will be capable of making use of fixed-connectivity patterns induced by LUT configuration as well. Of course, LUT input permutations can be used to swap rows in the LUT's truth table, making the configuration more favorable for routing using induced patterns. For one-level and two-level multiplexers, which currently dominate FPGA architecture, many LUT configurations would lead to short circuits in the multiplexers that share the configuration bits with them. However, since switching to fully encoded multiplexers would reduce the amount of SRAM used in the first place, and since contrary to SRAM, the area of encoded multiplexers continues to improve in the latest technologies [Sch22], this problem could be easily avoided in the future.

11.3.4 Turning Strict Design Rules into an Advantage

In Chap. 4, we have seen how the strict gridded nature of FinFET layouts enables development of precise, yet remarkably simple layout models. We also mentioned in the conclusion of that chapter that these severe restrictions of layout engineer's freedom can be leveraged for more than modeling—in the limit, we could create a fully integrated design automation flow, which would not only define a programmable interconnect architecture using approximate layout models, but also generate the entire layout itself. Besides enabling more informed decisions about architectural design, a great advantage that this could bring would be the ability to lose the tileability constraint. Once manual layout effort is eliminated, the primary reason to adopt tileable architectures is removed as well. Such increased flexibility for introducing heterogeneity of routing resources could bring further benefits to the performance of reconfigurable architectures. For example, different parts of a user circuits could be mapped onto regions of the reconfigurable fabric that have different connectivity, thus matching the needs of the circuit more closely. In a sense, this would be analogous to heterogeneous CPUs, comprising multiple different cores [Nay22]. Alternatively, it could be seen as integration of an entire family of custom programmable interconnect architectures, like suggested by Betz and Rose [Bet95], on a single chip.

Besides a fully integrated design flow that would remove the need for manual layout, to make this approach beneficial, algorithms capable of adequately partitioning user circuits and mapping the pieces to the most appropriate fabric regions are also required. From our prior experience [Nik19], an effective approach to solving this problem could be obtained by adapting graph-similarity measures used in other scientific fields, such as social network theory and computational biology, where comparison [Spa93] and alignment [Kin12] of graphs with size comparable to that of a typical circuit mapped onto FPGAs are regularly required [Zha20].

11.3.5 Are We Solving the Right Problem?

Once the tileability constraint is removed, it is useful to consider whether we should still be choosing prefabricated channel wires and switches that connect them while designing a programmable interconnect architecture. A reason for doubt lies in the abstract definition of the programmable interconnect design problem presented in Sect. 3.1, which is much broader. Once more, the programmable interconnect architecture is essentially just a network of stored-select multiplexers, in which multiple different circuit graphs can be embedded. That edges of this network can be interpreted as channel wires in an island-style FPGA is merely an implementational detail. Rather than constructing wire and switch sets as has traditionally been done, why not approach the problem directly? Can we automatically synthesize a minimal (in terms of total size, average configured depth, or some other precisely defined metric) multiplexer network that can implement all circuits from some set Γ? We believe that with the help of the aforementioned scalable network alignment algorithms, satisfactory solutions for this problem could be efficiently found. This approach would be especially appealing if the reconfigurable architecture under design is intended to be highly customized for a specific set of applications that can be captured by a small set of concrete circuits. However, much like when designing a classical island-style FPGA, making the set of circuits used in synthesis large and varied enough would ensure that, in practice, CAD tools will be able to map an arbitrary circuit of interest onto the produced architecture, albeit perhaps with a lower performance. By enforcing existence of certain paths in the architecture, as in Sect. 6.11, it may even be possible to prove some form of universality of the architecture.

11.4 Final Remarks

In this book, we have described the historical evolution of FPGA architecture design along with the challenges caused by recent changes in technology scaling that triggered a need for changing the design approach as well. We postulated that design automation is the most promising way to allow reconfigurable architectures to overcome these challenges and developed several algorithms for solving different, hard combinatorial problems that arise in the process of programmable interconnect design. Generality of these algorithms gives us hope that they will find practical use, perhaps in augmented and somewhat modified form, and contribute to producing high-performance reconfigurable architectures in many years to come. Of course, it is impossible to know whether this will truly be the case or the proposed algorithms will be quickly supplanted by other, still better ones. However, we believe that the lasting effect of this book will be the demonstration that automation of programmable interconnect design is feasible, despite the inherent difficulty of the combinatorial optimization problems that have to be solved along the way, or the various constraints imposed by practical layout and CAD tool limitations that the produced solutions must respect. Looking once more at the historical perspective, this should not come as a surprise, however. Although programmable interconnect design has a number of interesting and challenging peculiarities, difficult problems have been solved in every aspect of ASIC design automation for decades. The main difference between ASIC and programmable interconnect design automation is that there has been a great need for the former for more than half a century, whereas a real need for the later has only started to appear in the last few years. We hope that the results of Chap. 4 will be convincing enough for the reader to appreciate that the day has finally come when design automation for programmable interconnect is not only feasible, but also important. Finally, we hope that the attempt which we made in Chap. 3 to twist the formulation of the problem from how it is customarily presented will convince the reader that design automation for programmable interconnect is not only feasible and important, but also interesting. We admit that listing and counting wires and switches probably appear a bit boring—especially in comparison with designing a brand new embedded AI acceleration block—which may drive many prospective researchers away from this intriguing field. But, who can resist the appeal of aligning networks and searching for minimal common supergraphs?

References

[Bet95] Betz, V., & Rose, J. (1995). Using architectural "Families" to increase FPGA speed and density. In *Proceedings of the 1995 ACM Third International Symposium on Field-Programmable Gate Arrays*. Monterey, CA (pp. 10–16).

[Chr20] Chromczak, J., Wheeler, M., Chiasson, C., How, D., Langhammer, M., Vanderhoek, T., et al. (2020). Architectural enhancements in Intel®Agilex™FPGAs. In *Proceedings of the 2020 ACM/SIGDA International Symposium on Field-Programmable Gate Arrays*. Monterey, CA (pp. 140–149).

[DeH95] DeHon, A. (1995). *Entropy, counting, and programmable interconnect*. Transit Note 128. University of CA.

[Gai19] Gaide, B., Gaitonde, D., Ravishankar, C., & Bauer, T. (2019). Xilinx adaptive compute acceleration platform: Versal architecture. In *Proceedings of the 2019 ACM/SIGDA International Symposium on Field-Programmable Gate Arrays*. Seaside, CA (pp. 84–93).

[Gan16] Ganusov, I., & Devlin, B. (2016). Time-borrowing platform in the Xilinx UltraScale+ family of FPGAs and MPSoCs. In *2016 26th International Conference on Field Programmable Logic and Applications (FPL)*. Lausanne (pp. 1–9).

[Kap17] Kapre, N., & Gray, J. (2017). Hoplite: A deflection-routed directional torus NoC for FPGAs. *ACM Transactions on Reconfigurable Technology and Systems (TRETS), 10*(2), 1–24. ISSN: 1936-7406.

[Kap12] Kaptanoglu, S. (2012). Pathfinder: A negotiation-based performance-driven router for FPGAs. In *FPGA and reconfigurable computing hall-of-fame endorsement*.

[Kin12] Kingsford, C., & Patro, R. (2012). Global network alignment using multiscale spectral signatures. *Bioinformatics, 28*(23), 3105–3114.

[Koc21] Koch, D., Dao, N., Healy, B., Yu, J., & Attwood, A. (2021). FABulous: An embedded FPGA framework. In *The 2021 ACM/SIGDA International Symposium on Field-Programmable Gate Arrays*. Virtual Event (pp. 45–56).

[McM95] McMurchie, L., & Ebeling, C. (1995). PathFinder: A negotiation-based performance-driven router for FPGAs. In *Proceedings of the 1995 ACM Third International Symposium on Field-Programmable Gate Arrays*. Monterey, CA (pp. 111–117).

[Mur20] Murray, K. E., Petelin, O., Zhong, S., Wang, J. M., Eldafrawy, M., Legault, J.-P., et al. (2020). VTR 8: High-performance CAD and customizable FPGA architecture modelling. *ACM Transactions on Reconfigurable Technology and Systems (TRETS), 13*(2), 9:1–60.

[Nay22] Nayak, A., Chen, H., Mair, H., Lagerquist, R., Chen, T., Rajagopalan, A., et al. (2022). A 5 nm 3.4 GHz Tri-gear ARMv9 CPU subsystem in a fully integrated 5G flagship mobile SoC. In *2022 IEEE International Solid-State Circuits Conference (ISSCC)* (Vol. 65, pp. 50–52). San Francisco, CA.

[Nik19] Nikolić, S., Zgheib, G., & Ienne, P. (2019). Finding a needle in the haystack of hardened interconnect patterns. In *2019 29th International Conference on Field Programmable Logic and Applications (FPL)*. Barcelona (pp. 31–37).

[Sch22a] Schmit, H., & Denton, M. (2022). Multi-input serial adders for FPGA-like computational fabric. In *Proceedings of the 2022 ACM/SIGDA International Symposium on Field-Programmable Gate Arrays*. Virtual Event (pp. 35–41).

[Sch22] Schor, D. (2022). *IEDM 2022: Did We Just Witness The Death Of SRAM?* WikiChip. https://fuse.wikichip.org/news/7343/iedm-2022-did-we-just-witness-the-death-of-sram/. Accessed on 24 April, 2023.

[Spa93] Sparrow, M. K. (1993). A linear algorithm for computing automorphic equivalence classes: The numerical signatures approach. *Social Networks, 15*(2), 151–170. ISSN: 0378-8733.

[Ye06] Ye, A., & Rose, J. (2006). Using bus-based connections to improve field-programmable gate-array density for implementing datapath circuits. *IEEE Transactions on Very Large Scale Integration (VLSI) Systems, 14*(5), 462–473. ISSN: 1557-9999.

[Zha20] Zhang, S., & Tong, H. (2020). Network alignment: Recent advances and future directions. In *Proceedings of the 29th ACM International Conference on Information & Knowledge Management*. Virtual Event (pp. 3521–3522).

Index

© The Editor(s) (if applicable) and The Author(s), under exclusive license to Springer Nature Switzerland AG 2025
S. Nikolić, *Modern Programmable Interconnect Design*,
https://doi.org/10.1007/978-3-031-80629-2

MIX
Papier aus verantwortungsvollen Quellen
Paper from responsible sources
FSC® C105338

If you have any concerns about our products,
you can contact us on
ProductSafety@springernature.com

In case Publisher is established outside the EU,
the EU authorized representative is:
Springer Nature Customer Service Center GmbH
Europaplatz 3, 69115 Heidelberg, Germany

Printed by Libri Plureos GmbH
in Hamburg, Germany